Science 21

Science 21

Kahni Burrows

BICENTENNIAL 1807 WILEY 2007

jacaranda
jaconline.com.au

First edition published 2007 by
John Wiley & Sons Australia, Ltd
42 McDougall Street, Milton, Qld 4064

Offices also in Sydney and Melbourne

National Library of Australia
Cataloguing-in-publication data

Burrows, Kahni.
Science 21.

Includes index.
For secondary school students.
ISBN 978 0 7314 0419 3.

1. Science — Study and teaching (Secondary) — Queensland.
I. Title.

500.712943

Front cover image: © William Whitehurst/CORBIS/APL

Cartography by MAPgraphics Pty Ltd, Brisbane and the Wiley Art Studio
Illustrated by Steve Hunter and the Wiley Art Studio
Anniversary logo design: Richard Pacifico

Typeset in India by TechBooks
Layout by Wiley Composition Services

Printed in Singapore by
Craft Print International Ltd

10 9 8 7 6 5 4 3 2 1

All activities have been written with the safety of both teacher and student in mind. Some, however, involve physical activity or the use of equipment or tools. All due care should be taken when performing such activities. Neither the publisher nor the authors can accept responsibility for any injury that may be sustained when completing activities described in this textbook.

In this book, the word 'Aborigine' rather than 'Koori' is used when referring to indigenous Australians. The issues raised are not unique to the indigenous people of New South Wales and so the Australia-wide reference has been maintained.

It is recommended that teachers should first preview resources on Aboriginal topics in relation to their suitability for the class level or situation. It is also suggested that Aboriginal parents or community members be invited to help assess the resources to be shown to Aboriginal children. At all times the guidelines laid down by the Department of Education should be followed.

Contents

Preface

'Equipped with his five senses, man explores the universe around him and calls the adventure Science.'

— Edwin Powell Hubble, *The Nature of Science*, 1954

'You cannot teach a man anything; you can only help him find it within himself.'

— Galileo Galilei

To the average person, the word *scientist* conjures up images of old eccentrics in white coats surrounded by a maze of glassware filled with bubbling solutions, while *science* itself is perceived as being made up of a vast body of arcane formulae and complicated equations. What most people fail to realise is that we all started out as scientists. Whenever a child asks 'Why is the sky blue?', 'Where does the wind come from?', 'Why do we have five fingers and not four?' or even 'Will I die if I eat this crayon?', they demonstrate the curiosity about the world that underlies the work of the scientist. As the child grows up, the questions may change to something more like 'How did that red hair dye turn my hair khaki green?', 'Will I really get brain cancer from talking on my mobile for 3 hours a day?' or 'Why does my urine smell strange after I eat asparagus?', but the essential nature of the hidden scientist is still there.

Encouraging students to maintain this curiosity about their world and to find out things for themselves by *doing* science rather than just learning about scientific facts is the challenge thrown down to all science educators. The introduction of the subject Science 21 forms part of the Queensland Studies Authority's response to this challenge. Superseding Multistrand Science in the senior school, Science 21 has been designed as a viable option for students who are interested in studying a general science course beyond year 10 rather than focusing on a specific area of science such as physics, chemistry or biology. Through the introduction of this subject, which uses real-world science applications, it is hoped that the misconception that science has relevance only to professors in laboratories and has very little impact on the life of the average high school student may be overturned.

This textbook has been written to support the Queensland Studies Authority subject Science 21 and has been designed to confront the student who demands, 'Yeah, but how does all that science stuff affect me?' Each chapter is essentially a stand-alone module addressing a particular scientific aspect of the modern world in which our students live. Some of these chapters such as 'What's your poison?' and 'Survive that drive' address issues that have immediate relevance to the young adults that we teach. Others such as 'Life, the universe and everything', 'The restless earth' and 'Weapons through the ages' ask students to reflect on how our past has shaped our present, while 'The brave new world of genetic engineering', 'Nuclear energy — Australia's energy future?' and 'Colonising other worlds' introduce students to aspects of science that will shape their future world.

Each chapter is organised in such a way as to expose students to the idea that science is not a collection of stale facts that must be memorised with little direct application to their lives; rather, it is a way of thinking that allows them to find out things for themselves. At the beginning of the chapter, a question is posed for the consideration of the students, and the material that is included is intended to guide them to their own answer. While there is a large amount of information to fill students in on the basics of each specific area, there is plenty of scope for extension beyond the supplied material.

Investigations and **Activities** are included, where appropriate, as a way of allowing students to experience hands-on science techniques or demonstrating scientific principles. These, in turn, may lead the student to further investigation of some aspect of the observed phenomena.

Eye on the media sections feature articles from a variety of newspapers, journals, novels and magazines that are relevant to the issue under examination, followed by a series of questions based on the article. In this way, students are exposed to a variety of genres in which science may be presented as well as being encouraged to be critical rather than passive readers.

Internet activities direct students to websites where they may be required to collect research data, engage in experimental simulations or use interactive applications to further their understanding.

Discussion sections suggest topics for class consideration either in small groups, larger forums or debate.

End of chapter questions allow students to check their understanding of the basic principles discussed in each chapter.

At the end of the text, **Appendix A** addresses the fundamentals of the scientific method. Here, students can see how to develop an experiment under controlled conditions, whether it is a short experiment or a longer project. The structure of an extended report, the basics of data collection and presentation, error considerations, and suggestions on how to select an independent experiment topic are also considered. This appendix also shows students how to present material in a variety of forms, such as a research essay, an argumentative essay and a media file.

While this text may be used as the basis for an entire course, it can easily be used in conjunction with other topics that teachers may develop as being appropriate for their own class or school circumstances. In either case, it is hoped that this text may be a resource that helps students to develop an appreciation of the science that surrounds them everywhere.

Kahni Burrows
October 2006

Acknowledgements

The authors and publisher would like to thank the following copyright holders, organisations and individuals for their permission to reproduce copyright material in this book.

Images

• © Andrew Macrae: 1.6/using data from Harland, WB, Armstrong, RL, Cox, AV, Craig, LE, Smith, AG & Smith, DG 1990, *A geologic time scale 1989*, Cambridge University Press, New York; Gradstein, F & Ogg, J 1996, 'A Phanerozoic time scale', *Episodes*, vol. 19, no. 1&2, pp. 3–4 • © Andrew Smith: p. x • © Atomic Archive, www.atomicarchive.com. All rights reserved: 5.38 • Austral International: p. 128 • AAP Images: 5.30/AP Photo/Nick Ut; p. 160/AP Photo/Advanced Cell Technologies; 6.12/AP Photo/APTV • AKG London: 9.2/akg-images/Paul Almasy • ANTPhoto.com.au: 1.14/MWF Tweedie; 2.34/John & Elenor Baker • Australian Picture Library: 2.30/Corbis/Roger Ressmeyer; 3.1/NASA/Roger Ressmeyer; 3.15/Corbis/James Marshall; 3.24/Album; 3.26; 5.17/Corbis/Carl & Ann Purcell; 5.23/Corbis; 6.11(a)–(b)/Corbis/Robert Dowling; p. 190/Corbis/Tim Wright; 8.8/Corbis/Mediscan; 8.10/Scott Houston; 8.13/Corbis Sygma/Alan Lewis; p. 242/Corbis/Neal Preston • © Gary Clark, www.swamp.com.au: 6.19 • Commonwealth Copyright Administration: 8.20/© Commonwealth of Australia. Reproduced by permission • Corbis Images: 5.37/© Corbis Corporation • © David Malin Images: 3.22/Akira Fujii • © Digital Stock: p. 37, 7.13(a)/Corbis Corporation • © Digital Vision: p. 216 • Evidential Art, www.evidential-art.co.uk: 9.5/© Janet Richardson • Getty Images: 3.16/Time Life Pictures/NASA; 5.24/Hulton Archive; 5.29/Time & Life Pictures/Time Magazine/Larry Burrows; 7.4/Stone • © John Wiley & Sons Australia, Ltd: 2.35/Kari-Ann Tapp; 7.10/Jo Patterson; 7.13(b)/Joshua Quiroga • © Judith Kinnear: 1.12 • The Kobal Collection: p. 1/Touchstone/Spyglass Entertainment; p. 72/Paramount • Lonely Planet Images: 5.12/Bethune Carmichael • Mary Evans Picture Library: 1.13(a)&(b) • Milos Kalab: 9.3, 9.23(b)/Custom Medical Stock Photo • MAPgraphics Pty Ltd, Brisbane: 2.23, 2.36, 4.12 • © Neil O Hardy: 9.16 • © New Scientist: 1.27, 4.2 • © Newcastle Region Library: 2.27 • Newspix: 2.38, p. 104, 5.22; 3.2/AFP; 7.11/Angelo Soulas; 8.6/AFP Photo/Ramzi Haidar • Panos Pictures: 4.14/Stefan Boness • Peter Rozanski: A.3 • © Photodisc: 1.2, 5.1, 6.14(a)&(b), 6.17 • Photolibrary: 1.3/SPL/AATB/Royal Observatory; 1.11, 8.11/SPL/Sinclair Stammers; 1.18(a), 8.15/SPL; 1.20(a)/SPL/Vaughan Fleming; 3.3, 3.4, 8.19(a)–(e)/SPL/NASA; 3.12/SPL/John Sanford; 5.28/SIPA Press; 6.4/SPL/Manfred Kage; 6.5/SPL/Zephyr; 6.6/SPL/Dr P Marazzi; 6.9(a), 9.12/SPL/Mauro Fermariello; 6.9(b)/SPL/Kevin Curtis; 6.10/Photonica, Inc./Danielle Weil; 6.14(c)/Mauritius die Bildagentur GmbH; 6.15/SPL/Eurelios/Makoto Iwafuji; 8.5(a)/Brookhaven National Laboratory/Science Photo Library; 8.5(b)/SPL/Pascal Goetgheluck; 8.7/SPL/Gusto; 8.12/SPL/James King-Holmes; 8.14(a)/SPL/Arthur Glauberman; 9.1/SPL/Dr Jurgen Scriba; 9.23(a)/SPL/Dr GD Danilatos • Picture Media: 1.20(b)/Corbis/Reuters;

4.13/Corbis/Reuters/Vladimir Repik • The Picture Source: 7.7/Terry Oakley • © Queensland Police Service: 9.6, 9.7, 9.8 • RAAB Collection: A.1/Courtesy of The RAAB Collection • Robert Bosch GmbH: 7.2 • © The State of Queensland (Queensland Transport): 7.12 (five images) • Stock Photos Pty Ltd: 7.3 • SKM: 2.39(a)&(b)/© 2004 DigitalGlobe, Inc. All rights reserved • Victoria Forensic Science: 9.19, 9.20/© Victoria Police Forensics Services Department; • © The William Rafti Institute: 8.9/William Rafti

Text

• Advertiser Newspapers Pty Ltd: p. 275/David Washington/*The Courier-Mail*, included with permission of Advertiser Newspapers Ltd • Auspac Media: p. 8/© *Australasian Science* magazine. Distributed by Auspac Media • Laura A Huxley: p. 185/Excerpt from *Brave New World* included with permission of Laura A Huxley • Graeme Lofts: p. 278 ff./from *Jacaranda Physics 1 Second Edition* by John Wiley & Sons Australia, Ltd. Included with permission of the author • John Louie, PhD: p. 53/compiled by John Louie PhD, www.seismo.unr.edu/~louie/ • © *New Scientist*: pp. 22, 32, 52, 55, 60, 62, 67, 79, 95, 97, 116, 122, 155, 169, 177, 200, 225, 226, 248, 266 • Queensland Newspapers: pp. 142, 231/*The Sunday Mail*; p. 180/Jeff Sommerfeld/*The Courier-Mail*. Included with permission of Queensland Newspapers • Queensland Transport: p. 197; p. 199/© 2005 The State of Queensland (Queensland Transport), reprinted in *The Courier-Mail*, 29 June 2005. Included with permission of Queensland Transport and Queensland Newspapers • RACQ Magazine: pp. 195, 207/Reproduced courtesy of the RACQ's magazine, *The Road Ahead* • © *Scientific American*: pp. 28, 243 • © Time, Inc.: p. 76/All rights reserved. Reprinted from *Popular Science* magazine with permission

Every effort has been made to trace the ownership of copyright material. Information that will enable the publisher to rectify any error or omission in subsequent editions will be welcome. In such cases, please contact the Permissions Section of John Wiley & Sons Australia, Ltd, who will arrange for the payment of the usual fee.

1

Life, the universe and everything

Most cultures have their own beliefs about how the world was made and when humankind came into existence, and the scientific world is no different. For centuries, scientists have sought evidence of our beginnings and proposed theories that might explain how the universe and everything in it came into being. Over the millennia, these theories have shifted to accommodate new evidence and observations that have arisen from technological advances. So what are the current theories on it all and what proof do we have?

In the beginning . . .

Formation of the universe

The **Big Bang Theory** proposes that all of the matter and energy presently in the universe began as a single super-dense mass that exploded and expanded outwards.

A **billion years** is equal to 1 000 000 000 years.

While there are several different theories explaining the formation of the universe, the theory that is the most widely accepted is known as the **Big Bang Theory**. According to this theory, all of the matter and energy that exists now was originally compressed into an extremely small 'super-mass' under enormous pressure and extremely high temperature. About 16 **billion years** ago, this supermass erupted in less than one-millionth of a second, blasting matter in all directions. This early matter gave rise to the simplest atomic components — protons, neutrons, electrons — and other, more exotic particles such as neutrinos and particles of antimatter. As these particles cooled, they gave off energy and formed the simplest of the elements: hydrogen.

Nuclear forces are the forces that arise from the interaction of the particles that make up an atom. They are called nuclear forces because the most influential of the particles are found within the nucleus (or centre) of the atom.

Nuclear fusion is the process by which smaller nuclei combine to form larger nuclei, releasing a large amount of energy.

Under the influence of gravity and **nuclear forces**, the hydrogen atoms started to attract each other to form dense clouds of hydrogen gas. Over time, as more and more atoms combined, these clouds became very dense and very massive. The forces acting on these atom clusters became so massive that they collapsed in on themselves. This started a nuclear process (known as **nuclear fusion**) in which hydrogen atoms fuse to form heavier elements, releasing huge amounts of light and energy (see figure 1.1). These reacting clusters were the first stars.

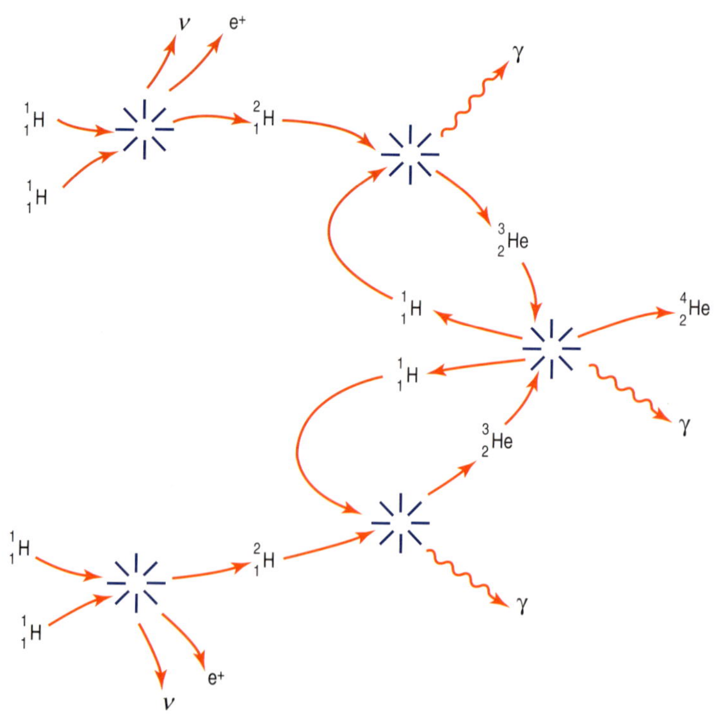

Figure 1.1:
In the process of nuclear fusion, the nuclei of hydrogen atoms combine to form helium and other heavier nuclei.

So long as there is hydrogen in the star, the energy created by the fusion process stops the star from collapsing. Eventually, over billions of years, a star runs out of the hydrogen that acts as its fuel and starts to collapse. The forces generated in the star's collapse ignite its outer layers and the star

A **nova** is a star that has depleted its supplies of hydrogen fuel and then collapsed, causing its gaseous envelope to ignite in a massive explosion that scatters the gases into a cloud surrounding the remains of its core. The nova's brightness is much greater than that of the original star.

A **nebula** (plural: nebulae) is a glowing cloud of gas resulting from the debris of stars that have exploded.

explodes as a **nova**, sending hydrogen, helium and other heavier nuclei out into space. The gaseous debris forms glowing clouds known as **nebulae** (see figures 1.2 and 1.3). Over time, the atoms in these nebulae will start to move together under the forces of gravity to form new atomic clusters that will give rise to new stars, and the process starts all over again.

Figure 1.2:
The Crab Nebula was formed when a star went nova. This exploding star was so bright that it was visible in the day sky on Earth in the early 11th century.

Figure 1.3:
Nebulae have been described as 'star nurseries' because new stars are formed inside them.

Another theory for the origin of the universe is the Steady State Theory. Find out as much as you can about both this theory and the Big Bang Theory. Compare the two theories by considering the following:

1. Describe how the theories state that the universe began.
2. When were the theories proposed and by whom?
3. What supporting evidence is there for each theory?
4. What are the things that contradict each theory?

A star is born

Like other stars, our sun was born from the material ejected by older stars. It is believed that our sun and its solar system first started to form about 4.6 billion years ago, when the gases in the solar nebula mixed with interstellar matter and formed clumps made of hydrogen, helium, iron, carbon and other elements. Under the effect of gravity, the clumps drew together and collapsed inwards. This inward motion and collapse caused the entire mass to begin rotating. This rotation in turn caused the dense gas cloud to flatten so that it resembled a disc with a bulge in the middle (see figure 1.4). At the dense centre of the disc, the inward collapse of the particles caused the temperature to rise until it was hot enough to start the nuclear fusion process and the sun started to burn.

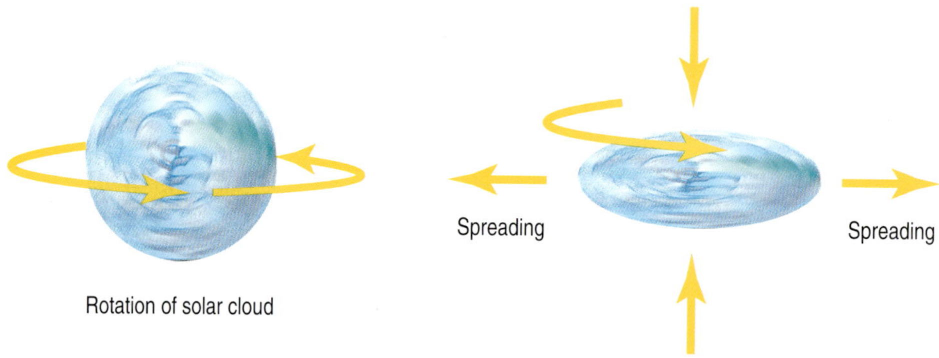

Figure 1.4:
As matter moved inwards, the solar cloud flattened into a disc.

Very close to the new sun, temperatures were very high but decreased with increasing distance from the centre of the solar nebula. In the region closest to the sun, the temperature was too high for lighter elements such as hydrogen and helium to exist in anything other than a gaseous form. However, the temperature was just low enough for heavier elements such as iron and nickel to condense from gases into liquids and solids. As a result, small chunks of rocky matter formed in this inner region. These chunks combined with dust still present from the original solar nebula to form **planetismals**. In a process known as **accretion**, the planetismals collided, broke apart and re-formed into larger chunks that looked a lot like asteroids. These, in turn, collided to form bigger masses that became the four inner rocky planets of our solar system — Mercury, Venus, Earth and Mars. It has been theorised that these inner planets would have been nearly molten when they first formed due to the extreme heat generated by the collision of the planetismals that created them.

Planetismals are chunks of material formed in the accretion disc of a solar nebula that combine to form planets.

Accretion is the process by which a star or planet gains mass when material is pulled into it by gravitational attraction.

Not all of the planetismals formed planets; some remained as asteroids. The **asteroid belt** between the orbits of Mars and Jupiter is made up of millions of these chunks of planetismal matter, ranging in size from specks of dust to chunky lumps that are hundreds of kilometres across (see figure 1.5). These chunks were unable to accrete into a fifth rocky planet because they were constantly being pulled apart by Jupiter's high gravity.

The **asteroid belt** is the belt of rocky debris that surrounds the sun between the orbits of the planets Mars and Jupiter.

When the sun was ignited, the lighter gases were blown further out in the accretion disc. In the regions beyond the terrestrial planets, the temperature was low enough to allow these gases to condense, forming the **gas giants** — Jupiter, Saturn, Uranus and Neptune. These planets, while made mainly of hydrogen, helium and methane gas, have large masses and therefore strong gravitational fields that pull smaller particles of dust, rock and ice into orbit around them. As a result, all of the gas giants have systems of rings made of this orbiting debris. Only Saturn's rings are large enough and bright enough to be visible through binoculars from Earth.

Gas giants are large planets composed almost entirely of gases. Jupiter, Saturn, Uranus and Neptune are the gas giants in our solar system.

Beyond the orbit of Neptune are millions of ice chunks and dust particles. This region is known as the **Kuiper Belt**. Some of the larger objects from the Kuiper Belt are in orbit around the sun. Pluto is one such object, and in fact it is believed to be a captured comet rather than a true planet. There is evidence to suggest that there are other masses in the Kuiper Belt that are both larger than Pluto and moving in a very slow orbit around the sun.

The **Kuiper Belt** is a region of icy bodies orbiting the sun beyond the orbit of Neptune.

Figure 1.5:
The asteroid belt is made of rocky material that was unable to accrete due to the gravity of nearby Jupiter.

Stages in the Earth's development

Having formed from chunks of rocky matter orbiting a new sun, the Earth slowly cooled and passed through a number of stages before it became the world that we know today.

Measuring geological time

Considering the vast periods of time that cover the Earth's history, scientists have found it convenient to break this time up into smaller units called **eons**. The four eons have been designated (from earliest to most recent) the Hadean, Archean, Proterozoic and Phanerozoic eons. The first three eons together span a four-billion year period of time called the Precambrian.

The Phanerozoic, the eon in which life became abundant, is divided into three **eras** — the Paleozoic, Mesozoic and Cainozoic eras. Eras are further divided into **periods**. These mark the most important changes in the kinds of fossils that are found in the rock records that form the basis of our understanding of what happened in the distant past. Finally, the periods are divided into **epochs**. Eras, periods and epochs divide up Earth's

Eons are the four major units into which the time that has passed between the Earth's birth and the present time is divided.

Eras are the three subunits into which the Phanerozoic eon is divided.

Periods are subdivisions of an era that signal changes in the fossil or geological record.

Epochs are the subdivisions of periods.

history into periods of time that are uneven in length, and take their names from Latin or Greek terms or from the places where notable samples were discovered showing a change in either the environment or the types of organisms that existed at that time.

A **geological timescale** is a chart that depicts the Earth's history as found in the geological and fossil records of its past.

Figure 1.6 is a chart showing an outline of Earth's history that is called a **geological timescale**. While there are a number of ways that these charts can be drawn, it is usual for Earth's earliest history to be shown at the bottom and its recent history at the top. As you can see, we are living in the Holocene epoch, which forms the last 10 000 years or so of the Quaternary period, which covers the last 1.5 million years of the Cainozoic era, which is in the Phanerozoic eon.

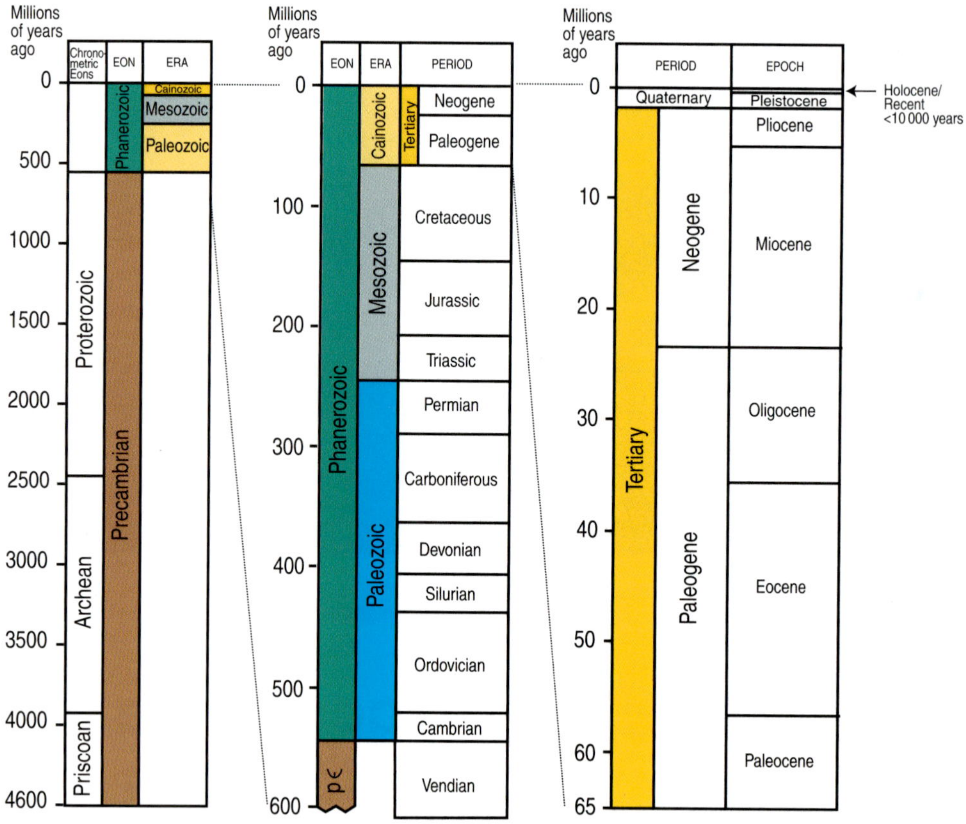

Figure 1.6: A geological timescale

The geological timescale is so large that it often describes events that happened millions of years ago, or MYA. Now that we have a better idea of the sheer immensity of the timescale that we are dealing with when considering where *we* fit into the picture, let us return to when the Earth first formed and how it developed from there.

The Hadean eon (4600–3900 MYA)

The word **Hadean** comes from the word *Hades*, which is another word for Hell.

The **Hadean** (or Pregeologic) eon is the period during which the Earth was transformed from a gaseous cloud into a solid body and its crust was formed. The atmosphere of the Earth at this time was short-lived. Made up of leftover interstellar matter, hydrogen and helium gas, the early atmosphere was both very light and very reactive, so it would have either floated off into space or reacted with other substances to form compounds.

In Earth's early molten state, the various materials were churned up quite heavily. As the outer layers of material cooled, they formed a crust. This crust was constantly broken up and redissolved by the hot rock that plumed from the interior. Eventually, after about 500 million years, the whole mass had cooled sufficiently for this pattern to stabilise. Rather than being constantly churned together, the elements started to separate out. The heavier elements such as iron sank towards the middle of the planet while the light rocky silicates floated towards the surface and hardened, forming the beginning of the Earth's present crust.

For the first 700 million or so years, the enormous amounts of rocky debris present in the solar system meant that all of the planets were under constant and intense asteroid bombardment. On the young Earth, these meteorites punched holes in the soft crust, allowing magma to ooze to the surface and effectively refill the holes. As a result, there is very little evidence of meteorite damage from this time.

Figure 1.7:
An artist's impression of Earth in the Hadean eon

Eye on the media

Wet Days in Hell

Following the news that the Hadean era of Earth's evolution had a wet climate and long, calm interludes (*AS*, June 2005, p. 11), it now appears that the formation of continents at the time worked much as it does today.

Last year Professor Mark Harrison of the Australian National University's Research School of Physical Sciences announced that zircons found in the world's oldest-known rocks indicated they had been formed in the presence of water. This suggested that the early Earth had oceans much like today.

The Hadean era was named after the ancient Greek underworld because this period of intense asteroid bombardment was considered hellish, but Harrison's work suggests that most of the time the environment may have been quite reasonable.

Further results obtained by Harrison suggest that the Earth was not a water world at the time, with continents forming much as they have ever since.

'We have evidence that massive amounts of continental crust were produced almost immediately upon the Earth's formation,' Harrison said. 'The Hadean Earth may have looked much like it does today rather than our imagined view of a desiccated world devoid of continents.'

Harrison's conclusion is based on the ratio of two isotopes of hafnium, one of which is formed by the radioactive decay of lutetium. The ratios found suggest that a reservoir of lutetium and hafnium had appeared prior to the formation of the zircons. Such reservoirs are consistent with modern continent formation.

Harrison admits that the picture he is drawing is a radical contrast to the perception most scientists hold of the early Earth. 'But these ancient zircons represent the only geological record we have for that period of Earth history, and thus the stories they tell take precedence over myths that arose in the absence of observational evidence,' he says.

Source: *Australasian Science*, January/February 2006, p. 5.

Clarification questions

1. What evidence is mentioned in this article for the Earth having a wet climate during the Hadean eon?
2. How does the Hadean Earth described in this article differ from the present theories about its appearance?
3. On what basis does Harrison make his claims?
4. When does Harrison believe the continents formed? How does this differ from present theory?

The Archean eon (3900–2500 MYA)

Volcanic outgassing is the process by which gases are released into a planet's atmosphere from its molten interior by volcanic activity.

The main characteristic of the Archean eon was the formation of Earth's first real atmosphere and the majority of the continental landmasses. The early atmosphere seems to have been made up of water vapour, carbon dioxide and nitrogen with small amounts of ammonia, carbon monoxide, sulfur dioxide and hydrogen sulfide. These gases were the product of the volcanoes that covered the Earth's surface. The release of these gases from the interior of the planet is called **volcanic outgassing**.

Figure 1.8:
An imaginary reconstruction of an Archean scene about 3.5 billion years ago

There was no gaseous oxygen in the atmosphere at this stage. If we were to travel back to that time, we would suffocate — if we didn't get roasted first! One aspect of this lack of oxygen is that there would also have been no ozone layer (the protective layer at the top of our atmosphere that filters out most of the deadly UV radiation from the sun).

As more gases accumulated in the atmosphere, the air pressure increased and caused the water vapour in the atmosphere to condense into liquid rain. This water gathered in the natural basins that had formed in the Earth's crust, creating the early oceans. Slowly, over millions of years, the water cycle was established (see figure 1.9). In this cycle, water from the oceans evaporates and rises to form clouds of water droplets. The clouds move over the land, where the droplets fall to the ground as rain. The rain forms watercourses that allow the water to flow back to the ocean, where the process continues.

With the water cycle established, the carbon dioxide gas in the atmosphere was dissolved out of the atmosphere by rain and entered the oceans, leaving nitrogen as the most common gas in the Earth's atmosphere. Even today, nitrogen gas makes up 79 per cent of our air. So where did the oxygen come from?

Figure 1.9:
The water cycle

The first life on Earth

It so happens that the origin of atmospheric oxygen is tied up inextricably with the origins of the earliest life on our planet. About 3500 million years ago, tiny micro-organisms called **microbes** appeared on Earth. The exact origin of these microbes is still a subject of much argument among scientists. One theory favours the idea that the microbes were carried to Earth from somewhere else in the universe, tucked into the deep fissures of meteorites (see figure 1.10). When the meteorites hit the Earth's surface, the microbes found a new home here. This is referred to as extraterrestrial seeding or **panspermia**. The more widely held view is that the simple organic molecules already present in the Earth's oceans gradually, over millions of years, combined to form larger and more complex molecules that eventually resulted in the microbes.

A microbe is an organism so small that you need a microscope to see it.

Panspermia is the theory that the earliest microbes were brought to Earth inside meteors from space.

The earliest microbes used hydrogen or methane in order to turn their food — other organisms or organic matter — into energy. This process (called **chemosynthesis**) was a form of **anaerobic respiration** where no oxygen is used to metabolise food into energy. Even today, microbes that live around volcanic vents in the deep ocean floor where there is no light and no oxygen still use this process. The early microbes would have eventually run out of food and become extinct if a new kind of microbe had not developed somewhere between 3.5 and 2.7 million years ago. This new microbe could produce its own food by using the energy of light from the sun in a process called **photosynthesis**. In this process, the hydrogen atoms are stripped from water molecules and combined with carbon dioxide to form starch and oxygen.

Chemosynthesis is the process by which organisms make organic compounds from elements other than oxygen.

Anaerobic respiration is the process by which food is turned into energy without the use of oxygen.

Photosynthesis is the process by which sunlight is used to convert carbon dioxide and water into oxygen and carbohydrate.

There was plenty of carbon dioxide available to these new microbes in the oceans, where carbon dioxide from the early atmosphere had been deposited by the rain. Over millions of years, the microbes (which were the simplest and earliest plants) made a large amount of oxygen as a result of the photosynthesis process. This oxygen took a while to appear in the atmosphere because, when it first started being made, it would combine with the free iron atoms present on the Earth's surface and form iron oxide. Eventually, about two billion years ago, the iron atoms had stopped mopping up the oxygen produced by the microbes and, at last, oxygen started to accumulate in the atmosphere. Sunlight reacted with the oxygen in the outer layer of the atmosphere to form ozone, eventually creating the ozone layer and shielding the Earth's surface from UV radiation.

Figure 1.10:
Did early life hitch a ride here?

Figure 1.11:
These fossilised remains of bacteria-like microbes from an ancient structure called a stromatolite are 3500 million years old. Stromatolites are layered rock-like structures formed by bacterial activity.

The Proterozoic eon (2500–542 MYA)

During the Proterozoic eon, plate tectonics began to prevail over other processes in shaping the Earth's crust while the nitrogen–oxygen atmosphere stabilised. The crust had formed into a series of moving sections called *plates* that carried the small raised continents over the liquid magma of the interior as if on rafts. As the interior of the Earth was hotter than it is today, the magma was more liquid and moved more quickly. This in turn caused the continents to move faster and collide more often. At this time, the face of the Earth experienced a continuous process of change. We will look at the process of plate movement in more detail later in this textbook.

The development of life on Earth

We saw that, during the Archean eon, microbes developed that were able to produce their food by using the process of photosynthesis. These were the earliest ancestors of plants on this planet and they produced the oxygen in the Earth's atmosphere. At about the time that free oxygen started appearing in the atmosphere (two billion years ago), a new form of microbe developed that used oxygen for its metabolic processes. These microbes were the earliest ancestors of animals.

Emergence of multicellular organisms

For many millions of years, the single-celled microbes were the sole forms of life on Earth. The next stage of development came when these single-celled organisms started to form into colonies, or groups. While many of these colonies formed only temporarily when individual cells found survival difficult, some formed permanent clusters of cells. Over time, cells in these clusters began to have specific functions. Some cells adapted to form structures that would move the colony; others were in charge of reproduction or digestion and so on. These clusters of specialised cells began to work as **multicellular** organisms rather than as colonies of identical cells. Scientists believe that this occurred about 800 million years ago.

Multicellular means 'made up of more than one cell'.

Diversification of organisms

The multicellular organisms continued to change over time, adapting to their particular environments. Fossils found in the Ediacara Hills (South Australia) indicate that the earliest multicellular **animals** appeared about 600 million years ago during the Cambrian period. These early animals were soft-bodied ocean dwellers. The development of animal life coincides with the increase in atmospheric oxygen.

Animals are multicellular organisms that use other organisms as a food source that they digest internally.

Approximately 400 million years ago, plants began to colonise the landmasses. These simple plants had no true root systems, leaves or flowers. Over time, however, the land plants became larger and more complex. Around 360 million years ago, the first land animals appeared and started to differentiate. Table 1.1 shows the approximate timescale over which these changes occurred.

Table 1.1: Key events in evolution		
Age (Myr: millions of years)	**Animals**	**Plants**
> 600	• Soft bodied animals	
Cambrian (540)	• Animals with exoskeletons	
Ordovician (505)	• First jawless vertebrates	
Silurian (440)	• First fish with jaws	• First true plants
Devonian (410)	• First insects • First ammonites • First sharks • First bony fish	• First gingkos • First ferns • First conifers
Carboniferous (360)	• First land-dwelling animals (amphibians) • First reptiles	• First giant club mosses
Permian (286)	• Extinction of trilobites	
Triassic (245)	• First pleisiosaurs • First ichthyosaurs • First dinosaurs	
Jurassic (208)	• First mammals • First birds	• First flowering plants
Cretaceous (146)	• Extinction of dinosaurs • Extinction of ammonites	
Tertiary (65)	• Diversification of mammals	• Diversification of flowering plants
Quaternary (2)	• First human species	

Figure 1.12: Fossils of trilobites (early sea crustaceans)

Evolution — the mechanism of change

These changes did not occur suddenly. Over the history of the Earth, the forms of life that arose were enormously varied. From the fossils left behind, we can see evidence of life forms that no longer exist as well as those that closely resemble organisms that are still alive today. We can also see that the variety of the organisms found in fossils seems to increase over time. So where did the missing organisms go? Where did the new ones come from?

While there are many different theories that explain these findings, the theory that best fits the facts at present is the **Theory of Evolution**. The central concept underlying evolution is that species change (or 'evolve') over time. The mechanism for evolution is **natural selection**, an idea expressed by the naturalists Charles Darwin and Alfred Russel Wallace (see figure 1.13).

Natural selection can be explained in the following way: in a particular group of organisms, variations occur over time as the result of mutation.

The **Theory of Evolution** describes the way in which organisms change in appearance, structure and function over long periods of time.

Natural selection is the process by which organisms exhibiting characteristics that increase their chance of survival in their environment pass these traits to their offspring, thereby increasing the incidence of that characteristic in the organism group.

Figure 1.13:
Charles Darwin (a) and Alfred Wallace (b) independently came up with the idea of natural selection as the means by which evolution occurs.

Figure 1.14:
The two variants of the English peppered moth

Some of these variations will give the organisms that have them a better chance of survival in their environment. These organisms live long enough to breed and pass on the useful characteristic to their offspring, while organisms without the characteristic die out. Over a period of time, all of the surviving organisms have the characteristic. Natural selection is often referred to as the 'survival of the fittest' because it is the organisms that are best suited (or 'fitted') to their environment who survive. The example of the peppered moth illustrates this (see figure 1.14).

In 19th century England, the English peppered moth generally had a light-grey speckled colour that allowed it to blend in when resting on the trunks of the trees during the day, thereby avoiding predators. Occasionally, moths were born that were black. These stood out on the pale tree trunks and so were quickly snapped up by birds and other predators.

However, over the course of the century, England's Industrial Revolution saw the building of more and more factories that poured smoke into the atmosphere. Over the years, the trunks of the trees became darker as the soot accumulated on the bark. Now, the black moths had the advantage because they blended into the darkened bark while the paler moths were more easily seen by predators and were eaten. The black moths survived to breed, passing the dark colouration on in their genes to their offspring. This happened over a relatively short period of time: the first black peppered moth was seen in Manchester in 1850 and, 45 years later, 98 per cent of the English peppered moths in Manchester were black. The black moths survived because they were the best suited for the new environment.

Speciation

Speciation is the process by which one species gives rise to another.

When Darwin and Wallace were writing about natural selection, the science of genetics had not yet come into existence. As a result, while they understood that some characteristics suddenly arose in a group of organisms and such characteristics could be passed down from parent to offspring, it was not until the middle of the 20th century that the way in which new species arose was understood. The process in which one species develops from another is called **speciation**. Organisms that are of different species are unable to mate and produce fertile offspring.

One of the things that trigger speciation is geographic separation. The members of a single population of organisms share a common pool of genes. Many of these allow the organisms to be well suited to their environment. As a population grows, it could spread out over a large area where, instead of remaining as a single group of organisms, it might divide into three or four isolated groups. Each of these groups is subjected to different environmental factors such as the number of predators, the climate, the availability of food and so on. In each of these groups, different characteristics arise as the result of random mutation and genetic variation. Some of these characteristics will be favourable and so will be passed down, while unfavourable characteristics will become less common in the gene pool. Over time, due to different environmental pressures and the randomness of genetic variation, the members of each group will share a different set of genes from the members of the other groups. Eventually, the groups will be so different genetically that they will be unable to breed with each other. They will have become different species (see figure 1.15).

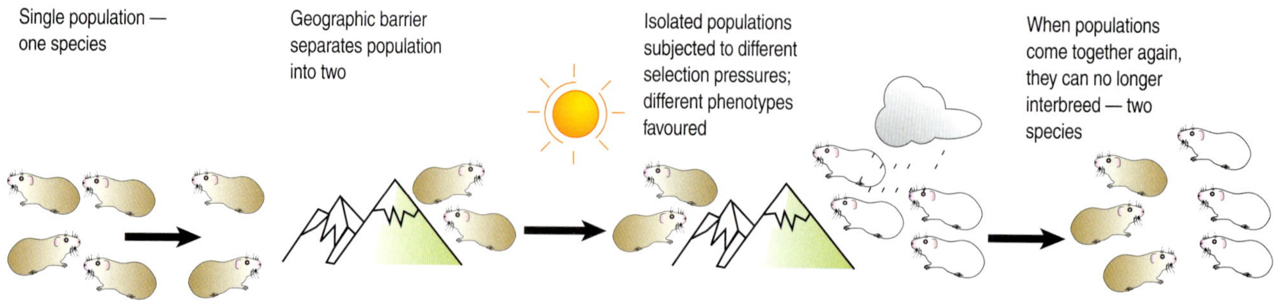

Figure 1.15: Natural selection can, over time, cause two different species to arise from a single population of organisms when subjected to different survival conditions.

Consider this example: a particular group of rodents spreads out so that some of the population settles in territory on the other side of a mountain range. While the original environment was warm and contained few predators, this new environment is very cold and full of predators. Over time, the characteristics of the rodents who survive there change by the process of natural selection. Eventually, the rodents who live on the far side of the mountain range become quite different from the original species: they have a thicker coat (to protect from the cold), a white colouration (to make it harder for predators to spot them in the snow) and they are smaller in size.

The process by which a single species changes over time to give rise to several new species is called divergent evolution.

Internet activity

To see adaptation at work as a simulation, go to the website for this book and click on the Evolution lab weblink (see Weblinks, page 309).

By changing the mutation rate and the selection strength, you can influence the adaptation of a small blue creature in its ability to reach its prey.

The average blue creature has a reach of two units. Due to mutation, sometimes creatures turn up that have reaches of one, three or even four units. Longer reach is not always an advantage. If a characteristic gives a creature a survival advantage, then it will pass the characteristic on to its offspring. In this simulation, mutation rate indicates the probability that a mutation will occur in a new blue creature, while the accessibility of the prey is determined by the selection strength.

Run the simulation for the following values of mutation rate and selection strength for 250 turns. Enter the data obtained after this time in a copy of the table below. While doing the simulations, also observe at what turn the creatures with a reach of one and two units can be said to be extinct.

Mutation rate	Selection strength	Average phenotype for reach of species (after 250 turns)	Turn when phenotype 1 is extinct	Turn when phenotype 2 is extinct
0.2	0.2			
0.2	0.4			
0.2	0.6			
0.2	0.8			
0.4	0.2			
0.4	0.4			
0.4	0.6			
0.4	0.8			
0.6	0.2			
0.6	0.4			
0.6	0.6			
0.6	0.8			
0.8	0.2			
0.8	0.4			
0.8	0.6			
0.8	0.8			

Questions

1. Under what conditions (if any) does the species average of two remain unchanged (in other words, no evolution occurs)?
2. Which characteristic (mutation rate or selection strength) has the greatest influence on the speed with which adaptation occurs in this simulation?
3. Compare your results with other students. Did you all get similar results? Explain any differences between your results.
4. In some simulations, the population of one type of creature would fall to zero but then suddenly reappear. How is this possible?
5. Were there any situations in which creatures with a reach of 4 actually died out?

The rise of humankind

As you can see from table 1.1, humankind has existed for only a short period of time compared to the Earth's long history. Contrary to popular belief, humans did not evolve from apes. While humans belong to the group of mammals known as *primates* — a group that also includes lemurs, gorillas, chimpanzees and orangutans — we are more like cousins than descendants. All of the primates evolved along different lines from a small tree-dwelling ancestor that existed about 65 million years ago. Over the intervening years, different types of primate evolved from this common ancestor and continued to follow different developmental pathways. The approximate time over which these separate pathways or divergences occurred is shown in figure 1.16.

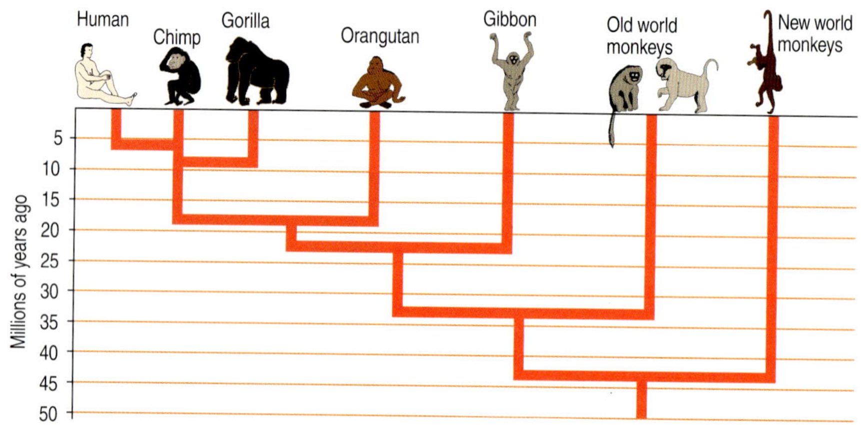

Figure 1.16:
Inferred evolution of the primate groups

As you can see from figure 1.16, the hominid line (which is the family of humans) diverged from the African ape line around six million years ago (as determined by Vincent Sarich and Allan Wilson in the 1960s). Fossils of the earliest hominids (the Australopithecines) have been found dating from around 3.9 million years ago. Over this period of time, the early hominids underwent a series of evolutionary changes:

- brains and skulls became larger (see figure 1.17)
- faces became shorter and flatter
- body size increased
- jaws and teeth became smaller
- changes to the hand allowed the fashioning and use of tools.

These changes marked the beginning of a new type of hominid around 2.4 million years ago: the *Homo* series of species that eventually gave rise to *Homo sapiens* (modern humans like us).

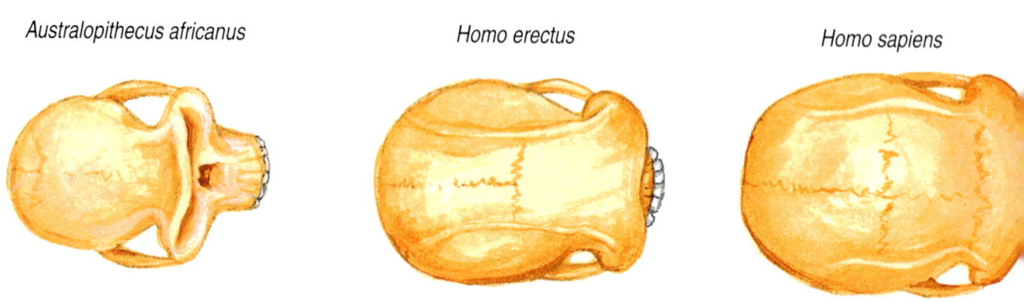

Figure 1.17:
A top view comparison of the skulls of a member of the Australopithecines (*Australopithecus africanus*) and a modern human (*Homo sapiens*). What differences can you see?

While organisms have a common name that is used in general conversation (such as kangaroo, horse or rose), these may be misleading. Koala 'bears' are not members of the bear family, for instance. To avoid confusion, scientists use a scientific name that consists of two parts when describing an organism. These scientific names (which are usually Latin words that describe a characteristic of the species) indicate how organisms are related to each other. Just as members of your family have the same surname, all species that are members of the same family have the same genus name. For example, the red wallaby (*Macropus rufus*), the Eastern grey kangaroo (*Macropus giganteus*) and the Western grey kangaroo (*Macropus fulginosis*) all have the genus name *Macropus* (meaning 'big footed'). The genus name is written first and starts with a capital letter. The second part of the scientific name is the specific name and describes a particular member of the genus. In the examples above, *rufus* means 'red' while *giganteus* means 'very large'. The specific name is written in lower case. The modern human's scientific name is *Homo sapiens*; *Homo* meaning 'man' and *sapiens* being Latin for 'thinking'. One of our earlier ancestors was *Homo erectus* ('upright man'). *Homo sapiens* and *Homo erectus* are both members of the Hominid family but belong to different species.

How did we get from *Australopithecus africanus* to *Homo sapiens* in 3.9 million years? Form into groups and use your library or the Internet to find when each of the following species is believed to have arisen:

- *Australopithecus africanus*
- *Australopithecus afarensis*
- *Australopithecus robustus*
- *Australopithecus boisei*
- *Homo habilis*
- *Homo ergaster*
- *Homo erectus*
- *Homo neanderthalensis*
- *Homo sapiens.*

Use this information to draw a timeline of human evolution. Note that some of these species became evolutionary dead ends and did not give rise to any further species. Make sure that you identify these and come up with a way of representing these on your timeline.

The cycle of life

It is important to realise that the process of evolution is continual. As conditions on Earth changed, species either adapted to the change or died out. Sometimes conditions on the Earth became so extreme that most of the life on the planet was wiped out. These are called **mass extinctions**.

Mass extinctions occur when large numbers of different species die within a short period of time, often as the result of a single catastrophic event.

The largest mass extinction occurred at the end of the Permian period 251 million years ago, when huge volcanic eruptions in what is now Siberia caused the amount of carbon dioxide in the atmosphere to be 10 times its normal level. About 90 per cent of marine organisms and 70 per cent of land organisms were wiped out as a result. The species that survived these conditions evolved to form new ones that slowly repopulated the land and the sea.

Some 65 million years ago, the dinosaurs died out when the ash and debris ejected into the atmosphere by the impact of a 10-kilometre-wide asteroid caused a massive decrease in the amount of heat and light reaching the Earth's surface. This disrupted the photosynthesis process, causing plant life to die off and the larger life forms to starve to death.

Just as species arose in the past and either died out or gave rise to new species, new species will arise in the future. As *Homo sapiens,* we could be the last stop in the evolution of the human race. Alternatively, we might evolve into a new species in the coming millennia.

Imagine a future in which the ozone layer has depleted to the extent that the surface of the Earth is flooded with UV radiation and cosmic rays. Describe what you think the humans in this future would be like in terms of functional, behavioural and structural adaptations.

Eye on the media

Our brains they are a-changing

New variants of two genes that control the development of our brains have swept through much of the world's population during the past several thousand years. In short, our brains are still evolving, and the implications of that are far from clear.

The evolution of a large complex brain has been the defining feature of the human lineage, although the size of our brain hasn't changed over the past 200 000 years. We don't yet know whether the new genetic adaptations discovered in our brains have any effect on brain size, or help to make us any more intelligent.

What's more, not everyone has the new gene variants, potentially inflaming the already controversial debate about whether the brains of groups of people from different parts of the world function differently.

'Whatever advantage these genes give, some groups have it and some don't. This has to be the worst nightmare of people who believe strongly there are no differences in brain function between groups,' says anthropologist John Hawks of the University of Wisconsin in Madison.

Generally speaking, genes that play important roles in the brain are highly conserved through evolution, with even distantly related species showing little difference in such DNA sequences. This could be because the brain is so complex and crucial to survival that the vast majority of mutations in related genes throw a monkey-wrench in the works.

Yet during the few million years of hominid evolution leading to humans, brain size tripled, even though a larger brain uses far more energy and makes childbirth dangerous. Recent genetic studies comparing humans with chimpanzees and other species have also shown that brain-related genes have evolved faster and are more active in the human lineage.

'Brains are very costly, so a large evolutionary increase suggests very strong selection on the brain,' says evolutionary neurobiologist Todd Preuss at Yerkes National Primate Research Centre in Atlanta, Georgia. But all the changes discovered so far occurred millions of years ago. Could this trend be continuing today?

Two genetic studies published in *Science* this week say it may be so. Geneticist Bruce Lahn of the University of Chicago and colleagues analysed the sequences of two genes active in the brain, one called *Microcephalin* and the other *ASPM.* Both specifically regulate brain size, as people carrying a non-functioning mutant copy of either of these genes suffer microcephaly, that is, they have brains that are structured normally but are much smaller than usual.

First the researchers sequenced the *Microcephalin* gene found in 89 ethnically diverse people, and in a single chimpanzee. The team found dozens of variants, or alleles, of the gene, but one particular set stood out. These alleles all carry a specific mutation that changes the protein the gene codes for, and are now in the brains of about 70 per cent of people, half of whom carry exactly the same version. The results suggest the mutation arose recently and spread quickly due to selection, rather than random genetic drift.

By looking at the variations in this gene, the researchers estimated that the distinctive *Microcephalin* variant arose between 14 000 and 60 000 years ago, with 37 000 years their best estimate. The new mutation is much more common among those from Europe, the Middle East and the Americas than those from sub-Saharan Africa (vol. 309, p. 1717).

The team also sequenced the *ASPM* gene from the same group, and again, among dozens of variants, found a stand-out mutation that alters the corresponding protein. The team estimate that this variant appeared between 500 and 14 000 years ago, with a best guess of 5800 years. It is present in about a quarter of people today, and is most common in Europe and the Middle East (vol. 309, p. 1720).

'The evidence for selection is compelling,' says population geneticist Rasmus Nielsen of the University of Copenhagen in Denmark. Yet it remains unclear how these genes work. Many researchers doubt there is any mechanism that selects for greater intelligence today, because human culture has effectively blocked any action natural selection might have on our brains. 'It seems to me that there is no simple positive relationship between heritable components of intelligence and the number of children people choose to have,' Nielsen says. The Chicago team's findings challenge this idea, though it is still a mystery what advantage the gene variants give.

Lahn and his colleagues are now testing whether the new variants give any cognitive advantage. Natural selection could have favoured bigger brains, faster thinking, different personalities or lower susceptibility to neurological diseases, Lahn says. Or the benefits might be counterintuitive. 'It could be advantageous to be dumber,' Lahn says. 'I highly doubt it, but it's possible.'

However, Lahn points out that since many genes are involved in shaping the brain, knowing what versions of *Microcephalin* or *ASPM* you possess will probably tell you little about how your brain works.

Nonetheless, Hawks predicts that the more we sequence active genes in the brains from people worldwide, the more we'll have to grapple with the meaning of the differences.

Source: Mason Inman, *New Scientist*, 17 September 2005, pp. 8–9.

Clarification questions

1. What studies are mentioned in this article as supporting the claim of brain evolution?
2. What disadvantages are involved in a larger brain size?
3. What functions do the genes *Microcephalin* and *ASPM* seem to have in the brain?
4. How common are these genes in the population?
5. What evidence is given that the mutation form of the *Microcephalin* gene arose fairly recently?
6. Does this article suggest that we are becoming smarter? Explain your answer.

Considering the evidence

The vast majority of the events that we have looked at so far happened long before there was anyone around to leave a record of what happened. So how do we know that any of this occurred at all, let alone *when* it occurred?

The fossil record

A fossil is the preserved remains of an organism that lived a long time in the past (see figure 1.18). The remains can be in the form of actual parts of the organism; remains in which the organic matter has been replaced with minerals through **petrification**; or impressions left behind in the rocks when the actual creature died. By studying fossils, we are able to reconstruct the appearance of the organism that formed the fossil.

Petrification is the process by which organic matter is replaced with minerals that are harder than the original material.

(a)

(b)

Figure 1.18:
(a) The fossilised bones of a dinosaur (b) The reconstructed appearance of the dinosaur

Generally, fossils are found in sedimentary rocks that formed when sediments accumulated at the bottom of bodies of water. As crust movements cause land profiles to change and water levels move up and down over long periods of time, layers are formed when new sediments settle on top of the old. The composition of these sediment layers allows geologists to determine a great deal about the climate, vegetation and atmosphere of the Earth at the time the sediment was deposited. As the layers of sediments accumulate on top of one another, their weight causes compression in the lower layers and forms rock. These rock layers are known as **strata**.

Strata (singular: stratum) are layers of rock that have formed from the compression of sediment.

Most fossils form when the remains of an organism are covered shortly after its death. The remains can be covered by ash or mud, or they can settle at the bottom of a body of water, such as an ocean or lake, when the remains of a land animal are swept into the water or when a marine animal's remains sink after death. New sediments are deposited over the

1. A dinosaur dies and is quickly covered by sediment.

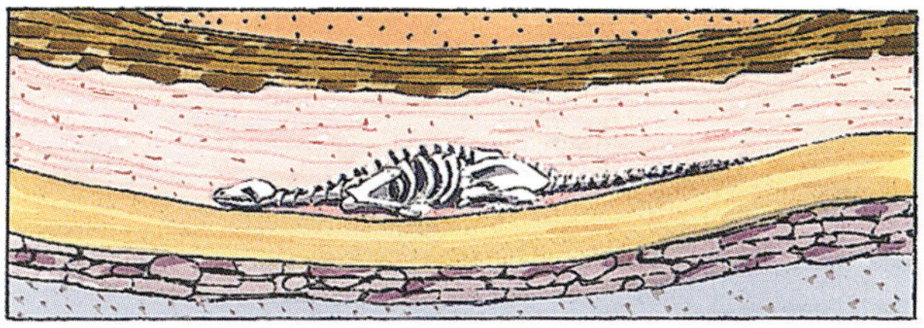

2. Over time, the sediment turns into rock. The remains of the dinosaur turn into a fossil.

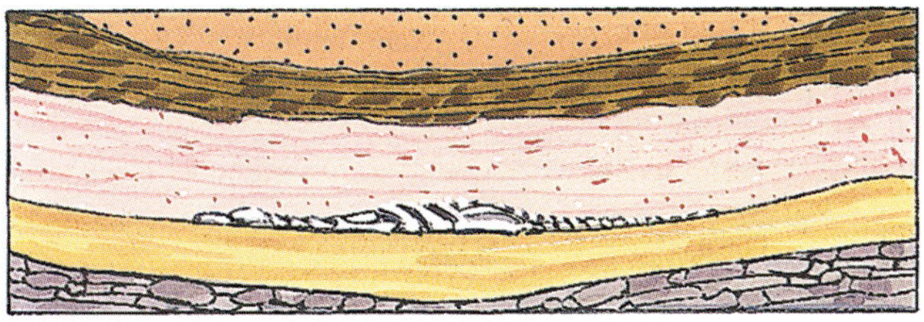

3. The fossil is flattened by the layers of rock.

4. The rock is folded and eroded and the fossil can be seen on the surface.

Figure 1.19: Formation of a fossil

covered remains. When these sediments turn into rock as the result of pressure, the remains are compressed between the rock layers. At this point, one of two things can occur:

- the remains can be preserved intact, or
- the organic matter can deteriorate, leaving a cavity in the rock (a mould). Minerals leach into the cavity over time and solidify, forming a cast (a copy of the original relic).

Fossils can also be created when small organisms are trapped in the sap of trees and later petrify to form **amber**, or when organisms are trapped in ice. This is shown in figure 1.20.

Amber is a transparent yellow, orange or greenish material that has formed from petrified tree sap. It is often used in jewellery.

Figure 1.20: (a) The remains of a woolly mammoth frozen during an Ice Age (b) An insect preserved in amber

Investigation

Creating fossils

Aim: To simulate the formation of fossils.

Materials: Plasticine; plaster of Paris; a piece of cardboard; plastic cup; ice cream stick; water; seashells, leaves or small plastic toys to use as your originals.

Method:

1. Roll the plasticine into a ball approximately 10 centimetres in diameter. Place the ball on the cardboard and flatten it with the palm of your hand until it is about 3 centimetres deep.
2. Press your selected object into the plasticine then carefully lift it out again. This should leave a clear impression of the object behind. (The impressions forms a negative of the original, called a mould.)
3. Half fill the plastic cup with plaster of Paris. Add a quarter of a cup of water and stir with the ice-cream stick until the plaster is the consistency of thickened cream. Add more water as needed.
4. Pour the plaster into the mould and allow to it to set for about 30 minutes. The plaster will heat up as it starts to set.
5. Once the plaster has cooled and hardened, carefully remove the mould.
6. Compare the plaster replica (called the cast) with the original.

To learn more about fossils in Australia go to the website for this book and click on the Fossils in Australia weblink (see Weblinks, page 309).

Eye on the media

Full splendour of 'lost' *Archaeopteryx* revealed

The world's best preserved *Archaeopteryx* fossil has surfaced after many years in a private collection. The exquisite specimen reveals that the bird was even more dinosaur-like than we originally thought.

'It's a phenomenally preserved specimen,' says Peter Makovicky of the Field Museum of Natural History in Chicago. The new specimen's foot has the hyper-extendible second toe of 'raptor' dinosaurs — the killer-talon of *Jurassic Park's* infamous *Velociraptor*.

But the specimen also reveals that the first toe was not fully reversed as in modern birds. A key skull bone also looks more dinosaurian than birdlike. In other specimens these details are either missing, entombed in the supporting rock, or obscured by cracks, says team-leader Gerald Mayr of the Senckenberg Research Institute in Frankfurt (*Science*, vol. 310, p. 1483).

The team used ultraviolet light to enhance the fossil's details. This makes minerals in the fossilised bone fluoresce, revealing the skeleton in more detail.

The story of how the specimen came to light is a mysterious one. Like the other nine *Archaeopteryx* fossils to come to light, this one came from a limestone quarry at Solnhofen in Bavaria, Germany. Its first known owner was a Swiss fossil collector who died in the 1970s, leaving behind a large collection of Solnhofen artefacts. It lay unrecognised until around 2000, when his widow tried to sell it to the Senckenberg Institute.

The institute failed to raise the money and so contacted Burkhard Pohl, fossil collector and founder of the Wyoming Dinosaur Centre in Thermopolis, Wyoming. Pohl arranged for the fossil to be sold to an anonymous private collector, who has pledged it will remain on display. After the Senckenberg has made a cast, the fossil will go to the Wyoming Dinosaur Centre on long-term loan, where it will have the distinction of being the only *Archaeopteryx* in America.

Source: Jeff Hecht, *New Scientist*, 10 December 2005, p. 18.

Clarification questions

1. What have scientists learned by examining this fossil?
2. Why was the fossil examined under ultraviolet light?
3. Why was this fossil able to tell scientists more than the other nine *Archaeopteryx* fossils that have already been found?

The study of the structure of the Earth is called **geology**.

Fossils tell us what organisms in the past looked liked, but determining when these organisms lived requires scientists to piece the past together much like a detective puts together the sequence of events in a crime based on the evidence. In this case, the evidence comes from the rocks in which the fossils are found or the rocks found adjacent to the fossils. Scientists are able to determine the timescale over which these events occurred by using radiometric dating combined with stratigraphy (the study of rock layers), paleontology (the study of life existing in former geological periods) and **geology** (the study of the Earth and its rocks). To understand radiometric dating, we must first understand how radioactive decay occurs.

Process of radioactive decay

Isotopes are different forms of the same element that have the same number of protons but a different number of neutrons.

All matter is made of atoms. There are currently 109 known types of atom, a few of which only exist for a fraction of a second under experimental conditions. Most atoms occur in more than one form. These different forms of atoms are called **isotopes**. Isotopes of the same element have the same number of protons in their nucleus but different numbers of neutrons. For example, hydrogen-1 and hydrogen-2 (also called deuterium) are isotopes of the element hydrogen. Hydrogen-1, the more common form, has only one proton in its nucleus but hydrogen-2 has one proton and one neutron in its nucleus.

Radioactive decay is the breakdown of unstable isotopes into more stable nuclei by the release of particles.

Half-life is the time taken for half of a group of unstable nuclei to decay.

In larger atoms, there are usually more neutrons than protons present in the atom's nucleus. If there are too many neutrons compared to the number of protons, then the atom or isotope becomes unstable and starts to break up into smaller, more stable atoms. When unstable atoms break up, they release smaller atomic particles and energy in the form of heat and radiation. This process is called **radioactive decay**. (A more detailed explanation of nuclear stability is given in the discussion of nuclear energy in chapter 4.)

Different materials decay at different rates, depending on their level of instability. This rate of decay is related to the **half-life** of the isotope. The interesting thing about the half-life is that it does not matter how much of the material is left — it will always take the same amount of time for half of the remaining amount to decay. Consider this example: Polonium-210 has a half-life of 138 days. If we start with a jar containing 100 grams of polonium-210 then, after 138 days, half of those 100 grams will have decayed to leave 50 grams of polonium-210. The 50 grams polonium-210 that have decayed have not simply disappeared into thin air; they have broken down into smaller atoms (lead-206 atoms in fact), releasing heat energy and radioactive particles in the process. After 138 days, the jar will contain 50 grams of polonium-210 and about 49 grams of lead-206, the other gram having been lost as heat and radiation.

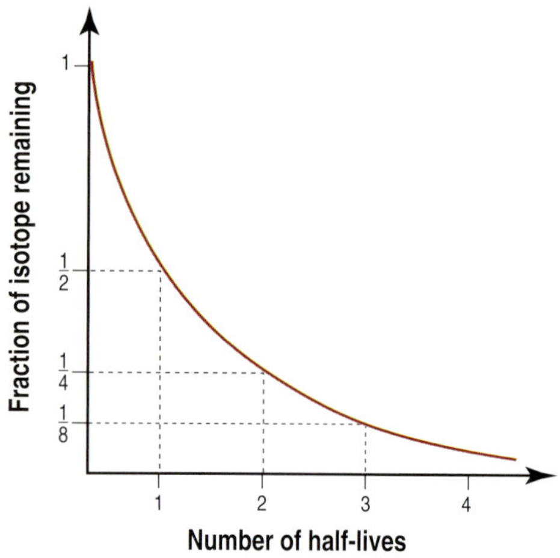

Figure 1.21: The decay of an isotope over time

After a further 138 days (a total of 276 days from when we started), half of the remaining 50 grams of polonium-210 will have decayed, leaving 25 grams of polonium-210 in the container and about 73.5 grams of lead-206. If we were to come back every 138 days, we would find less and less polonium-210 in our container and more lead-206. So long as there is polonium-210 present, no matter how small an amount, it will continue to decay. Figure 1.21 and table 1.2 show the decay of our 100-gram sample of polonium-210 over a period of time.

Table 1.2: The decay of polonium-210

Time elapsed (days)	Number of half-lives elapsed	Amount left (grams)	Proportion of sample remaining
0	0	100	1
138	1	50	$\frac{1}{2}$
276	2	25	$\frac{1}{4}$
414	3	12.5	$\frac{1}{8}$
552	4	6.25	$\frac{1}{16}$
690	5	3.12	$\frac{1}{32}$
828	6	1.56	$\frac{1}{64}$
966	7	0.78	$\frac{1}{128}$

If you look at the table, you will notice something interesting about the second and fourth columns of numbers. It becomes clearer if we rewrite our fourth column as shown in table 1.3.

Table 1.3: The decay of polonium-210, reworked

Time elapsed (days)	Number of half-lives elapsed	Amount left (grams)	Proportion of sample remaining	
0	0	100	1	$\frac{1}{2^0}$
138	1	50	$\frac{1}{2}$	$\frac{1}{2^1}$
276	2	25	$\frac{1}{4}$	$\frac{1}{2^2}$
414	3	12.5	$\frac{1}{8}$	$\frac{1}{2^3}$
552	4	6.25	$\frac{1}{16}$	$\frac{1}{2^4}$
690	5	3.12	$\frac{1}{32}$	$\frac{1}{2^5}$
828	6	1.56	$\frac{1}{64}$	$\frac{1}{2^6}$
966	7	0.78	$\frac{1}{128}$	$\frac{1}{2^7}$

Can you see the relationship now? The fraction of sample left is a function of how many half-lives have elapsed. We can therefore come up with a relationship that shows this a little more formally:

$$\frac{\text{Amount of sample left}}{\text{Original amount of sample}} = \frac{1}{2^n}$$

where n = number of half-lives that have elapsed, that is:

$$n = \frac{\text{Period of time elapsed}}{\text{Half-life of material}}$$

What this equation allows us to do is work out how much of a sample will be left after a certain period of time provided that we know its half-life. More importantly to us here, the equation allows us to determine how much time has elapsed given the fraction of sample remaining.

Let us see how this works: opening a sealed cabinet that is used to store radioactive material, a lab technician finds that only 5 grams of polonium-218 (which has a half-life of 1.3 months) is left in a container that originally contained 40 grams when it was put in the cabinet. We can work out how long the polonium has been in the cabinet by using our equations.

First, we work out the fraction of polonium-218 that is left compared to the original amount:

$$\text{Fraction left} = \frac{5 \text{ grams}}{40 \text{ grams}} = \frac{1}{8}$$

We know that this fraction will have to be related to how many half-lives have elapsed:

$$\frac{1}{8} - \frac{1}{2^n} \text{ and so } 8 = 2^n$$

As we know that $2^3 = 8$, this must mean that $n = 3$. In other words, three half-lives have elapsed. As the half-life of polonium-218 is known to be 1.3 months then:

$$\begin{aligned}\text{Period of time elapsed} &= 3 \times 1.3 \text{ months} \\ &= 3.9 \text{ months}\end{aligned}$$

Therefore, we know that the polonium-218 had been in the cabinet for nearly four months. In the next section, we will look at how scientists use all of this information to determine the age of rocks and fossils.

Investigation

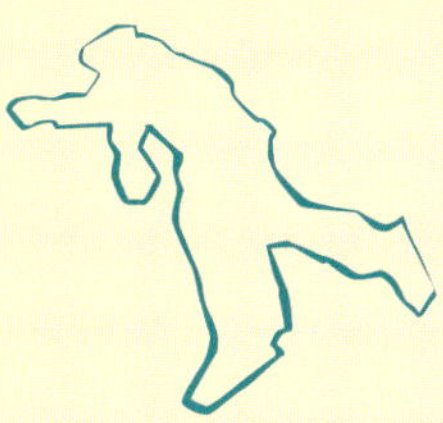

Simulating radioactive decay

Aim: To use a simple model to explain the process of radioactive decay

Materials: 100 M&Ms, clean sheet of white paper, large plastic container

Method:

1. Lay out all the M&Ms on the sheet of paper with the 'm' side down. This represents all of the atoms in the parent isotope before radioactive decay occurs.
2. Place all of the M&Ms into the plastic container and shake them up thoroughly. The M&Ms should be able to move around freely in the container, so shake them well. Pour the M&Ms onto the paper again and spread them out *without changing which side up each one has landed.*

(*continued*)

3. Separate the ones that have landed with the 'm' facing up. These represent the atoms that have decayed into a new isotope. Count them and enter the amount in a copy of the data table below. Place these to one side.
4. Collect the remaining M&Ms (the ones that landed 'm' side down) and return them to the container. Shake the container, pour out the M&Ms and repeat step 3.
5. Continue in this way until all of the M&Ms have been removed. You may need more rows in your data table than have been shown here.

Turn	Number of M&Ms in container at start of turn	Number of M&Ms that 'decayed' this turn	Total number of M&Ms 'decayed' since the start	Percentage of M&Ms that have decayed since the start	Percentage of M&Ms remaining from the start
1	100				
2					
3					
4					
5					
6					

6. Plot your results as a line graph showing the turn number on the horizontal axis and the percentage of 'undecayed' M&Ms remaining at the end of the turn on the vertical axis.

Discussion questions

1. Compare your results with the rest of the people in your class. Did you all get similar results? Explain why there would be some variation.
2. Your teacher will combine the class results. Plot these results on a line graph using the same axes as you used in step 6 of the method above. Compare this graph shape to the one you drew in step 6.
3. In this simulation, identify how the following were represented:
   - parent atoms
   - daughter atoms
   - the decay process
   - the half-life.
4. Was it easier to guess how many M&Ms would emerge 'm' side up when there were a large number of M&Ms or when there were only a few? Give an explanation for this.

Internet activity

To see a simulation of radioactive decay, go to the website for this book and click on the Radioactive decay weblink (see Weblinks, page 309).

By adjusting the slider, you can specify the half-life of the isotope and then watch how quickly it decays. Use the simulation to explore the relationship between an isotope's half-life and the time taken for it to be considered completely gone. A copy of the data table at the top of the facing page may be helpful for this:

Half-life of isotope(s)	Time taken for complete decay(s)
5	
10	
15	
20	
25	
30	
35	
40	

1. Plot the values in the table as a graph, showing the half-life on the horizontal axis and the extinction time on the vertical axis. What relationship between the two quantities does the graph shape suggest?
2. Use the data in your table to determine what relationship there is (if any) between these two quantities and put it in the form of an equation. Use the letter *H* for the half-life and *E* for the extinction time.
3. Use either your graph or your equation to determine how long an isotope would take to reach extinction if it had a half-life of:

   (a) 7 s (b) 45 s.

Radiocarbon dating

All life on our small blue planet is based on the element carbon. While the majority of this carbon is in the form of the stable isotope carbon-12, some of the carbon in living things is in the form of radioactive carbon–14. Carbon-14 is made when cosmic rays interact with the nitrogen-14 in the atmosphere, causing it to effectively lose a proton from its nucleus. The carbon-14 atom combines with oxygen in the atmosphere to form radioactive carbon dioxide, which is taken in with normal carbon dioxide by green plants during photosynthesis. When animals eat the green plants, they also take in the carbon-14 as do the animals that eat these animals. In this way, all living things on Earth take in carbon-14 (see figure 1.22).

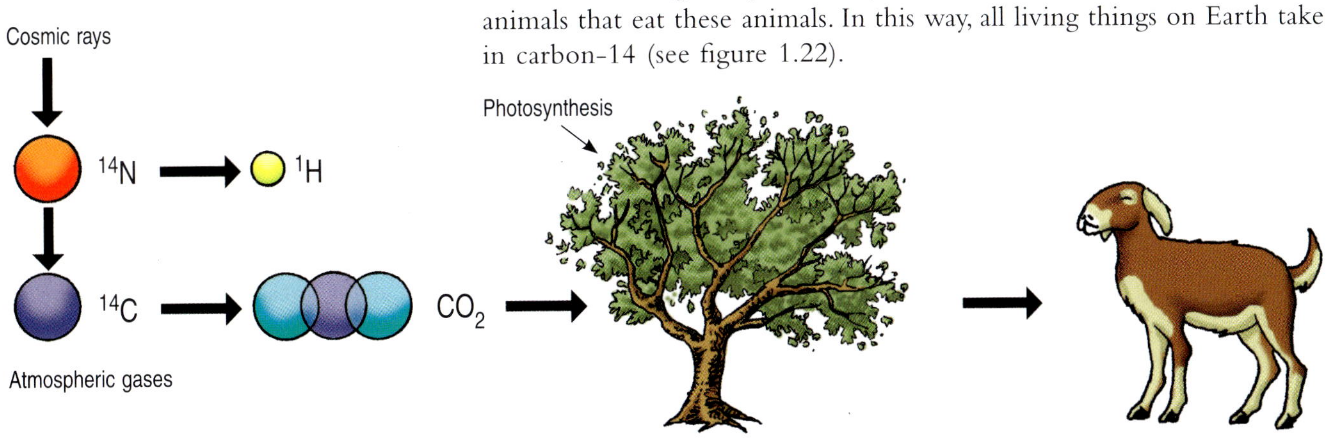

Figure 1.22
How carbon-14 is produced and enters the food chain

The amount of carbon-14 in living things is constantly being lost or decayed and then replenished, so it tends to stay at a fairly constant level while the organism is alive. When the living thing dies, however, carbon can no longer be taken in and the radioactive carbon that was present in the organism at the time of death begins to decay. By measuring the amount of carbon-14 in the organic remains and comparing it to the amount that would have originally been present, it is possible to use the known half-life of carbon-14 (5730 years) to determine how long the organism has been dead. This process can be used on any organic material or relics with organic content, including bones, parchment, wooden implements and fabric scraps.

For example, a bone is found in which the carbon-14 content is 1/16 of what it is expected would have been present originally. As was shown in the previous section, this allows us to determine the value of n (the number of half-lives that have elapsed):

$$\frac{1}{16} = \frac{1}{2^n} \text{ therefore } 16 = 2^n$$

This allows us to find that $n = 4$, so the organism which the bone belonged to has been dead for four half-lives, or roughly 23 000 years.

A **Geiger counter** is a device that measures the rate at which a material is undergoing decay. It does this by detecting the radioactive particles produced during the decay.

Accelerator mass spectrometers (AMSs) are devices that allow the number of carbon-14 atoms in a sample to be counted with high accuracy.

There are several different ways in which the amount of carbon-14 in a sample can be determined. Most of the time, samples are burned with oxygen in a sealed canister to form a gas. This gas is then cooled until it condenses into a liquid. A **Geiger counter** is used to detect the rate at which the carbon-14 is decaying, from which the number of carbon-14 atoms present can be determined. More recently, **accelerator mass spectrometers (AMSs)** are used. AMSs were originally used by physicists investigating subatomic particles, and they can determine the number of carbon-14 atoms in a sample by essentially counting them very quickly. This method is expensive, however.

There are limits to radiocarbon dating:

- It is not very accurate with samples that are more than 50 000 years old.
- Contamination of samples by carbon emissions from heavy industry or motorways can lead to inaccurate dates.
- It is assumed that cosmic radiation enters the atmosphere at a constant rate, causing the production of carbon-14 to occur regularly over time. This may not be the case.

Eye on the media

Cold war clues

Atomic tests allow carbon dating of baby boomers

Frustration was the mother of invention for Jonas Frisen of the Medical Nobel Institute in Stockholm, when he set out to map the ages of various body parts in living people. As a neuroscientist working toward regenerating brain tissue, he would have found it handy to know whether some or all of the human brain ever regenerates naturally and, if so, how often. 'I was pretty annoyed,' Frisen says, because the question was unanswerable for humans: the techniques used to tag cells and watch their life cycles in animals employ toxic chemicals that could not ethically be used on people.

Then Frisen learned of a natural tag unique to people born after 1955, when aboveground testing of nuclear weapons increased substantially. The explosions, which stopped after the 1963 Limited Test Ban Treaty, threw enormous amounts of carbon-14 isotope into the atmosphere that quickly diffused around the globe.

Plants incorporated the carbon-14 into their cells, animals ate the plants, and people ate both, absorbing the isotope into their own cells and creating a trail Frisen could follow.

He teamed up with Bruce A Buchholz of Lawrence Livermore National Laboratory, who had already used atomic-era carbon-14 traces to determine the age of plaques in brain tissue samples from Alzheimer's victims. Frisen's project was much more complex, however, because cells are chock-full of proteins that are continually being produced and degraded. To learn the true age of a cell, Frisen needed something that is formed at the moment of the cell's birth and remains stable throughout its life, which meant he needed to isolate and date its DNA.

By measuring the amount of carbon-14 incorporated into the DNA molecule at its creation, then correlating it with atmospheric carbon-14 levels, Frisen finally had a test that could give him answers. As it turns out, the team reported in the journal *Cell* this past summer, many parts of our bodies are much younger than the whole. Jejunum cells from the gut tissue of subjects in their mid-30s were less than 16 years old. Skeletal muscle from two subjects in their late 30s was just over 15 years old.

But the big surprises were in the brain, where Frisen tested cells from the cerebellum and occipital cortex. Neurons in both tissues dated to a period at or near the subjects' birth, indicating that those parts of the brain are formed at the beginning of life and do not normally turn over.

Frisen thinks it is too soon to know whether his finding means less hope for therapeutic approaches to regenerating damaged brain tissue. 'I wouldn't say so. Perhaps a little less hope than if there was ongoing vigorous regeneration,' he admits. But he is eager to do follow-up studies on brain tissue from stroke victims, for example, to see if those neurons do regenerate after injury. He also plans to study heart muscle and the insulin-producing beta cells of the pancreas, both tissues whose regenerative capacity is hotly debated and both of great therapeutic interest. Frisen hopes other researchers will employ the technique as well to study their favourite organs.

Bomb-era carbon-14 levels started dropping rapidly after 1965, sloping downward by 50 percent every 11 years. After 1990, according to Buchholz, the signal left in human cells is faint. Scientists can use this technique indefinitely to date stored tissue samples from the period, Frisen notes. And Buchholz is now employing the method for forensic work. Given the 'huge segment of the population' born during the relevant era, Buchholz-a baby boomer himself-says, 'the technique will certainly be very useful for the rest of my career.'

Source: Christine Soares, *Scientific American*, December 2005, pp. 11–12.

Clarification questions

1. Who or what is a 'baby boomer'?
2. Why do different cells have different ages in the same body?
3. Why do baby boomers have increased levels of carbon-14 in their bodies?
4. How is this increased level of carbon-14 of use to scientists who are trying to determine the true age of a cell?

Radiometric dating

Whereas radiocarbon dating can be done on an artefact that contains organic matter, other forms of radiometric dating allow the age of inorganic matter (such as rocks) to be determined. One of the most commonly used isotopes for dating rocks is potassium-40.

Potassium is a common element in the minerals that are found in igneous rocks. These rocks form in two ways:

- The molten magma under the Earth's crust cools slowly and becomes solid while still underground. Later earth movements and erosion cause the rock to be exposed. These rocks have large crystals of minerals.
- Molten magma flows onto the Earth's surface (through vents or volcanoes) as lava, which then cools quickly and solidifies. Rocks made this way have small crystals.

The majority of the potassium found in igneous rocks is stable potassium-39, but around 0.012 per cent is radioactive potassium-40. The half-life of potassium-40 is approximately 1.25 billion years. When potassium-40 decays, it forms the inert gas argon-40, which accumulates in the rock.

When dating a rock sample, the age is determined by comparing the amount of potassium-40 that was present when the rock was last molten (calculated from the amount of stable potassium-39 in the rock) with the amount that has decayed since that time (calculated from the amount of argon-40 that has accumulated in the rock). This allows scientists to determine how many half-lives have elapsed since the rock was last molten and therefore find the age of the rock.

By using radiometric dating, the world's oldest mineral crystals (found in Western Australia) were found to be about four billion years old. As a result, we know that the Earth was formed *at least* this long ago.

We have examined only a couple of dating methods. Divide into groups and research these other dating methods:

(a) fission track dating
(b) stratigraphy
(c) fluorine analysis of bone
(d) biostratigraphy
(e) dendrochronology
(f) paleomagnetic dating
(g) thermoluminescence dating.

Report your findings back to the class.

Using the geological timescale

The age of fossils can also be determined by looking at the rock layers in which they were found. When considering the way in which rock layers form, geologists interpret a set of strata in terms of three underlying principles:

- *Principle of uniformitarianism:* the geological processes that are observed working today in the Earth's crust are the same as those that were in operation in the crust in the past.
- *Principle of superposition:* sedimentary rock layers are deposited in chronological order, with the oldest rock layers being lower down than the younger ones at the top. In this way, a set of rock layers (called strata) that have not been disturbed by geological upheaval can be viewed as a timeline from the past to the present.
- *Principle of original horizontality:* sediments in rock layers are deposited in horizontal layers. Changes in the orientation of the layers are the result of movements in the crust.

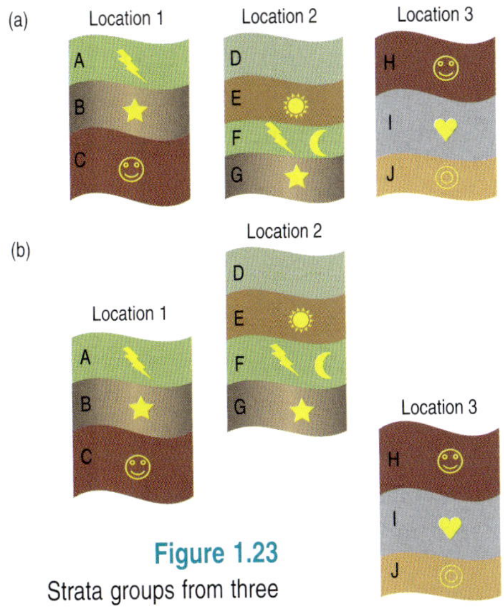

Figure 1.23 Strata groups from three locations with similar layers: (a) unaligned (b) aligned

In the 19th century, geologists noticed that certain fossils or groups of fossils occurred in distinct rock strata, even those sets of rock strata that were separated by large distances. By assuming that the fossil is approximately the same age as the rock that surrounds it, a timescale may be devised that allows the relative dating of the fossil. For example, *Archaeopteryx* fossils have only been found in rock layers that date to the middle of the Jurassic period allowing the radiometric dating of *Archaeopteryx* to be about 150 million years ago. Fossils found in layers above the *Archaeopteryx* will be younger than this, while those found in lower layers will be older. Also, the fossil of an unknown organism can be dated if it is found in the same rock layer as a fossil whose age we do know or if it is found in a rock layer that lies between strata containing known fossils.

Consider the three strata groups in figure 1.23. Each strata group came from a different location.

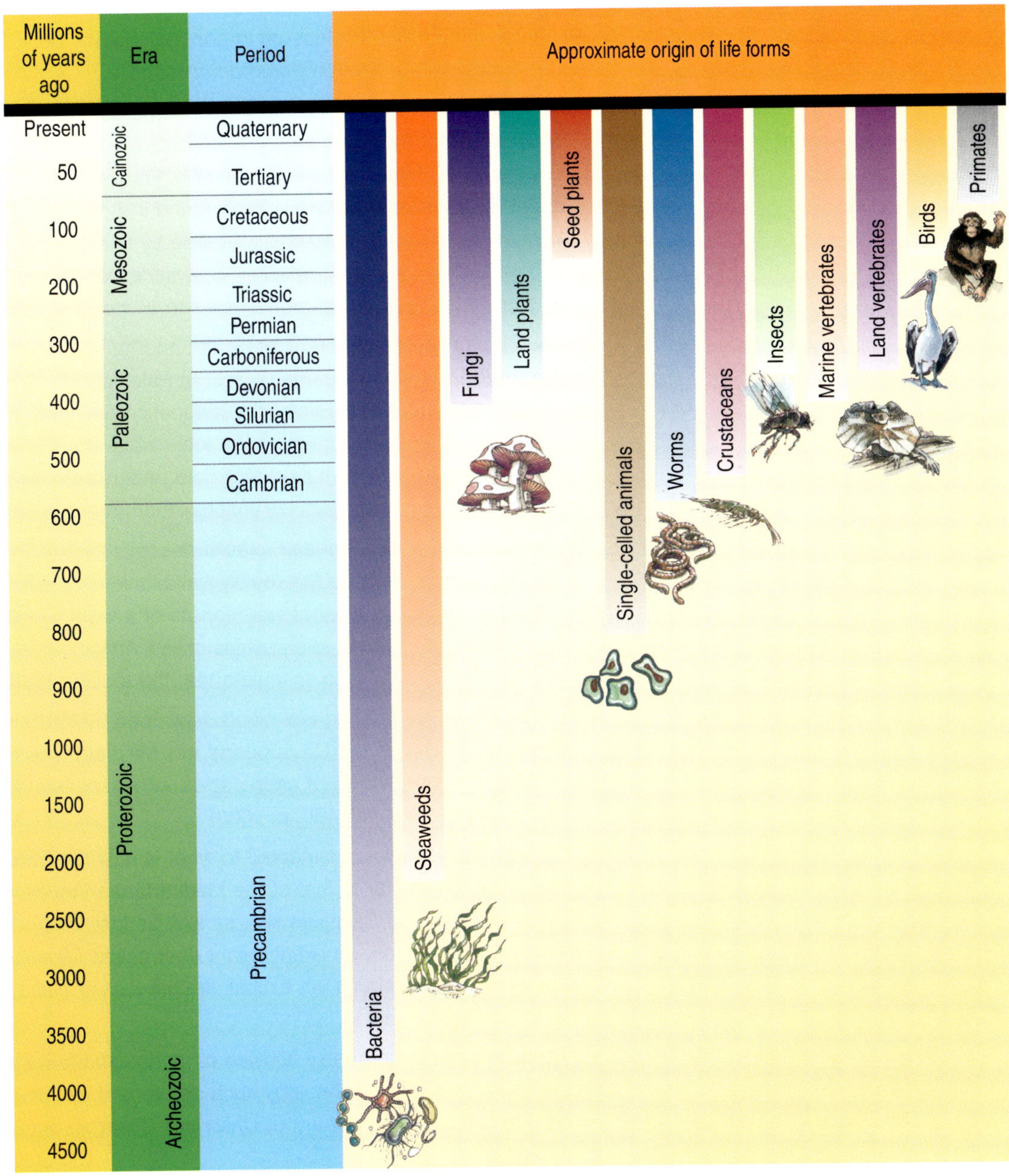

Figure 1.24 A simplified representation of the fossil record

If we realign the blocks so that similar rock types containing similar fossils are lined up, we can see that the youngest rock layer is layer D while the oldest is layer J. We can also see that layers A and F date from the same time period, while C and H are also from similar time periods. Note that layer F contains a fossil that layer A does not.

Figure 1.24 shows a simplified representation of the fossil record as collated so far.

Eye on the media

European history has a new chapter

Humans may have colonised northern Europe 200 000 years earlier than previously thought. Stone tools found in eastern England suggest that humans were there at least 700 000 years ago.

'We don't know for sure what species it was,' says team member Chris Stringer of the Natural History Museum in London, 'but my bet is it's an early form of *Homo heidelbergensis* or *Homo antecessor*.'

H. heidelbergensis is known to have been present in central Europe about 500 000 years ago. Bones were first discovered in 1907 near Heidelberg, Germany, and have since been found in France and Greece. Hominid remains about 800 000 years old have been found in Spain and Italy, indicating that early humans had colonised southern Europe by this time. These early humans have been classed as another species, *H. antecessor*, though arguments remain over whether it is a really separate species to *H. heidelbergensis*.

The 32 stone tools, made of black flint and many of them still sharp, were discovered by amateur archeologists at Pakefield, Suffolk. They have been dated using several methods. Firstly, the magnetic polarity of iron-containing minerals in the sedimentary rocks where the tools were found is aligned north–south, just as it is today. The Earth's magnetic field underwent a polarity reversal 780 000 years ago, so the site must be younger than that.

The tools were found beneath glacial deposits laid down during a period 450 000 years ago when the region was blanketed in ice, so they must be older than this. Also present were fossils of a water vole *Mimomys*, which was superseded by another vole species called *Arvicola* around 500 000 years ago. This leads the authors to speculate that the tools are around 700 000 years old.

A new amino-acid dating technique developed by Kirsty Penkman of the University of York in the UK supports this estimate. The method was used to measure the breakdown of amino acids within shells of a freshwater snail found at the site (*Nature*, vol. 438, p. 1008).

Back then Britain was connected to what is now the European mainland and had a climate similar to that of the Mediterranean today. The researchers found hippo fossils at the Pakefield site, as well as fossils of other warmth-loving species such as lion, an extinct giant deer called *Megaloceros dawkinsi* and *Palaeoloxodon antiquus*, an extinct straight-tusked elephant. Rhinos and hyenas also roamed the region.

The warm climate probably allowed early humans to migrate northwards without the need to develop technology such as fire and clothing or to adapt to a colder climate, says Anthony Stuart of University College London, who coordinated the project.

But the climate got the better of them eventually. 'People couldn't settle here long-term,' says Stringer. 'They would have been swept away by the cold stage that followed about 100 000 years ago.'

Source: Rowan Hooper, *New Scientist*, 17 December 2005, p. 11.

Clarification questions

1. What evidence has been cited for the belief in an earlier human colonisation of northern Europe?
2. What dating techniques were used to determine the age of the stone tools discovered?
3. How were early humans able to travel between Britain and Europe without boats?
4. What evidence was used to back up the claim that northern Europe 700 000 years ago had a climate 'similar to that of the Mediterranean today'?
5. What species were these early humans?

End of chapter questions

1. Define the following terms:
   (a) accretion
   (b) planetismal
   (c) nebula
   (d) Kuiper Belt.
2. What evidence exists for Earth being over 4.5 billion years old?
3. Describe how radiocarbon dating is able to determine the age of a fossil with organic matter.
4. One of the criticisms levelled at the movie *Jurassic Park* was that the majority of dinosaurs in it (*Apatosaurs, Tyrannosaurus rex, Triceratops* and *Velociraptors*) were not from the Jurassic period at all. What period do you think would have been a more accurate name to use for the park?
5. How long did the Hadean eon last?
6. Arrange the following units of time from longest to shortest:
   period epoch era eon year.
7. In what era are we living?
8. How is biostratigraphy different from stratigraphy?
9. For which of the following items, would it be possible to use radiocarbon dating methods? Give explanations in cases where radiocarbon dating would be inappropriate.
   (a) A dinosaur tooth
   (b) The remains of a stone spearhead
   (c) The skeleton of a murder victim buried wearing an Apple ipod
   (d) A medieval manuscript written on vellum (calfskin)
   (e) A piece of petrified wood
10. An archeology student is shown a cave painting by a local guide that has never been publicised before. The painting shows a scene of humans hunting a dinosaur with spears done in some kind of pigment. Explain how the student can tell that the cave painting is a forgery.
11. The following observations have been made on the Galapagos Islands:
    - Only one species of rat is found on these islands.
    - More than 13 species of finches differing in shape and size of their bills and in their preferred diets inhabit these islands.
    - Land iguanas with rounded tails and unwebbed toes live on these islands.
    - Seaweed-eating marine iguanas with flattened tails and webbed feet live on the rocky shores of these islands and nowhere else in the world.

    Given that these observations are true, assess whether the statements below are most likely true or false (explain your decision in each case):
    (a) Rat populations do not evolve.
    (b) Rats probably reached the Galapagos Islands before the first finches arrived.
    (c) Marine iguanas could not survive anywhere else in the world.
    (d) Marine iguanas evolved from ancestral land iguanas.

(e) At various times in the past, different species of finches migrated to the Galapagos Islands from South America until a total of more than 13 kinds had arrived.

(f) Flattened tails and webbed feet developed in iguanas as a result of their trying to swim.

12. Your younger cousin proudly announces that he has learned from a book that all humans are descended from monkeys and that is why we look like them. How would you explain the flaws in his understanding?
13. The rock strata in figure 1.25 have been disturbed by a fault. Using the diagram, determine:
    (a) the youngest layer of rock
    (b) the oldest layer of rock.

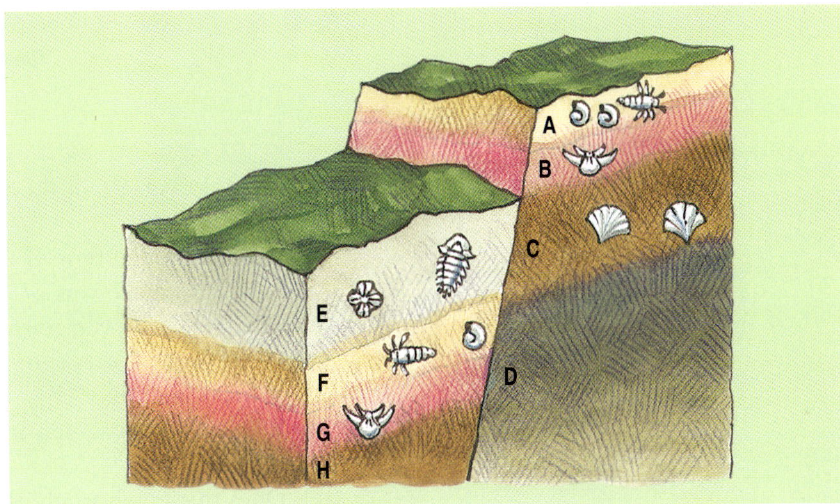

Figure 1.25
Rock strata on either side of a fault

14. Approximately what fraction of the original amount of radioactive carbon-14 would you expect to find in:
    (a) a wooden Aboriginal spear 11 000 years old?
    (b) a skull 23 000 years old found in a cave?
15. Why are the rocky planets found closest to the sun while the gas giants are much further away?
16. Describe the changes in the composition of the atmosphere during the Precambrian.
17. Figure 1.26 shows individuals from two very different parts of the world. The man in (a) comes from a hot area in Africa while the man in (b) is an Inuit from the cold Arctic regions. Describe how the populations that these two men represent have adapted to fit their environment.
18. What conditions need to occur for a fossil to form from the remains of an organism?
19. Why are more fossils of ocean organisms found than those of land organisms?
20. Figure 1.27 shows the average time period between critical events in history as a function of time.

Figure 1.26
Contrasting members of two populations living in different climates

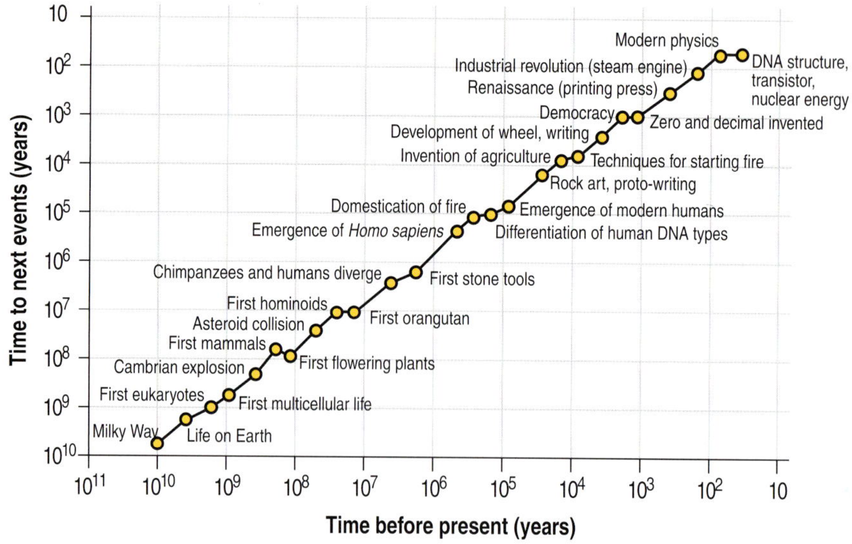

Figure 1.27

Critical events represented as a graph

(a) What do you think the values on the vertical axis (time to next event) signify?

(b) Why did the author of the graph have the values on the axes increasing in powers of 10?

(c) What was the average time interval between critical events:
- (i) 1 000 000 years ago?
- (ii) 100 years ago?

(d) When does this graph indicate that the domestication of fire occurred?

(e) Does this graph imply that there are now smaller intervals of time between critical events or larger intervals of time?

(f) What does this graph indicate about the rate at which human discoveries have been made over the last one million years?

(g) The original title of this graph was 'Growth of order and complexity'. Do you think that this is an appropriate title for this graph? Explain your answer.

2

The restless Earth

The Earth that we stand on seems very firm under our feet and it is very easy to forget that underneath us the seething interior is far from placid. It is only when we see footage of erupting volcanoes, or read in the newspaper of thousands who are dead or homeless as a result of terrible earthquakes, that we are reminded of just how fragile our existence is on the thin skin of the Earth. Where do these powerful forces come from?

Structure of the lithosphere

You may recall from chapter 1 that the Earth was molten in its early stages. As the Earth cooled and solidified, the materials from which it was made underwent a process of separation. Five million years after the Earth was formed by the accretion of rocky planetismals, heavier elements such as iron and nickel sank towards the centre of the planet while the light rocky silicates floated on the surface, hardening as they cooled to form the Earth's crust. The various materials separated to form four layers: the crust, the mantle, the outer core and the inner core.

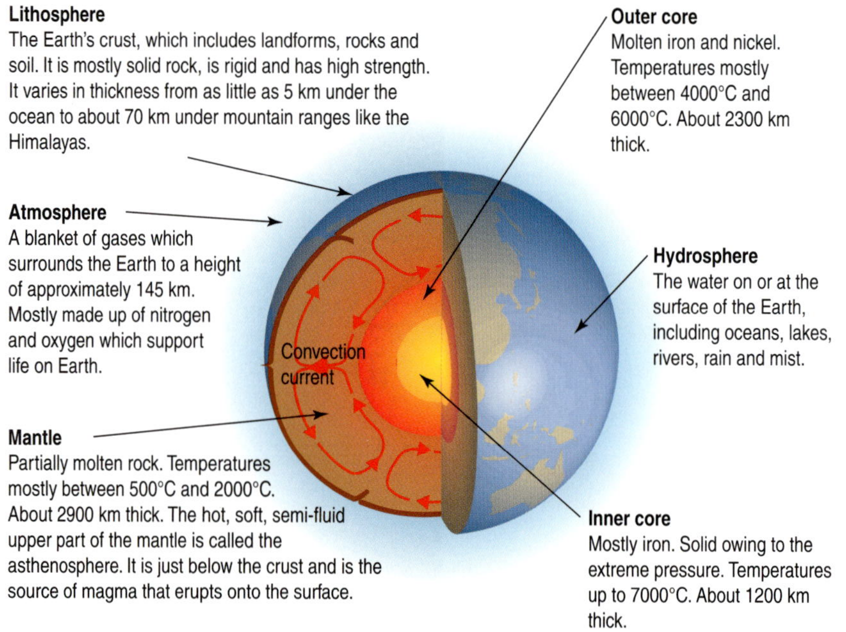

Figure 2.1:
The structure of the Earth

The Earth's crust

The Earth's outer crust is very thin — about 35 kilometres deep on average under the continents and only about 5–10 kilometres under the oceans. If you were to cut an apple in half and compare the thickness of the peel to the radius of the apple, then you would have a good model of how thin the crust is compared to the radius of the Earth.

There are two types of crust structure — **oceanic crust** and **continental crust**. These two types of crustal material are formed in very different processes and are composed of quite different materials. Oceanic crust is between 6 and 8 kilometres deep, and tends to be made of dense igneous rock such as basalt. The oceanic crust material is constantly being formed by magma, which oozes up between the cracks in the lower crust at the bottom of the oceans and forms 'pillows' of basalt. Continental crust, on the other hand, is much less dense than oceanic crust, and can be between 35 and 70 kilometres thick. It is made of a wide range of rock types including igneous rock such as granite, as well as sedimentary and metamorphic rock. Much of the continental crust was formed early in the Earth's development. As the name implies, it is this type of crustal material from which the continents are made.

Oceanic crust describes the thinner regions of the Earth's crust, made of dense igneous rock such as basalt. Oceanic crust is constantly being formed.

Continental crust describes the thicker regions of the Earth's crust, made of a variety of less dense rock types.

The **lithosphere** is a series of plates made up of the crust and the top layer of the mantle, forming the outermost skin of the Earth.

A **seismologist** is a scientist who studies earthquakes.

The **Mohorovicic discontinuity** (or '**Moho**') is the region where the bottom of the Earth's crust meets the upper layers of the mantle.

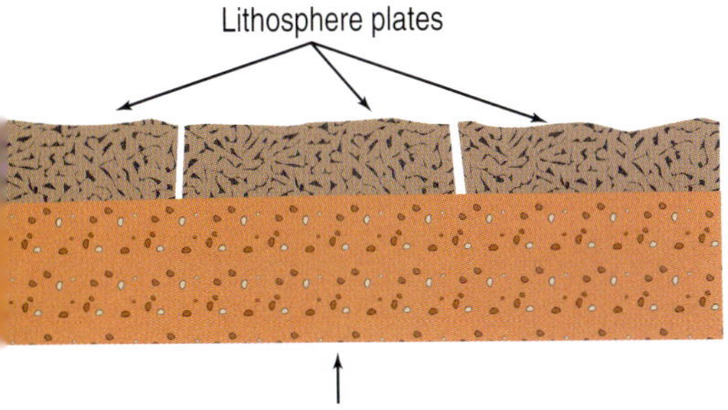

Figure 2.2:
The lithosphere plates

The **asthenosphere** is a section of the mantle on which the lithosphere plates move.

The lithosphere

The crust and part of the upper mantle together make up a structure known as the **lithosphere**. In Greek, *lithos* means 'rock'. Where the bottom of the crust material meets the upper mantle is a region that **seismologists** refer to as the **Mohorovicic discontinuity** (or '**Moho**'). Rather than being a continuous shell like the coating on the outside of a Jaffa lolly, the lithosphere is made up of 10 major and many minor separate plates much like a giant jigsaw puzzle (see figure 2.2).

All lithosphere plates contain some mantle materials at their bottom layers, but some have only oceanic crust over this while most have continental crust. However, due to the larger sizes of the oceanic plates, there is a greater proportion of oceanic crust than continental. The lithosphere varies in thickness from an average of 100 kilometres under the older ocean basins up to 300 kilometres under the continental shields.

The plates of the lithosphere float on a layer of magma that can be up to 200 kilometres thick and is called the **asthenosphere** (see figure 2.3). The asthenosphere is actually more of a solid than a liquid even though it is very hot — over 1500°C — simply because of the high pressure that the material is under. This high pressure squeezes the particles of the mantle much closer together than they would be at that temperature if they were on the Earth's surface where the pressure is much lower.

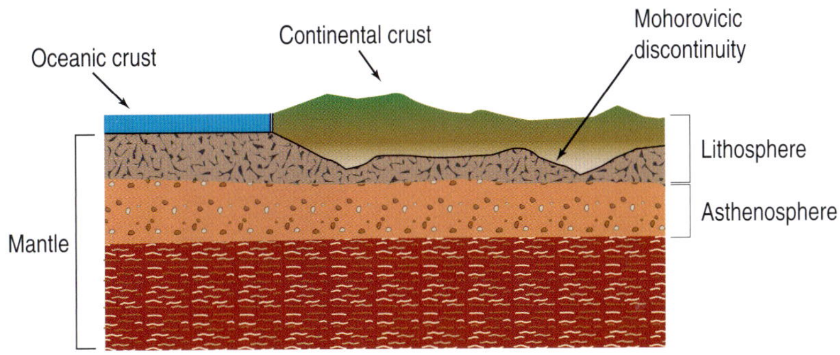

Figure 2.3:
The lithosphere and the asthenosphere

Pangaea is the name for the original supercontinent that contained all the present continents as a single mass. It broke up into Laurasia and Gondwanaland.

Continental drift

The *theory of continental drift* was first proposed in 1912 by a German meteorologist, Alfred Wegener, who theorised that the continents were not always in their current locations. This theory was developed when Wegener considered a number of geological facts. First, many of the continents were shaped like jigsaw puzzle pieces that could be fitted together. Second, there were matching rock types and fossils on these interlocking masses. In his paper published in 1915, Wegener theorised that the continents were once all part of a single massive supercontinent, which he named **Pangaea**. He thought that this huge mass had broken up over millions of years and that the smaller continents had drifted away from each other into their present positions. While many scientists at the time were openly critical of Wegener's theory, later dating techniques and geological methods led scientists in the last few decades to discover that the basic theory was valid.

Wegener's Pangaea existed about 250 million years ago. Over the next 70 million years, this giant landmass separated into two smaller masses:

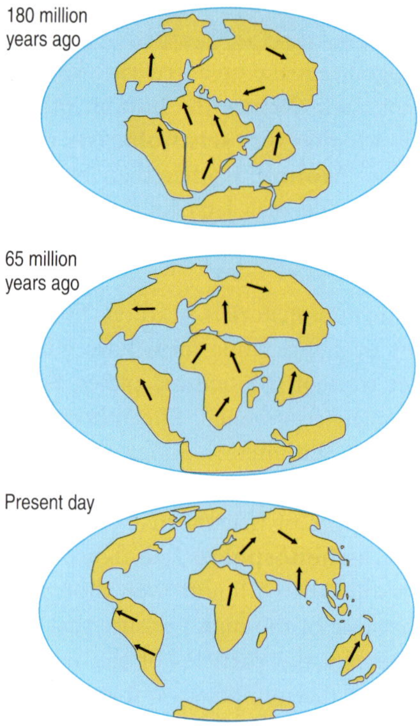

Figure 2.4:
The break-up of the continents

Laurasia (the northern mass comprising what is now Europe, Asia, North America and Greenland) and Gondwanaland (the southern mass made up of Africa, South America, Australia, Antarctica and India). These landmasses continued to break up into smaller continents and move until they assumed their present familiar locations (see figure 2.4).

So the position of the continents has changed over time and, in the future, they will change again. In fact, Australia is currently moving towards the Asian continent.

Figure 2.5:
The future location of Australia?

1. Photocopy figure 2.6, which shows the outlines of the southern continents that formed Gondwanaland.
2. Carefully cut out the continents along the dotted lines. These represent the edges of the continental shelves.
3. Paste the shapes onto cardboard.
4. Try to fit the continents together to approximate their positions when they were part of Gondwanaland. Use the following pieces of information to help you do this:
   - The east coast of South America fits into the west coast of Africa. Rocks of a similar age and structure are indicated.
   - Grooves left by glaciers are shown by arrows. The arrows show the direction in which the glaciers moved.
   - The locations of basalt outcrops formed about 100 million years ago are shown.

- Places where fossils of the extinct reptile *Mesosaurus* (which lived about 220 million years ago) have been found are shown.

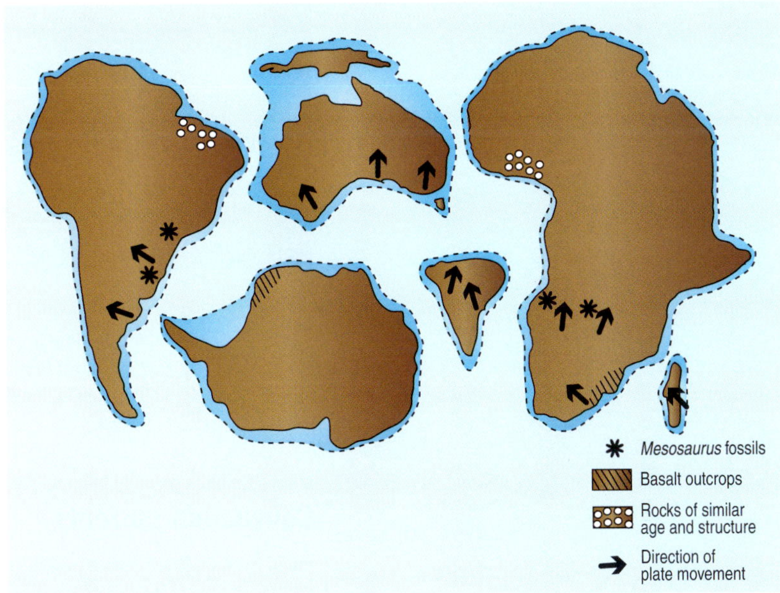

Figure 2.6: The continents of Gondwanaland

Plate tectonics

Continental drift formed the basis of the *theory of plate tectonics*. The word *tectonic* comes from a Greek word meaning 'to build'. The theory of plate tectonics explores the way in which the plates of the lithosphere interact with each other and how this movement affects the continents mounted on those plates. Plate tectonics provides the mechanism for continental drift that Wegener was unable, in his time, to explain.

Because the continents are made of lighter sections of continental crust floating on the top of the plates, they tend to lag behind as the tectonic plates are moved around by the convection currents. When oceanic crust rocks are subducted at an oceanic–continental plate boundary, the continental crust layers remain at the surface while the denser oceanic crust subducts, or slides, underneath.

Convection

Convection is the cycle in which hot, less dense material rises and cooler, denser material sinks.

While the plates of the lithosphere are fairly rigid, the asthenosphere on which they float is more fluid. The magma in this region is hot as a result of radioactivity lower down in the Earth. This heat is transferred through the liquid mantle by the process of **convection**. Hot magma lower in the mantle expands and becomes less dense, so it rises to the top of the mantle where the asthenosphere is. As it reaches these cooler upper layers, it loses its heat and starts to become cooler. This causes the magma to become denser again and it sinks towards the bottom of the mantle again (see figure 2.7). This cycle of heating and cooling gives rise to convection currents, movements in the magma that act much as sea currents do. These convection

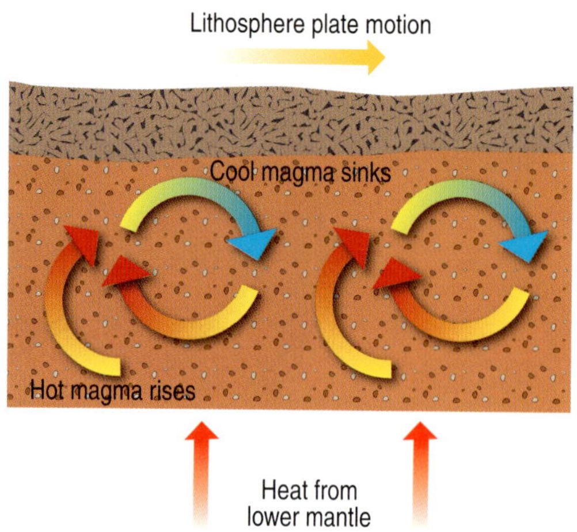

currents move the plates as if they were on a conveyer belt. The Earth's plates move at different speeds but plate motion is generally very slow — only a few centimetres per year. The plate on which Australia is located moves 5–6 centimetres per year and is the fastest moving of the lithosphere plates.

Figure 2.7: Convection currents

Investigation

Convection currents

Aim: To observe the process of heat transfer through convection and to observe the motion of convection currents

Materials: Incandescent desk lamp with 75-watt bulb; rheoscopic fluid; French square bottle; ring stand

Method:

1. Shake the bottle of rheoscopic fluid to ensure it is thoroughly mixed.
2. Fill the French square bottle completely with rheoscopic fluid, being careful not to leave any pockets of air. Screw the cap on tightly.
3. Place the bottle on the ring stand as shown in figure 2.8.
4. Place the lamp directly under the centre of the bottle as shown in the figure.
5. Turn on the lamp and observe the convection currents that begin to form from the side of the bottle and the top.

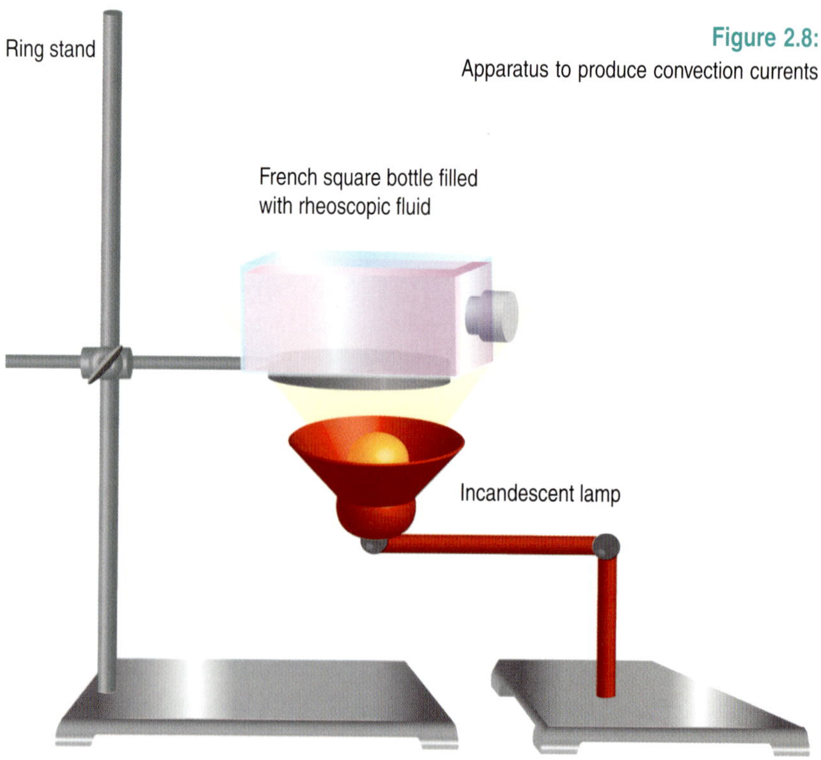

Figure 2.8: Apparatus to produce convection currents

Discussion questions

1. Draw a simplified diagram showing the motion of the convection currents from the side and when viewed from the top. How do you explain the differences in the motion shown in the two diagrams?
2. Do you think that the convection in the mantle would be faster or slower than those produced in this investigation? Explain your reasoning.
3. A material called rheoscopic fluid was used in this investigation. What other materials do you think could be used instead? Try them out and see which would be the best replacements.

A **plate boundary** occurs where one tectonic plate meets another.

As a result of convection, the plates move relative to each other. The region where one plate meets another is called a **plate boundary**, and it is along these boundaries that the forces exerted by plate motion become most apparent. The motion of the plates causes geological change throughout the crust. The three main types of plate interaction are divergence, convergence and translation.

Plate divergence

Divergence occurs when two plates are moving away from each other.

Divergence or 'spreading' occurs when two plates are being forced away from each other by the injection of magma from the upper mantle (see figure 2.9). This magma forms basalt at the edge of the plates, becoming new ocean floor. A ridge forms in the middle of the ocean floor where new material is being made. The youngest sections of the ocean bed are closest to the ridge while the oldest sections are up against the continental edges.

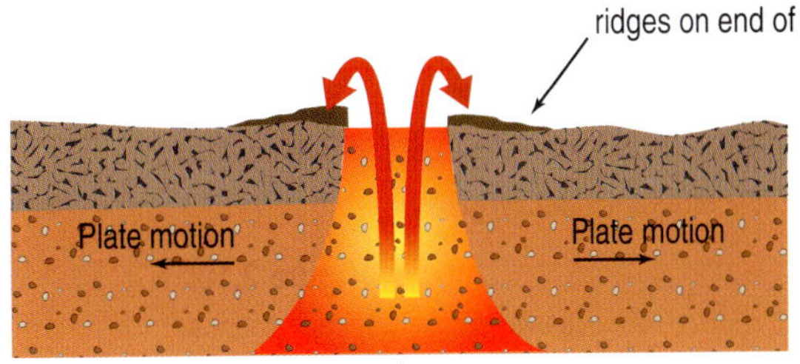

Figure 2.9: Divergence between two plates

One such ridge is the Mid-Atlantic Ridge, which is a submerged mountain range that extends from the Arctic Ocean to the south of the African continent. At the northern section of the ridge, the North American Plate and the Eurasian Plate are being pushed away from each other while, in the south, a similar thing is happening with the South American Plate and the African Plate. As a result of these motions, pillow basalt is being intruded at the Mid-Atlantic Ridge at a rate of 2.5 centimetres every year.

Plate convergence

Convergence occurs when two plates are moving towards each other.

Convergence, which is sometimes also referred to as 'collision', occurs when two plates are moving towards one another. The type of collision that takes place between plates depends on the kind of lithosphere plates involved. Convergence can occur between an oceanic and a continental plate, between two oceanic plates, or between two continental plates.

Oceanic–continental convergence

Subduction occurs when one plate moves downwards under the edge of another plate.

A **subduction zone** is the region where the subduction of one tectonic plate under another occurs.

When an oceanic plate and a continental plate move towards each other, the more dense oceanic plate will move under the less dense continental plate in a process called **subduction**. The section of the lower plate that is forced under the upper plate re-melts into the upper mantle, later rising as plumes through the overlying continental plate to form volcanoes (see figure 2.10).

The region where one plate slides under another is referred to as the **subduction zone**. Where an oceanic plate subducts beneath a continental plate, trenches up to 10 kilometres deep form in the ocean floor while mountain ranges form along the edge of the continent. This is the case where the Nazca Oceanic Plate is being subducted beneath the continental South American Plate, forming the Peru–Chile Trench off the coast of South America and the Andes Mountains along the edge of the continent.

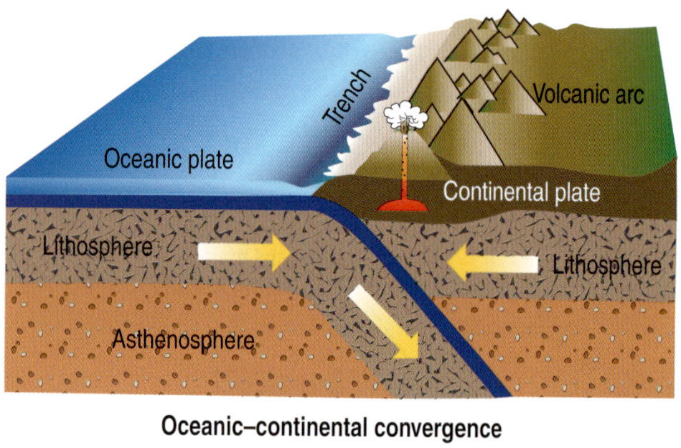

Figure 2.10: An oceanic–continental convergence

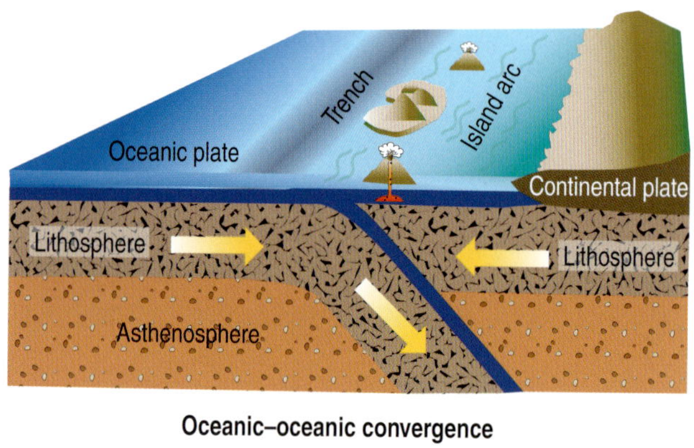

Figure 2.11: An oceanic–oceanic convergence

Oceanic–oceanic convergence

When two oceanic plates converge, one will usually subduct and form large trenches in the ocean floor (see figure 2.11). The Mariana Trench, which extends more than 10 kilometres into the Earth's interior, was formed as a result of the convergence of the Pacific and Philippine plates. Subduction in oceanic-oceanic plate convergence also results in the formation of volcanoes.

Continental-continental convergence

Where two continental plates converge, they are equally buoyant because they are both made of the same light continental material and therefore resist downwards motion much as corks floating in water. As a result, when one plate subducts under the other, they form a double layer of crustal material (see figure 2.12). The enormous Himalayas were formed by the convergence of the Indian and Eurasian plates. As the Indian Plate continues to burrow underneath the Eurasian Plate, the Himalayas will grow higher by a few millimetres every year (see figure 2.13).

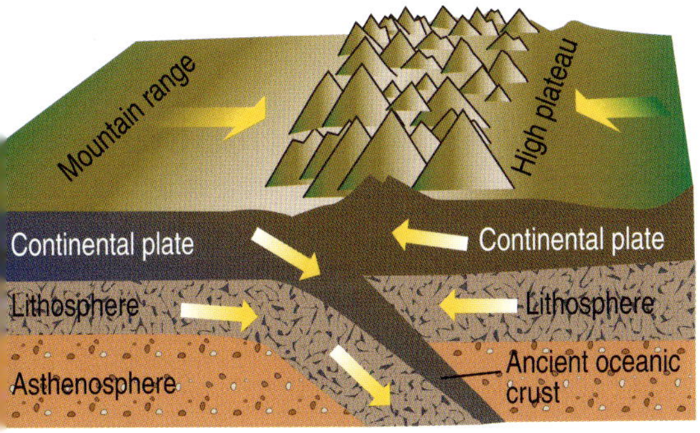

Continental–continental convergence

Figure 2.12: Continental–continental convergence

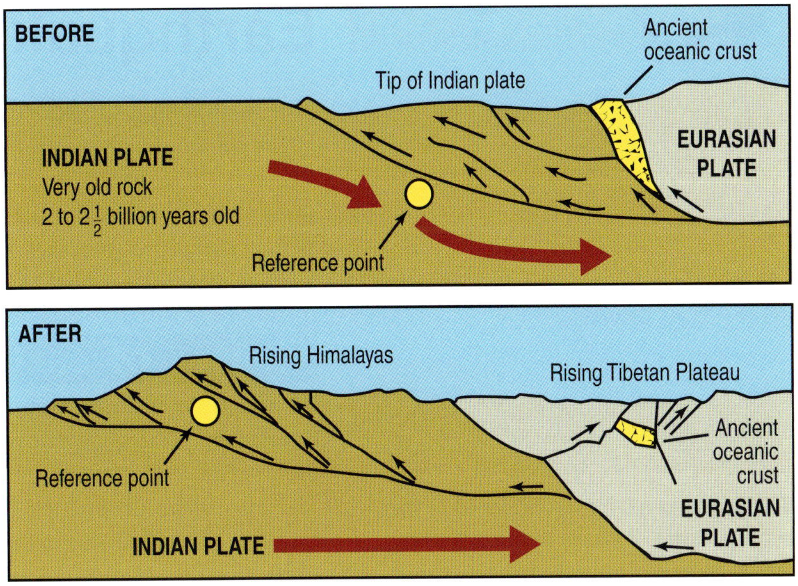

Figure 2.13: The formation of the Himalayas

Plate translation

Translation occurs when two plates slide past each other either horizontally or vertically without creating or destroying crust material.

Transform faults are areas of weakness and cracking in the Earth's crust that are caused by the forces generated by plate motion.

Translation occurs when two plates slide past each other (see figure 2.14). Crust material is neither created nor destroyed in this process, but the plates themselves can get 'stuck'. This causes enormous stress forces to be generated in the plates, which leads to the generation of weaknesses and cracks in the crust material called **transform faults**. Major earthquakes occur along these transform faults.

Transform faults are usually found on the floor of the ocean but there are cases where the faults occur on land. The San Andreas Fault is an example of this, caused by the grinding north-westerly motion of the Pacific Plate against the North American Plate. This fault is more than 1200 kilometres long and 10 kilometres wide in places (see figure 2.15).

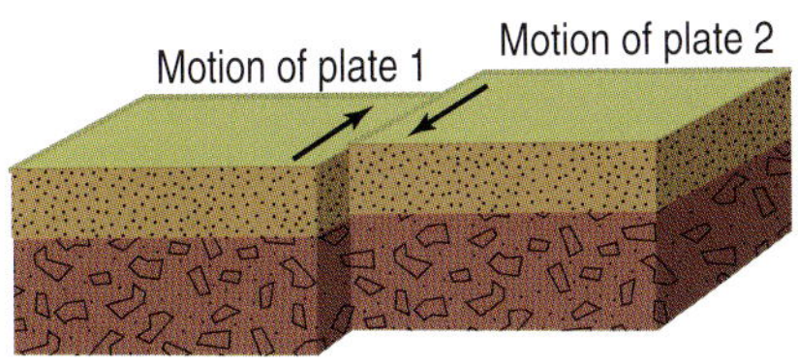

Figure 2.14: Translation of plates

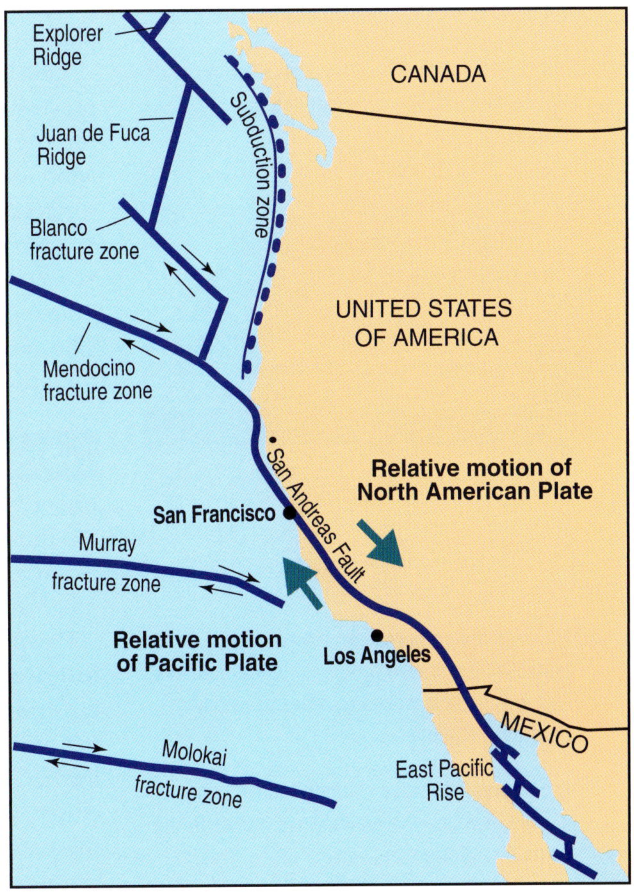

Figure 2.15: Translation along the San Andreas Fault

Earthquakes

Not surprisingly, the enormous forces exerted on and by the plates have an effect on the Earth's surface. While the movement of these plates is generally very slow, the way in which they interact can cause enormous pressures to build up in the mantle and at the interface of the plates themselves. When these pressures are released, the results can be sudden, catastrophic and — literally — earthshaking!

Causes of earthquakes

Seismic waves are the waves formed by the sudden release of energy that occurs when locked plates move suddenly.

An **earthquake** occurs when the stresses between locked plates and rocks are suddenly relieved, causing sudden land movement and the release of energy.

Intra-plate earthquakes are earthquakes that occur when the plate flexes in response to stressing forces on the edge of the plate.

The **focus** is the location inside the Earth of the fracture or faulting that causes an earthquake.

The **epicentre** of an earthquake is the point on the surface of the Earth that lies directly above the focus.

As we have already seen, convection currents in the mantle cause the lithosphere plates to move past each other, away from each other or underneath one another. However, this motion is by no means smooth, and sometimes the plates lock together and jam in position. As the convection currents are still trying to make the plates move, enormous stress builds up in these jammed plates. As the stress increases, the lithosphere plates become distorted. When the stressing forces finally get large enough to overcome the locking of the plates, the plates will suddenly snap into a new position. This rapid movement causes rocks in the plate to break and energy to be released. About 5 per cent of this energy travels through the Earth in the form of waves. These are called **seismic waves**, and it is the vibration of the Earth in response to the passage of these waves that we call an **earthquake**. Most earthquakes occur at the plate boundaries where the locking of plates occurs, but they can sometimes occur within a plate. On a global basis, **intra-plate earthquakes** are less common than those caused by movement along the plate boundaries. The 1989 Newcastle earthquake was an intra-plate event.

The location in the Earth's lithosphere where the rocks fracture in the process of releasing locked plates and generating the earthquake is called the **focus**. The seismic waves radiate away from the focus (see figure 2.16). Most of the time, the focus is located within 8 kilometres of the Earth's surface. Earthquakes can be classified according to the depth of the focus:

- *shallow focus:* less than 35 kilometres from the surface
- *intermediate focus:* between 35 and 300 kilometres from the surface
- *deep focus:* between 300 and 700 kilometres from the surface.

The point on the Earth's surface that is directly above the focus is called the **epicentre**.

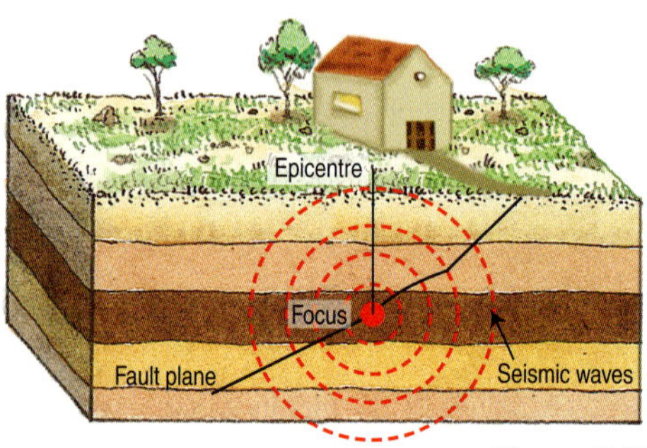

Figure 2.16: The focus and the epicentre of an earthquake

Seismic waves

P-waves are compression waves, which are the fastest of the waves produced in earthquakes and are the first waves to arrive at seismic recording stations.

A **compression wave** (or **longitudinal wave**) is a wave in which the disturbance moves in the same direction as the wave.

Energy released during an earthquake travels in the form of waves. There are three main types of wave that are generated by earthquakes: P-waves, S-waves and L-waves. These waves differ in terms of their orientation, their speed and the regions of the Earth through which they travel.

P-waves are compression waves. A **compression wave** (also called a **longitudinal wave**) is one in which the disturbance moves in the same direction as the wave. As the disturbance moves through the particles of the medium, it alternately pushes them closer together and then pulls them further apart. The positions where the particles are crowded together the closest are called *compressions* and those where the particles are spread furthest apart are referred to as *rarefactions* (see figure 2.17). Soundwaves move in this manner.

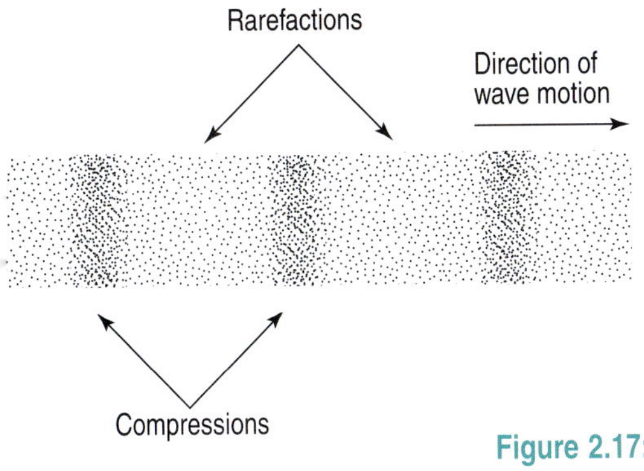

Figure 2.17: The movement of compression waves

S-waves are transverse or shear waves produced during an earthquake.

A **transverse wave** is one in which the disturbance caused by the transfer of energy acts perpendicularly to the direction of the wave itself.

P-waves are the fastest of the waves and are, therefore, the first to be detected by a seismometer. As a result, they were named primary waves (abbreviated to P-waves). The speed at which they travel depends upon which part of the Earth they are travelling through. In the crust, P-waves have a speed of 5 kilometres per second. In the top of the mantle they travel at about 8 kilometres per second, and at 14 kilometres per second in the bottom of the mantle.

After the P-waves, the second sets of waves to be detected are the secondary waves or **S-waves**. These travel in the form of transverse waves. A **transverse wave** is one in which the disturbance caused by the transfer of energy acts perpendicularly to the direction of the wave itself (see figure 2.18). The ripples on the surface of a pool when a rock has been dropped into it form transverse waves. The particles of the water itself move up and down as the wave moves outwards.

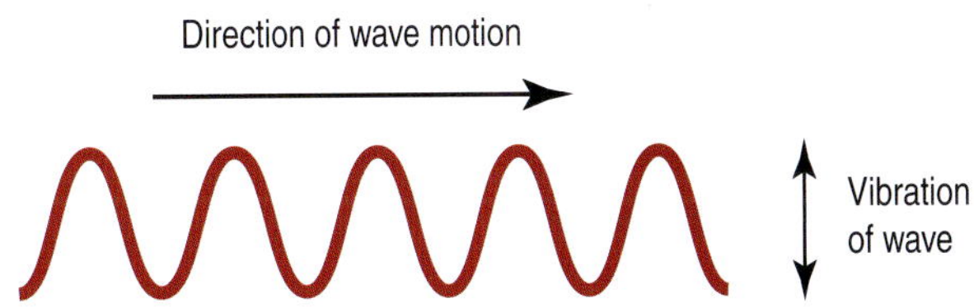

Figure 2.18: Transverse waves

As P-waves and the S-waves can travel through the interior of the Earth, they are said to be *body waves*. L-waves, however, can travel only on the surface of the Earth. They are therefore classed as *surface waves*. While they travel much more slowly than either the P-waves or S-waves, it is these surface waves that are responsible for the majority of an earthquake's destructive power. This is because all of the L-wave energy is distributed across the surface of the Earth rather than being dissipated in the Earth's interior, as in the case of P- and S-waves (see figure 2.19).

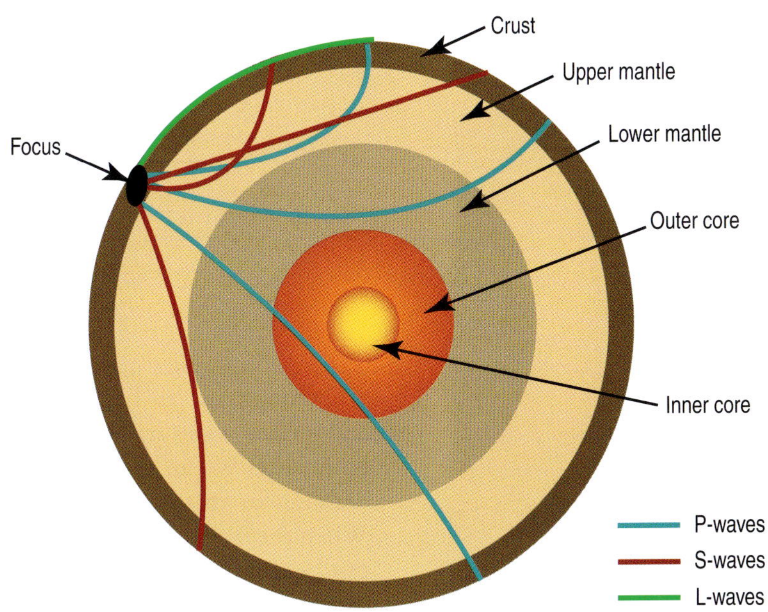

Figure 2.19: How the different waves travel across and through the Earth

Detecting earthquakes

A **seismograph** is made up of a seismometer, an amplifier and a system that records the readings of the seismometer.

A **seismometer** is a device that detects small horizontal and vertical movements of the Earth's surface.

The waves travelling from the focus can be detected using a device called a **seismograph**. The modern seismograph has four main components:

- a **seismometer** that detects horizontal and vertical motion in the Earth.
- an amplifier that increases the seismometer's signal strength
- a clock
- a recording device

The seismometer itself may be located thousands of kilometres away from the recording device, sending its data electronically. The frame of the seismometer is attached to a heavy base (often made of concrete and steel) that is anchored securely to solid rock. When the rock vibrates, the frame vibrates with it. A heavy mass — which can be up to 450 kilograms in large mechanical seismographs — is attached to the frame with a spring. When the frame moves, inertia causes the mass to remain steady. The movement of the frame relative to the mass is detected electronically and recorded.

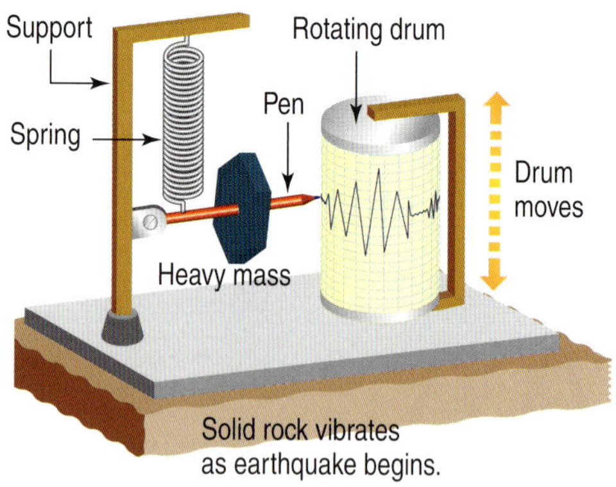

Figure 2.20: A simple seismograph

In many parts of the world, simpler types of seismographs are used where the weighted mass in the seismometer is attached to a pen that moves over a rotating drum of paper on which the traces of seismic waves are recorded as they occur (see figure 2.20). A big disadvantage of this set-up is that, in the event of an earthquake near the seismograph, both the device and the data are lost.

A **seismogram** is the record of the ground motions detected by a seismometer.

The record of seismic activity obtained from a seismograph is called a **seismogram**, and the waves on it appear as a continuous sequence of wriggles. The height of these wriggles indicates the intensity of the waves that are being recorded, with bigger amplitudes coinciding with more energetic waves. When no seismic waves are detected, the seismogram traces a flat line (see figure 2.21).

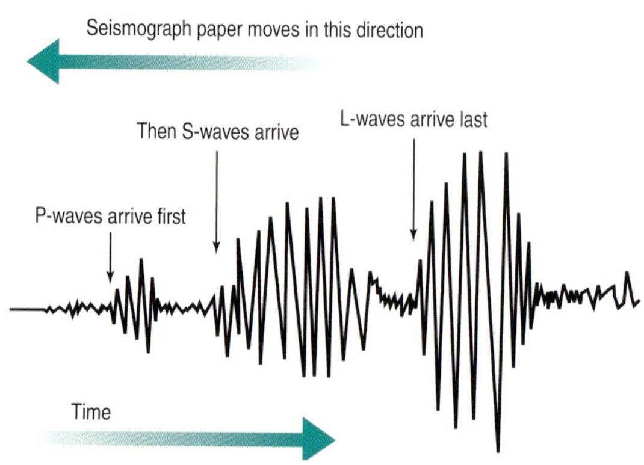

Figure 2.21: A seismogram

Seismograms can be quite complicated to interpret. As the body waves and surface waves travel around or through the Earth, they meet a variety of boundaries where there may be faults, or layers of rock that have different densities or that are under differing pressure and temperature conditions. When the waves strike these boundaries, their direction and type of motion can be disturbed: some of the P-waves can be converted into S-waves and vice versa. Experienced seismologists can interpret these variations in the seismograms and use them to learn more about the structure of the Earth itself.

Internet activity

Generate seismograms based on the seismic data received for a particular day during the last two years at any of the seismograph locations in Australia. Go to the website for this book and click on the Earthquake seismogram weblink (see Weblinks, page 309).

Investigation

Seismograph model

Aim: To make a simple seismograph
Materials: Retort stand, armature and boss-head; a strong spring; a 0.5–1.0 kg weight; a felt pen; stiff cardboard; a sheet of A4 paper; sticky tape
Method:

1. Set up your equipment as shown in figure 2.22.
2. Have one person move the cardboard (on which the paper has been taped) slowly past the pen. At the same time, thump the surface on which the seismograph is sitting. Do this a few times with different force as the paper is moved. Ensure that the pen remains in contact with the paper.
3. Examine the seismogram that you have generated.

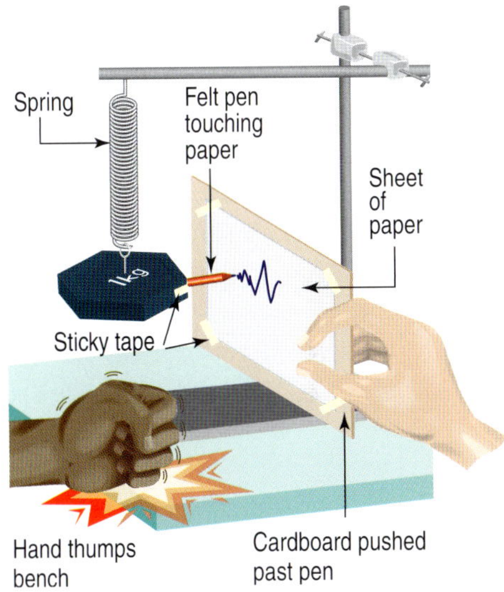

Figure 2.22: A model seismograph

Discussion questions

1. How clearly does your seismogram show the difference in the intensities of the thumps? If they are not clear, give possible explanations for why this is the case.
2. What effect would you expect to see on your seismogram if the paper had been pulled across at a faster rate?
3. How could you modify this set-up in order to:
   (a) record thumps over a half-hour period of time?
   (b) make the device more sensitive?
4. How does your model seismograph differ from the simplified mechanical one shown in figure 2.20?

If you feel like a challenge, there is a more ambitious plan for a model seismograph. Go to the website for this book and click on the Model seismograph weblink (see Weblinks, page 309)

Locating the epicentre

Have you ever wondered how seismologists can tell where the epicentre of an earthquake is located even if it is on the other side of the world or in the middle of an unpopulated region? The epicentre of an earthquake can be determined by comparing the time interval that elapses between the arrival of the P-waves and the S-waves at different seismograph locations. In order to do this, the readings from at least three different seismic stations must be obtained. This process is known as **triangulation**.

Triangulation is the process by which the location of an earthquake's epicentre can be determined based upon the difference in arrival times of seismic waves from three different seismograph stations.

Materials: A ruler; a pencil; a drawing compass; a photocopy of the map in figure 2.23; a sheet of A4 paper torn into strips that are 10 centimetres long and about 1 centimetre wide

Figure 2.23:
Map of the Pacific region

Scenario: The recording stations at Port Moresby, Noumea and Brisbane all detect the seismic waves from an earthquake. The time lag that each station observes between the arrival times of the P-waves and the S-waves are as shown below:

Recording station	Time lag between arrival of P- and S- waves (min)
Port Moresby	3
Noumea	4
Brisbane	4.8

Task: Use this information to determine the location of the earthquake's epicentre. To do this, the task must be broken into a number of steps.

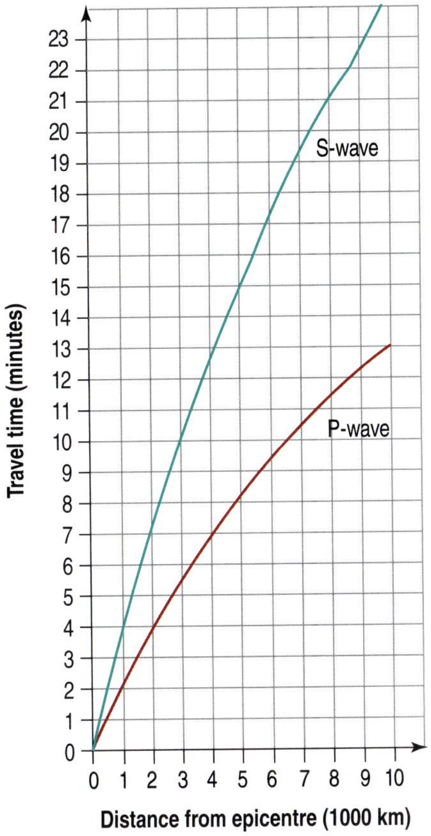

Figure 2.24: Earthquake S- and P- wave travel-time graph

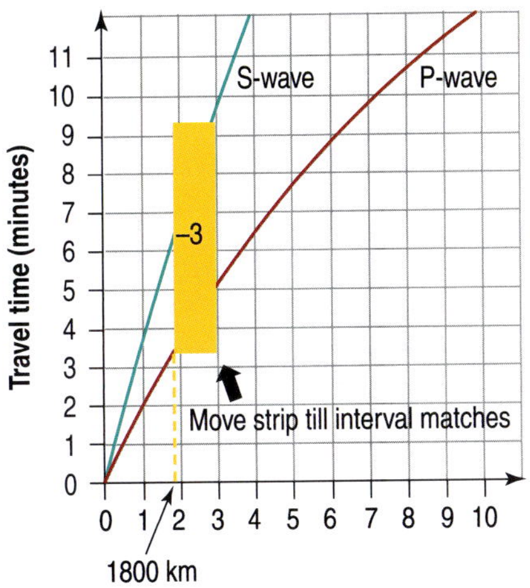

Figure 2.26: Reading the distance corresponding to the time lag

Part A: Determining the distance of the epicentre from each station

To do this, use the graph in figure 2.24, which shows the travel time of the P- and S- waves as a function of their distance from the epicentre. As you can see from this graph, P-waves travel much faster than S-waves. That is why they are always detected first regardless of how far away the epicentre is from the detector.

First, find the distance of the epicentre from Port Moresby.

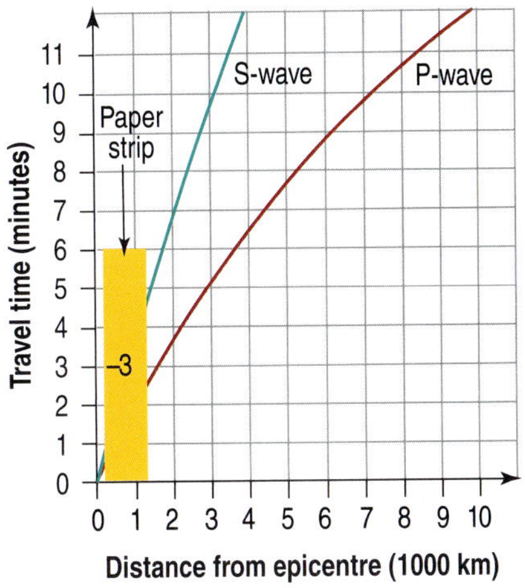

Figure 2.25: Marking the strip at the 3-minute position

1. Take a paper strip and place it along the vertical axis of the graph. Using a pencil, mark the strip of paper at the position corresponding on the axis to the time lag for Port Moresby (3 minutes). Make sure that your paper strip is straight and that its bottom corner is lined up with the corner of the graph (see figure 2.25).
2. Move this strip of paper so that it remains parallel with the vertical axis and place it so that the pencil mark lies on the S-wave line and the bottom corner touches the P-wave line.
3. Read down from the paper edge to the distance scale. You will find that the distance for Port Moresby station is 1800 kilometres. This means that Port Moresby is located 1800 kilometres from the epicentre (see figure 2.26).
4. Repeat this method to find the distances that correspond to the time lag at the other two stations.

Part B: Using triangulation to find the epicentre

By using the method in part A, you will have found the following distances of each of the three stations from the epicentre:

Recording station	Time lag (min)	Distance (km)
Port Moresby	3	1800
Noumea	4	2500
Brisbane	4.8	3300

Now we will use triangulation to find the epicentre.

1. Place the point of your compass at the 0 mark of the scale at the top of the map in figure 2.23. Adjust the compass width until the compass pencil is at the point on the scale where 1800 kilometres would be.
2. Without adjusting the compass width, place the compass point on the marker for Port Moresby and draw a circle around it. The epicentre will lie somewhere on this circle.
3. Repeat steps 1 and 2 for the other two stations, adjusting the width of the compass to correspond to the new distance each time. The epicentre will be located at the point where all three circles intersect.

You should find that the epicentre in this case is located north of the Solomon Islands and west of Nauru. Quite often when using this method, you may end up with a small triangle where the circles do not quite cross at exactly the same point. In this case, locate your epicentre in the middle of the triangle. The smaller this triangle is, the more accurate your method. If the circles intersected at a single point, well done!

Part C: Practising the technique

Using the procedures which you have just worked through in parts A and B, you will now locate the epicentre of the earthquake that produced the following data:

Recording station	Time lag (min)
Wellington	4.1
Noumea	2.0
Brisbane	4.3

While all of these calculations are now done by computers, spare a thought for the seismologists in the pre-computer era who did this all by hand just as you are doing now!

Eye on the media

How to recognise the 'Big One' the moment it begins

In places like Japan and California, where tiny tremors occur every day, it is not enough to know when and where an earthquake will hit — you also need to know just how serious it will be. Now seismologists think they can predict how bad an earthquake will be within seconds of an underground rupture beginning, perhaps giving vital seconds' warning of the main seismic waves. Their findings could provide an early-warning system to help limit earthquake damage.

Until now, seismologists believed you could only assess an earthquake's full force once the quake was over. Earthquakes were thought to progress unpredictably, with one rupture triggering the next as weak masses of rock give way along a fault line. But when Richard Allen at the University of California, Berkeley, began looking at earthquake records he realised that this was wrong. 'An earthquake essentially knows how big it will be within the first few seconds of rupture,' he says.

When an earthquake strikes, seismic waves radiate from the focus at about 4 to 6 kilometres per second. These fastest-travelling primary waves, which are weak and do little damage, can arrive more than 20 seconds before the slower, more destructive waves hit. Allen and his colleague Erik Olsen, at the University of Wisconsin in Madison, studied 71 earthquakes in Japan, Taiwan, California and Alaska and found that they could use the frequency of the waves arriving at sensors within the first 4 seconds to predict the magnitude of the entire quake (*Nature,* vol. 438, p. 212).

While 20 seconds might not sound long, it might be enough time for automated emergency systems to stop trains, move machinery into safe modes, shut down gas lines and sound alarms to warn people to shelter under desks or in doorways. And the emergency systems would only activate if a major quake is on its way — avoiding unnecessary panic.

Accurate magnitude predictions could lead to smart buildings that adjust their stiffness to cope with incoming shock waves.

Source: Zeeya Merali, *New Scientist,* 19 November 2005, p. 29.

Clarification questions

1. At what speed do seismic waves leave the focus of an earthquake?
2. How far ahead can seismologists predict the magnitude of an earthquake using the system in this article?
3. What are the advantages of these predictions?
4. How does the progress of an earthquake as theorised in this article differ from the present accepted theory?

Comparing earthquake strengths

Not all earthquakes cause buildings to collapse and the ground to open up like in the movies. In fact, earthquakes happen every day but some of them are too small to be noticed — they can often be mistaken for a passing truck on a nearby road. Seismologists have different ways of classifying the strength of an earthquake. The best known of these is the **Richter scale**.

The **Richter scale** is used to determine the magnitude or size of an earthquake and reflects the amount of energy released.

The Richter scale is an open-ended scale developed by Charles F. Richter. It rates an earthquake as having a particular magnitude that reflects the amount of energy it has released. Richter noticed that the destructive potential of an earthquake was related to the amplitude of the shear waves that appeared on its seismogram. From this observation he was able to generate a calibrated system of measurement where the amplitude of the shear waves, and the delay between the arrival times of P-waves and S-waves at the seismometer, allowed an earthquake to be assigned a number called a **magnitude**.

Magnitude A **magnitude** is a number assigned to an earthquake that reflects the destructive energy it has released.

The Richter scale is not a linear scale. An increase in a magnitude of 1 on the Richter scale represents an increase in energy by a factor of 10. So an earthquake that has a magnitude of 5 on the Richter scale releases 10 times more energy than a magnitude 4 earthquake, and 100 times more energy than a magnitude 3 earthquake. This type of relationship is described in mathematics as a **logarithmic** relationship. The Richter scale is used to determine the extent of damage caused by an earthquake (see table 2.1).

A **logarithmic** is a relationship in which a single unit increase in the independent variable results in tenfold increases in the dependent variable.

Table 2.1: The Richter scale

Richter magnitude	Description of effects
0	• Smallest magnitude detectable by seismometer
1.0	• Undetectable by humans • Equivalent to the energy released by the detonation of 30 kilograms of TNT
2.0	• Smallest magnitude that can be noticed by humans • Tremors of magnitudes less than this are called microquakes
<3.5	• Noticeable by humans
3.5–5.4	• Often felt but rarely causes damage • Vibrations similar to those felt when a heavy truck drives past • Objects on shelves may rattle or fall down
<6.0	• Slight damage to well-designed buildings but can cause large cracks and material failure to poorly constructed buildings
6.1–6.9	• Can be destructive in areas up to about 100 kilometres across
7.0–7.9	• Major earthquake • Can cause serious damage over larger areas
≥8	• Great earthquake • Can cause serious damage in areas several hundreds of kilometres across

The **Mercalli intensity scale** is used to describe the effects of an earthquake in a particular region.

While the Richter scale is well known, it has a major limitation: while the magnitude of an earthquake can be calculated, the actual effects of the earthquake on an area vary enormously depending upon the area's distance from the epicentre, the stability of the ground in the region, how well the buildings are constructed, the presence of fault lines and so on. As a result, seismologists use the **Mercalli intensity scale** to indicate the extent of the destruction an earthquake causes. This classifies the earthquake intensity based upon what is observed in the earthquake rather than as the result of a calculation (see table 2.2).

Table 2.2: The Mercalli intensity scale

Scale	Classification	Description of effects
I	Negligible	• Detected by instruments only
II	Feeble	• Felt by some people • Suspended objects swing
III	Slight	• Vibration like passing truck • Parked cars may rock
IV	Moderate	• Can be felt indoors • Sensation like a heavy truck striking the building • Windows, doors and dishes rattle
V	Rather strong	• Felt by most people • Dishes and windows broken • Pendulum clocks may stop • Pictures on walls move • Sleeping people are woken up • Trees shake
VI	Strong	• Felt by all people • People experience difficulty walking • Furniture moves • Pictures fall off walls • Wall plaster cracks • No structural damage to buildings
VII	Very strong	• People have difficulty standing • Weak structures suffer moderate damage • Can be felt by people in moving cars • Loose bricks fall from buildings
VIII	Destructive	• Drivers have trouble steering cars • Weak structures suffer severe damage • Strong structures suffer slight damage • Heavy furniture is overturned • Tree limbs break
IX	Ruinous	• Total destruction of weak structures • Considerable damage to strong structures • Building foundations damaged • Underground pipes broken • Fissures open in ground
X	Disastrous	• Only the strongest buildings survive • Foundations ruined • Ground badly cracked • Railway tracks bent • Large landslides
XI	Very disastrous	• Few masonry structures remain standing • Wide fissures in the ground
XII	Catastrophic	• Total destruction • Ripples and waves visible in the ground

Eye on the media

One disastrous slip . . .

The devastating earthquake that levelled parts of northern Pakistan and neighbouring Kashmir on 8 October came as no surprise to geologists. While the region is geologically active, it has suffered relatively few large earthquakes in the last century, leaving it primed for a big one.

As *New Scientist* went to press, the quake had reportedly claimed more than 33 000 lives, and hopes of finding many more survivors were fading. The magnitude 7.6 earthquake struck at 0850 local time (0350 GMT), with the epicentre close to the city of Muzaffarabad and only 10 kilometres below the surface.

The affected region lies along the fault line where the northward-moving Indian tectonic plate plunges beneath the Eurasian plate, pushing up the Himalayan mountain range. Saturday's earthquake was the most powerful along the fault since 1950.

GPS measurements taken on opposite sides of the fault suggest that the plates are moving towards each other at a rate of around 2 centimetres a year. Records of earthquakes in the region examined in 2001 by Peter Molner and colleagues at the University of Colorado in Boulder showed that most of the stress built up by this convergence had not been released (*Science,* vol. 293, p. 1443). Particularly vulnerable were those parts of the fault where earthquakes had not occurred recently.

Earlier this year, Roger Bilham, also at the University of Colorado and Nicholas Ambraseys, an engineer at Imperial College London, predicted that the accumulated stress was equivalent to four quakes of magnitude 8.5 or larger (*Current Science,* vol. 88, p. 1658).

But it is impossible to predict the timing of such quakes. 'Can we say, "Tomorrow don't go to school"? No we are not at that stage,' Molner says. However, knowing which areas face the greatest threat should help governments prepare, by building structures that can withstand an earthquake and educating people about the dangers.

There is a risk that last weekend's quake will make other regions along the fault more prone to a slip, although again Molner says it is hard to make specific predictions. 'There is a possibility that the region between this earthquake and the one that occurred in 1905 would be at greater risk,' he says.

Source: James Randerson, *New Scientist*, 15 October 2005, p. 8.

Clarification questions

1. Why does this article state that geologists were not surprised by the size of the Pakistan–Kashmir earthquake?
2. How was the rate of motion of the Eurasian and Indian plates determined?
3. What benefits are possible by being able to assess which areas are more likely to experience earthquakes?

Earthquakes in Australia

Most earthquakes occur in the regions where plates meet. Australia is located in the middle of a plate, so you would think that we would experience very few earthquakes. In fact, our country experiences many earthquakes as the result of intra-plate flexing.

Australia is seismically active and earthquakes pose a substantial risk as demonstrated by the deadly magnitude 5.6 Newcastle earthquake of 1989 (see figure 2.27). When compared to plate margin regions such as California or Japan, the frequency of earthquakes is lower but relative to other intra-plate regions, Australia's earthquake activity is quite high. Table 2.3 lists some of the larger earthquakes that have occurred in Australia.

Table 2.3: Notable Australian earthquakes

Date	Location	Magnitude	Modified Mercalli intensity	Comments
1941 29 Apr.	Meeberie (WA)	7.2	VIII	• Largest onshore earthquake in Australia • Caused minor damage in Perth, 600 km away
1960 11 Nov.	Brisbane (QLD)	5.0	–	• Mt Glorious
1978 28 Nov.	Heron Island (QLD)	5.2	IV	• Widely felt in Gladstone and along coast from Rockhampton to Maryborough
1979 2 June	Cadoux (WA)	6.2	VIII	• Over $10 million damage
1988 22 Jan.	Tennant Creek (NT)	6.4 6.4 6.9	–	• Three earthquakes in the same day • Darwin–Alice Springs pipeline was shortened by a metre
1989 28 Dec.	Newcastle (NSW)	5.6	VII	• 13 deaths • 160 injuries • 60 000 buildings damaged • Felt in an area of over 200 000 square km • $4 billion damage
1994 6 Aug.	Cessnock (NSW)	5.2	VII	• Over $35 million damage
1996 18 Jul.	Brisbane (QLD)	2.9	IV	• In Acacia Ridge • Widely felt from southern Brisbane suburbs to Logan City in the south
1998 2 Nov.	Rockhampton (QLD)	4.7	–	• Widely felt on Heron Island • Felt in Rockhampton

Figure 2.27:
The Kent Hotel in Beaumont was severely damaged in the Newcastle earthquake.

Volcanoes

There are few more spectacular and fearsome sights than that of an erupting volcano. They are the central theme of many disaster movies and even a few Tarzan adventures as producers cash in on the public fascination for them. Unfortunately, the facts often tend to be set aside in favour of spectacle and drama, and many people have gained a few strange ideas about volcanoes, courtesy of Hollywood! Let us look at the reality.

How volcanoes form

Magma chambers are underground cavities, usually found beneath volcanoes, that contain magma.

A volcano forms when molten rock and gases from the mantle are able to escape onto the Earth's surface. This can occur when weak regions of crustal material give way as the result of pressure exerted from under the lithosphere or in regions near plate boundaries. When these weak areas collapse, a vent (tunnel) or a fissure (crack) is formed. Magma, made of molten rock combined with dissolved gases, moves upwards into these vents because it is less dense than the solid rock that surrounds it. When the magma first moves up through the weak region, it forms reservoirs in the Earth's crust. These are called **magma chambers**. From the vent, the magma is ejected onto the surface in an event called an eruption. There the molten rock starts to cool and, in the case of vent eruptions, form the typical cone-shaped mounds, while the flow from a fissure tends to create ridges on either side of the gap. Mid-oceanic ridges are fissure or vent eruptions (see figure 2.28).

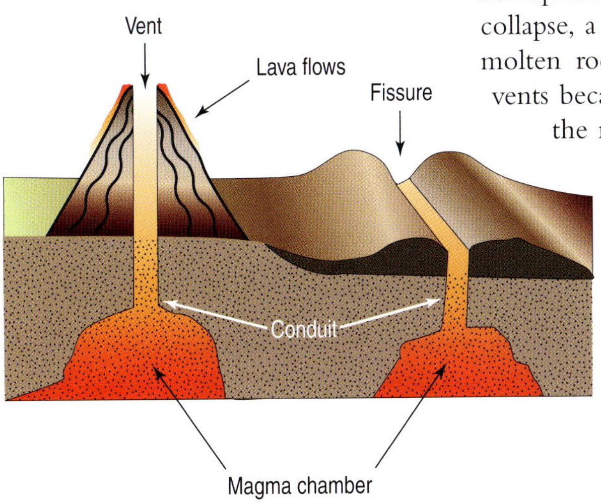

Figure 2.28:
Fissure and vent-type volcanoes

When a volcano erupts, magma flows from the magma chamber up into the vent or fissure (see figure 2.29). The eruption ends when the reservoir is depleted.

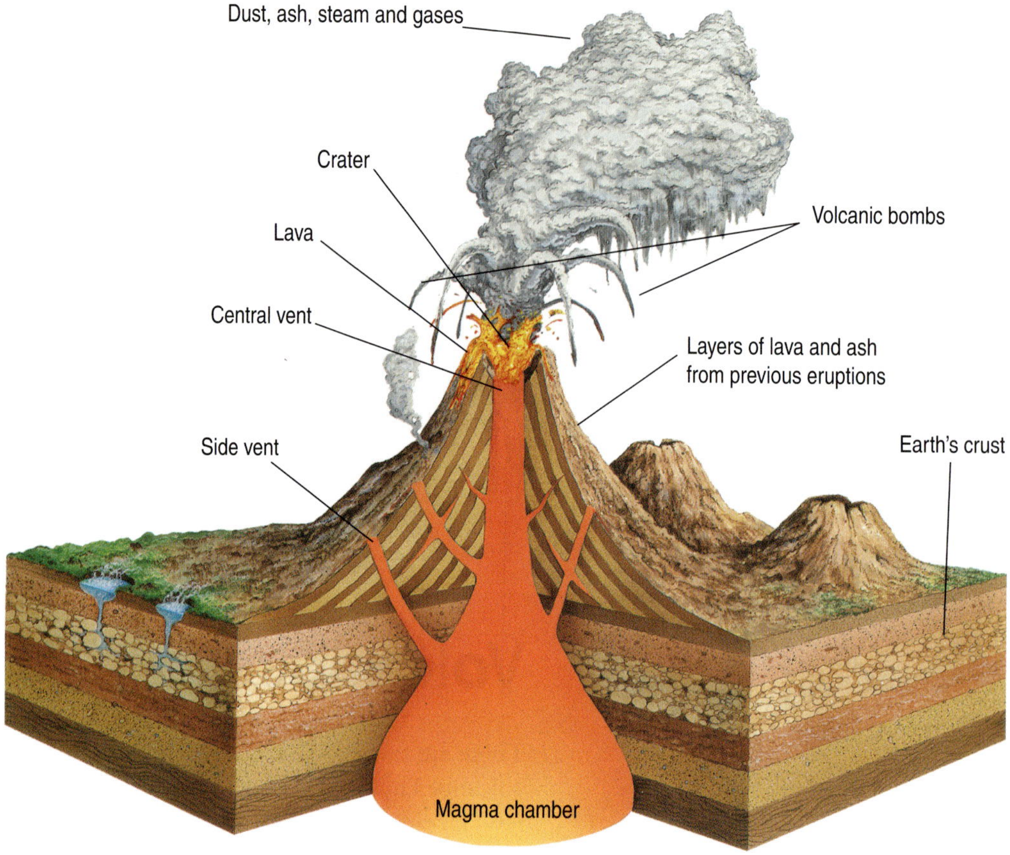

Figure 2.29:
The structure of a volcano

Materials produced by volcanoes

Depending upon the conditions in the magma chamber, volcanoes spit out a variety of different substances during an eruption.

Lava is molten rock that comes out of a volcano onto the Earth's surface, where it forms igneous rock when it solidifies.

Viscosity is a measure of the 'stickiness' of a fluid. A fluid with high viscosity (like honey) flows slowly, while a fluid with low viscosity (like water) flows easily.

Volcaniclastic material is a general term for the solids, liquids and gases that are ejected from a volcano during an eruption.

Lava is the term given to magma that has reached the surface of the Earth. The lava that flows from some volcanoes is very thick and travels very slowly, only a few metres every hour. The lava produced by the volcanoes in Hawaii is like this. At the other extreme, there have been cases where the lava was particularly liquid and travelled so rapidly that it consumed people who were unable to get out of its path in time. The behaviour of lava depends mainly upon its chemical composition. This affects the **viscosity** or 'stickiness' of the lava, which in turn affects its rate of flow.

Volcaniclastic material is made up of fragments of the volcano itself. These are usually produced when an eruption is explosive. This can occur when:

- Gases build up due to the high viscosity of the magma.
- A vent has been blocked by solidified magma, which is called a *plug*. The pressure under the plug increases as the forces acting on the magma chamber cause it to push more and more magma into the vent. Suddenly, just as the cork is blasted out of a champagne bottle, the fragmented plug

along with pieces of the cone itself is ejected from the mouth of the volcano along with hot gases, ash and dust.

- Water enters the vent. This turns immediately into superheated steam that rapidly expands.

The rock fragments thrown out from the volcano range widely in size from small particles up to boulders as large as a house! These larger chunks are called *volcanic bombs.*

Volcanic gases are those released by a volcanic reaction. The most common gas is superheated steam but carbon dioxide, carbon monoxide, sulfur dioxide and hydrogen sulfide can also be produced.

A **vulcanologist** is a scientist who studies volcanoes.

A **pyroclastic flow** occurs when hot volcanic material produced by an explosive eruption flows like a liquid.

Volcanic gases are considered by many **vulcanologists** to be the most dangerous of the materials that a volcano produces. While the most common gas produced is superheated steam, deadly vapours of carbon dioxide, carbon monoxide, sulfur dioxide and hydrogen sulfide are also produced.

Sometimes, material is ejected from a volcano in the form of a **pyroclastic flow** (see figure 2.30). This is a mass of hot gas, ash and rock that travels as if it were a liquid, and it is the most destructive aspect of an explosive volcano. These flows move at frightening speed close to the ground, moving away from the crater. The speed of a pyroclastic flow increases with the steepness of the slope and the mass of material involved. The gases in the flow can have temperatures up to 800°C and reach speeds of 150 kilometres per hour.

Figure 2.30:
A pyroclastic flow in action

Types of volcanoes

With each eruption from a vent, layers of lava, rock, gas and ash flow out over those left behind by previous eruptions, causing a cone to form. Cone volcanoes can be classified according to the steepness of their sides and the material from which they are made.

Shield cones

Shield cones are shaped like low domes and are made from layers of basalt that formed from very fluid, fast-cooling lava (see figure 2.31). This lava has a low viscosity, so gases are easily able to escape. As a result, shield

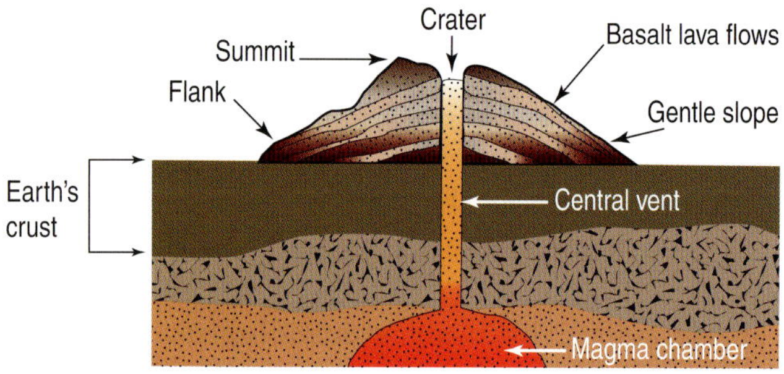

Figure 2.31: A shield volcano

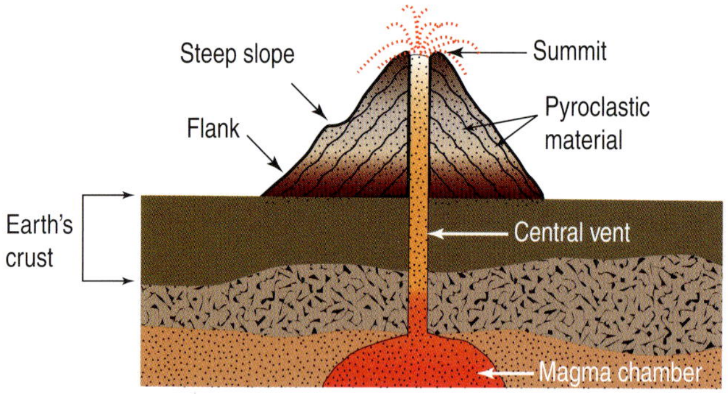

Figure 2.32: A cinder cone

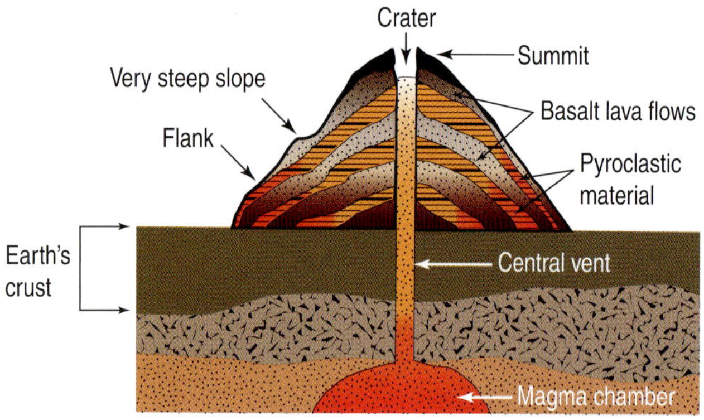

Figure 2.33: A composite cone

Dormant volcanoes are those that are not currently active but have the potential to erupt.

Extinct volcanoes are volcanoes that have not erupted for thousands of years and are effectively dead.

volcanoes tend not to have explosive eruptions. The sides of shield volcanoes are very gentle — forming shallow angles of between 2 and 10 degrees with the surface. Despite their subtle shape, shield cones can be very large. The largest volcano on Earth is Mauna Loa (in Hawaii). It is a shield volcano that is 96 kilometres long and 48 kilometres wide, and rises to a height of 9 kilometres above the ocean floor.

Cinder cones

As their name implies, cinder cones have a very regular pointed shape (see figure 2.32). Their slopes are usually around 30 to 40 degrees, so they are steeper than shield volcanoes but they do not get as large. Few cinder cones exceed 500 metres in height. They are made almost entirely of pyroclastic material such as cinders, ash, volcanic rocks and so on, and are formed during only a few periods of eruption.

Composite cones

Most volcanoes form composite cones. These have very steep sides and are made of alternating layers of lava and pyroclastic material (see figure 2.33).

Dormancy

While the Earth is covered in thousands of volcanoes, not all of them are capable of erupting. Only 500 or so volcanoes are considered to be *active,* which means that they have either erupted recently or show signs of future eruption. In fact, there are only about 50 volcano eruptions a year, lasting anywhere between a day and a month. About half of these eruptions are *explosive;* the rest, involving slow lava flow and gas emission, are said to be *effusive*. Volcanoes that have not erupted for 20 years or more are said to be **dormant** or 'sleeping'. These volcanoes may still become active again in the future. A volcano that has not erupted for thousands of years is classified as **extinct**. These volcanoes have no active magma chamber beneath them and their vents have solidified. All of the volcanoes that ever existed on the Australian mainland have been extinct for thousands of years.

Eye on the media

How a supervolcano covers a continent

When supervolcanoes blow they can cover entire continents with ash. But how this happens has been a puzzle because wind and the initial force of the eruption are not enough to carry the ash over such long distances. Now an examination of prehistoric eruptions has come up with an answer.

Supervolcanoes are classified as volcanoes that spew out more than a trillion tonnes of material when they erupt — equivalent to 30 Krakatoas. Such volcanoes cannot be studied directly as the most recent was Toba in Sumatra around 71 000 years ago.

So Peter Baines from the University of Melbourne, Australia, and Stephen Sparks from the University of Bristol, UK, used geological records of ash volume and magma chamber size to estimate the energy of past blasts and model the plumes they would have generated. From this they deduced that the Earth's rotation fans ash out into it giant spinning cloud up to 6000 kilometres wide within one day. 'It is a bit like a hurricane, but on a much larger scale,' Sparks says.

Unfortunately the findings don't offer a solution for surviving a future eruption, such as if the supervolcano underneath Yellowstone National Park in the US were to blow. 'I'm not sure what we could do, except stay underground,' says Sparks.

Source: *New Scientist*, 14 January 2006, p .19.

Clarification questions

1. What is a 'supervolcano'?
2. Why can't supervolcanoes be studied directly?
3. According to the theory in this article, how does the Earth's rotation affect the distribution of ash?
4. What are the limitations of this theory?

Queensland's volcanic past

While there are presently no active volcanoes in Queensland, it was a very different story 22 million years ago. The largest of these was the massive Tweed Volcano, which extended from Mt Tamborine in Queensland down to Byron Bay in Northern New South Wales. The vent of this volcano was over 1900 metres above sea level and had a diameter of 30 kilometres. When the volcano became extinct, the hard rock plug that formed in the vent was exposed because the softer surrounding rock was eroded by wind and rain over the millions of years that have since passed. Left behind is a circular ridge with a 40-kilometre diameter (referred to in tourist brochures as the Scenic Rim) that was once the outer wall of the volcano, and the remains of the plug, which is now known as Mt Warning (see figure 2.34).

Figure 2.34: Mt Warning

To the north of Brisbane are the Glasshouse Mountains (see figure 2.35). These are the remains of volcanic plugs from a cluster of volcanoes that once covered all of South-East Queensland. The volcanic soil in this region is particularly fertile due to the good drainage properties of the coarse volcanic particles.

Figure 2.35:
Glasshouse Mountains

Eye on the media

Ol Doinyo Lengai

Bigger isn't always better when it comes to volcanoes. Mount Vesuvius in Italy and Mount St Helens in Washington state are hardly giants, but for the scale and destructive power of their eruptions they are hard to beat. Ol Doinyo Lengai, some 120 kilometres north-west of Arusha, Tanzania, in east Africa's Rift Valley can pack a punch too. Three times during the last century it propelled boulders and dense columns of ash several kilometres into the air.

But this is not what gets geologists most excited. Ol Doinyo Lengai (the name translates as Mountain of God) is the only volcano on Earth with alkali-rich carbonatite lava. Unlike the basalt lavas of other volcanoes, this is rich in sodium and potassium carbonates and low in silicon. With a temperature of around 550°C, it is also relatively cool — a good 500°C cooler than basalt lavas and not hot enough to glow in daylight. And it's as fluid as oil.

The carbonatite emerges looking like black or dark-brown mud, but turns quickly to grey and then white as it reacts with moisture in the atmosphere. The slopes of Ol Doinyo Lengai look a bit like dirty snowfields streaked with oil slicks. And since 1983, it has been in a state of near-constant eruption.

The volcano is not short of scientific visitors. The low temperatures mean researchers can get close to the lava flows, channels, vents and lakes without protective equipment. And the low viscosity of the lava appears to make it particularly sensitive to changing atmospheric conditions. A recent study by Josh Gordon and Fred Belton at Middle Tennessee State University in Murfreesboro found that Ol Doinyo Lengai's activity increased when atmospheric pressure was falling or at a minimum. The researchers suggest the falling pressure might trigger an eruption by causing a slight expansion of the gases in the lava. They also noted more eruptions when the moon was closest to the Earth, possibly because of the stronger tidal pull. But they stress they need a lot more data to support these theories.

There is another reason why researchers get excited about the Mountain of God: it could offer a hint about what happens on the surface of Venus. Our sunward neighbour has several lava channels on its surface, some up to 7000 kilometres long. Only a low-melting-point magma like carbonatite could have stayed liquid long enough to flow such a long way.

Fancy a visit to Ol Doinyo Lengai? First check whether the volcano is in a particularly eruptive phase. It can turn explosive without warning, as it did in 1917, 1940 and 1966, when the whole summit was destroyed. 'One should always be prepared to flee,' advises Charles Frankel, whose book *Worlds on Fire,* describing the solar system's most spectacular volcanoes, is published this week by Cambridge University Press. 'Camping at the summit is not a good idea,' he warns.

Source: *New Scientist*, 17 December 2005, p. 52.

Clarification questions

1. What's so special about the Ol Doinyo Lengai volcano?
2. How does atmospheric pressure appear to affect the volcano?
3. Why do the slopes of the volcano look 'a bit like dirty snowfields streaked with oil slicks'?
4. Where is Ol Doinyo Lengai located?
5. How is the basalt of this volcano different to that of other volcanoes?

The Ring of Fire

The **Ring of Fire** is a seismically active region that mainly follows the outline of the Pacific Plate.

The majority of active volcanoes lie along the perimeter of the Pacific Plate, forming a circuit that passes through Japan, Indonesia, New Zealand, Hawaii and Alaska, and then down along the eastern coasts of North and South America. Due to the enormous amount of seismic and volcanic activity that occurs around this circle, it is called the **Ring of Fire** (see figure 2.36). The boundary of the Pacific Plate forms subduction zones and transform faults, producing tectonically unstable regions.

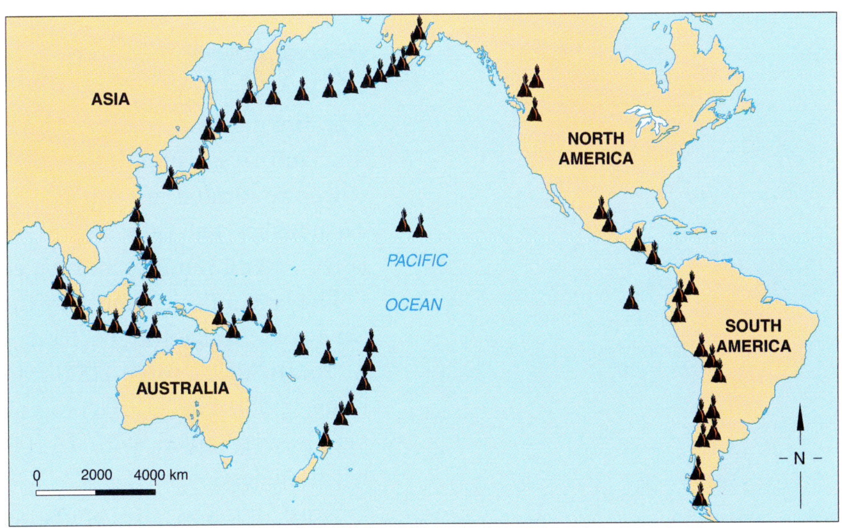

Figure 2.36: The Ring of Fire

Tsunamis

Tsunamis are giant sea waves produced when water is displaced by underwater earthquakes, landslides or volcanic eruption.

Tsunamis are ocean waves that are generated by the sudden displacement of large volumes of water. While these displacements can be caused by landslides or volcanic eruption, they are usually the result of movement of the sea floor by earthquakes. Rather than being a single wave, a tsunami is actually a series of waves that travel across the ocean at speeds of up to 800 kilometres per hour.

How tsunamis form

Amplitude is the height of a wave.

When a large volume of water is suddenly displaced, it travels outwards from the disturbance much as ripples move outward from the point where a rock is dropped in a pond. In deep water, the waves can be separated by as much as a hundred kilometres, and they can have a small **amplitude** — only a metre or so — which is not noticeable to observers on ships travelling through them or those on aircraft above them.

As the waves approach the shallower water along a coastline, the waves slow down. Waves further back start to catch up to those in front, which causes the waves to increase in height as they bunch up. These waves can reach heights of 30 metres before they break onto the land. The massive body of water moves inland, causing severe flooding and considerable loss of life and property (see figure 2.37).

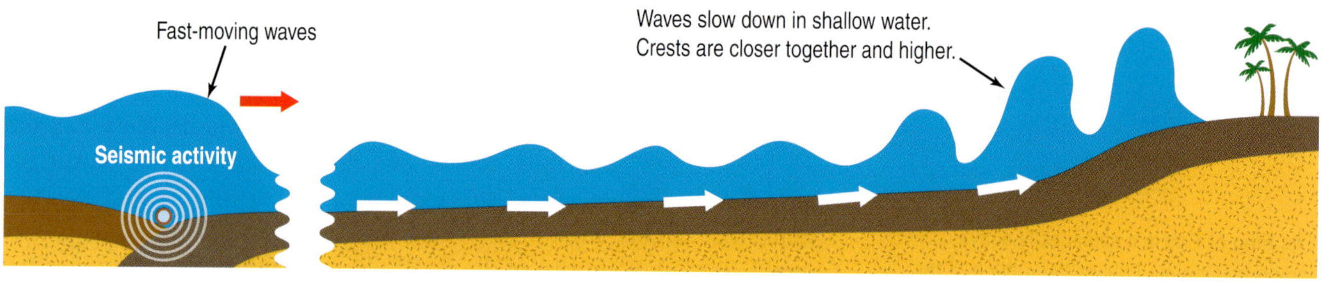

Figure 2.37: How tsunamis form

Case study: Boxing Day 2004

The Boxing Day 2004 Tsunami (also referred to as the Sumatra–Andaman Tsunami) was caused by an earthquake of magnitude at least 9.0, with an epicentre located just north of Simeulue Island in Indonesia. Whereas most earthquakes last only a few seconds, the Sumatra–Andaman earthquake lasted several minutes and triggered secondary earthquakes in Alaska.

The tsunamis generated by the earthquake hit the coastlines of Indonesia, Sri Lanka, India and Thailand — and even as far away as Port Elizabeth in South Africa, some 8000 kilometres from the earthquake epicentre (see figure 2.38). The US Geological Survey reported that as many as 283 000 people were killed, with 14 100 people still missing and over one million left homeless.

Worst hit was the Sumatran city of Banda Aceh (see figure 2.39). A 20-metre tsunami pushed a 3-metre wall of water, mud and debris a distance of 10 kilometres inland from the coastline.

Path of destruction ... how the tsunami exacted its deadly toll

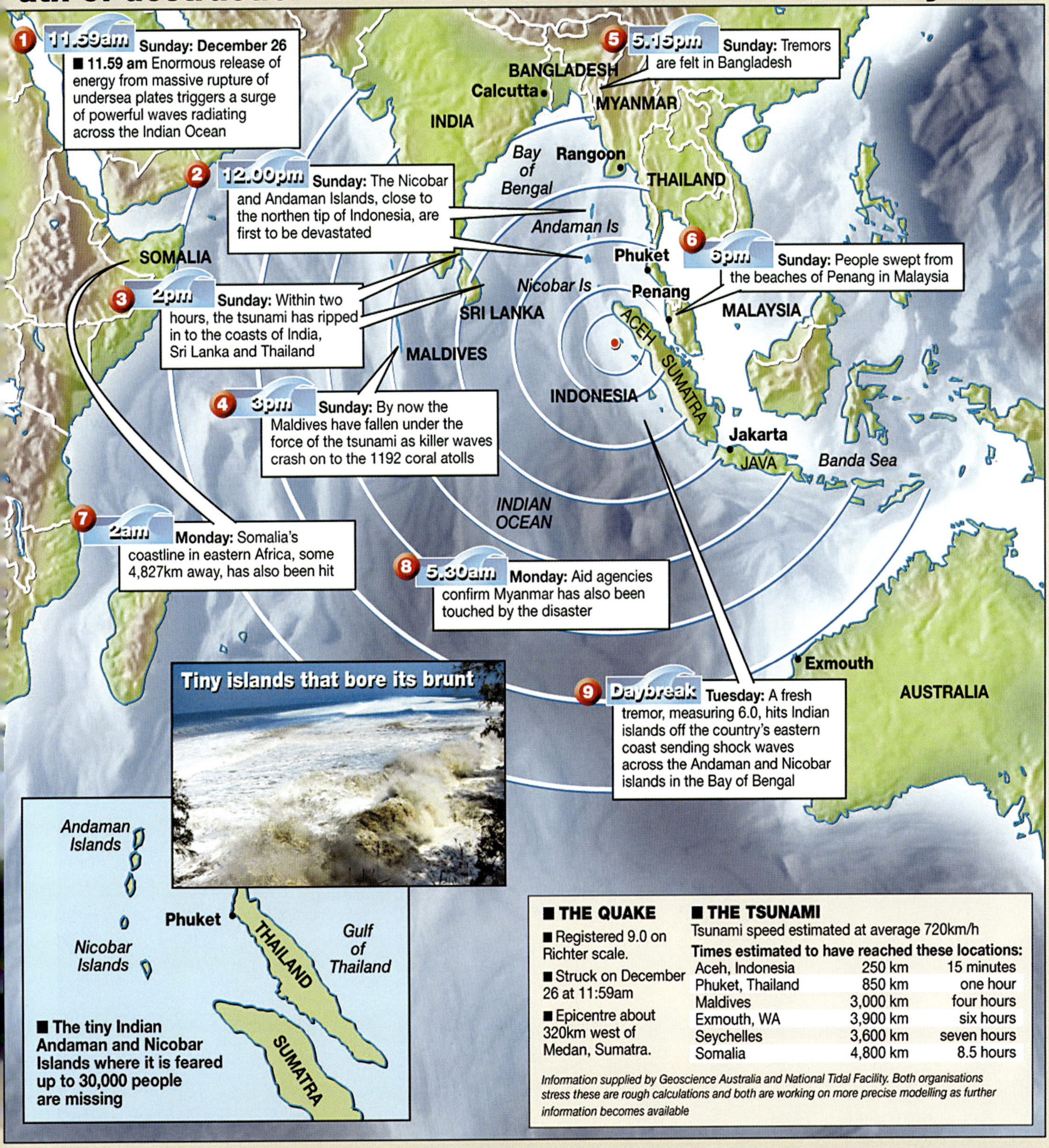

Figure 2.38:
Countries most directly affected by the 2004 Sumatra–Andaman earthquake

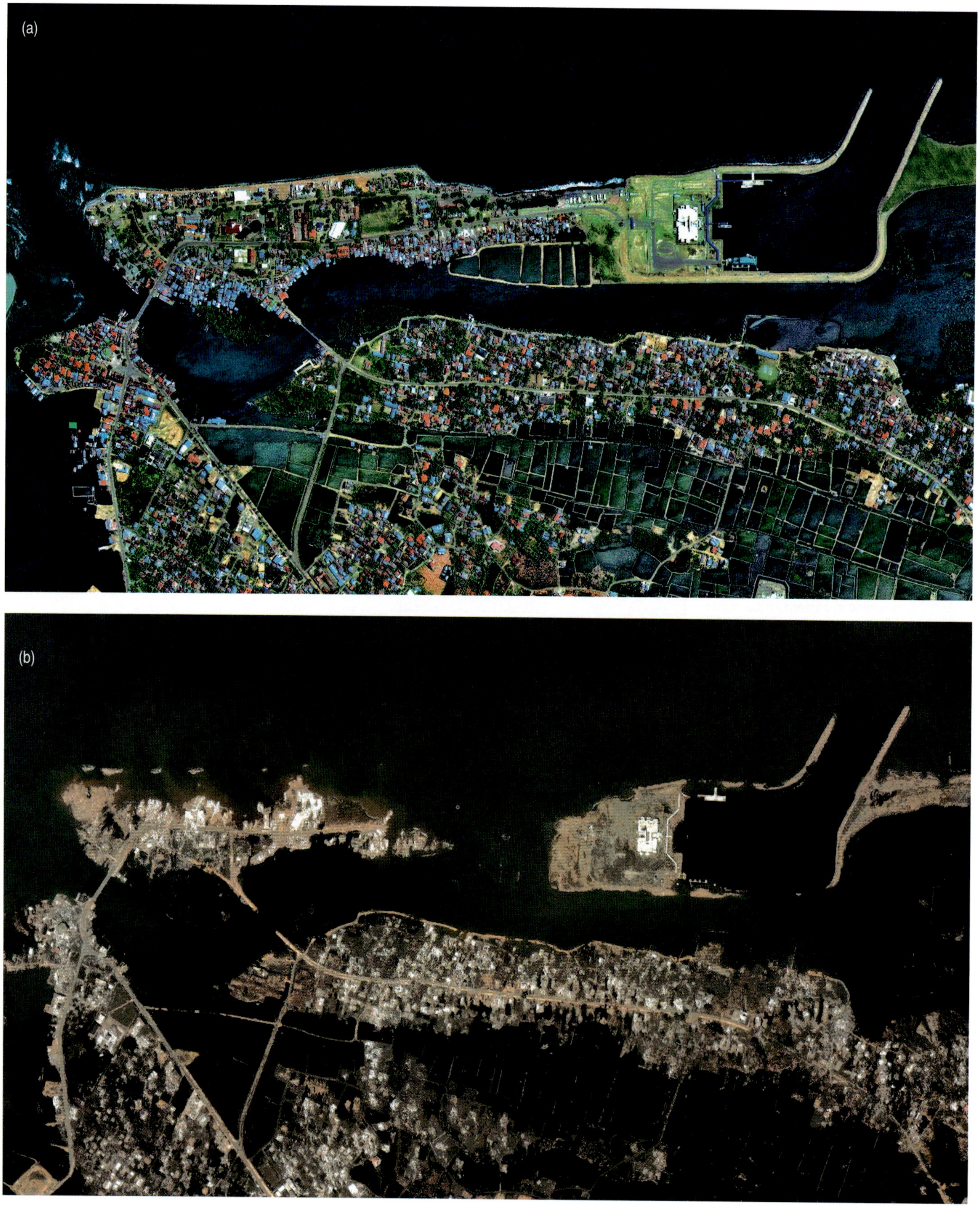

Figure 2.39:
Banda Aceh before (a) and after (b) the tsunami.

Eye on the media

Tsunami waves 'shot along' mid-ocean ridges

The Asian tsunami of 26 December last year sent waves ricocheting around the world. But the pattern of the waves was puzzling. Beaches in Brazil, on the other side of the Earth from the underwater earthquake, were hit by larger waves than the shores of the Cocos Islands just 1700 kilometres away. Now there is an explanation: underwater structures such as mid-ocean ridges and continental shelves funnelled the waves across massive distances.

Vasily Titov of the US National Atmospheric and Oceanic Administration's laboratory in Seattle and his colleagues collected data from tide gauges around the world and satellite data of sea-level changes that day. They then used computer models to analyse how the tsunami spread around the globe.

They found that two main factors affected the reach of the tsunami: the direct effect of the earthquake's force and how the topography of the sea floor guided the waves. Close to the epicentre the waves were controlled mostly by the shape of the fault and the long, thin body of water the earthquake displaced. 'The Cocos Islands lie to one side, so they didn't receive much direct energy,' says Titov. Tide gauges on the islands recorded waves about 50 centimetres high, while gauges in India and Sri Lanka at a similar distance from the epicentre recorded waves nearly 10 times that.

Waves further afield were influenced by the sea floor. 'The energy shot along mid-ocean ridges and continental shelves, to reach locations like Peru and Mexico,' says Titov. Some beaches there recorded waves more than 50 centimetres high (*Science,* DOI:10.1126/science.1114576).

The computer model also explains why some nearby islands, such as Nias, suffered little initially, but were hit by a large wave hours later. 'Nias was close to the source, but it lay to the side of the main energy beam. It received its largest wave around four to six hours later, reflected back from the shores of Sri Lanka,' says Titov.

Now the researchers plan to use their model to predict the impact of potential tsunamis originating elsewhere, such as along the large underwater fault running up the US west coast.

Source: Kate Ravilious, *New Scientist*, 3 September 2005, p. 13.

Clarification questions

1. What was puzzling about the pattern of the waves of the Boxing Day Tsunami?
2. What explanation for this pattern is put forward in the article?
3. What evidence is there to support this explanation?
4. According to this model, what was the effect of the sea floor on the tsunami waves?
5. What does the word 'topography' mean as used in this article?
6. How do the scientists who developed this model plan to apply it in the future?

Is Brisbane in danger from a tsunami? Discuss this after considering the following questions:

1. How far is Brisbane from a plate boundary?
2. Have there been any earthquakes that have an epicentre in the Pacific Ocean near the South-East Queensland coastline? If so, what magnitude were these earthquakes?
3. What magnitude does an earthquake have to be to generate a tsunami?
4. Is there a clear path for a tsunami to travel from the open ocean to the coastline?
5. How close is Brisbane to the coastline? What proportion of the city is within 10 kilometres of the coast?

End of chapter questions

1. Describe the main types of plate interaction.
2. How is continental crust different from oceanic crust?
3. What is the lithosphere?
4. State whether each of the following statements about earthquakes is true or false.
   (a) A subduction zone occurs where one plate is forced into the mantle by another plate.
   (b) Continental crust is usually older than oceanic crust.
   (c) Earthquakes are more likely to occur in New Zealand than in Australia.
   (d) P-waves move slower than S-waves.
   (e) The place within the crust where rocks fracture is called the epicentre of an earthquake.
5. What causes tsunamis?
6. Vulcanologists have stated that dust in the atmosphere is a major factor in causing changes in the Earth's climate. Suggest some ways in which volcanic dust blasted high into the atmosphere could disturb climatic patterns.
7. Explain the theory of continental drift.
8. How does the Richter scale differ from the Mercalli intensity scale?
9. Table 2.4 shows the number of people killed in some the major earthquakes in recent years.

Table 2.4: Earthquake death tolls

Year	Location	Number of deaths	Richter magnitude
1997	Jahalpur, India	38	5.8
1998	Papua New Guinea	2183	7.0
1999	Chi-Chi, Taiwan	2400	7.7
2000	Napa, USA	0	5.0
2001	El Salvador	844	7.7
2001	Coastal Peru	75	8.4
2001	Gujarat, India	20085	7.7
2003	Bam, Iran	40000	6.6
2004	Sumatra	213000	9.0
2005	Kashmir, Pakistan	76000	7.6

   (a) Which two earthquakes best show that the Richter scale measurement does not indicate the death rate in earthquakes?
   (b) What factors, apart from the Richter scale magnitude, affect the likelihood of deaths in an earthquake?
10. Figure 2.40 demonstrates how to use Richter's original method to measure a seismogram for a magnitude estimate in Southern California.

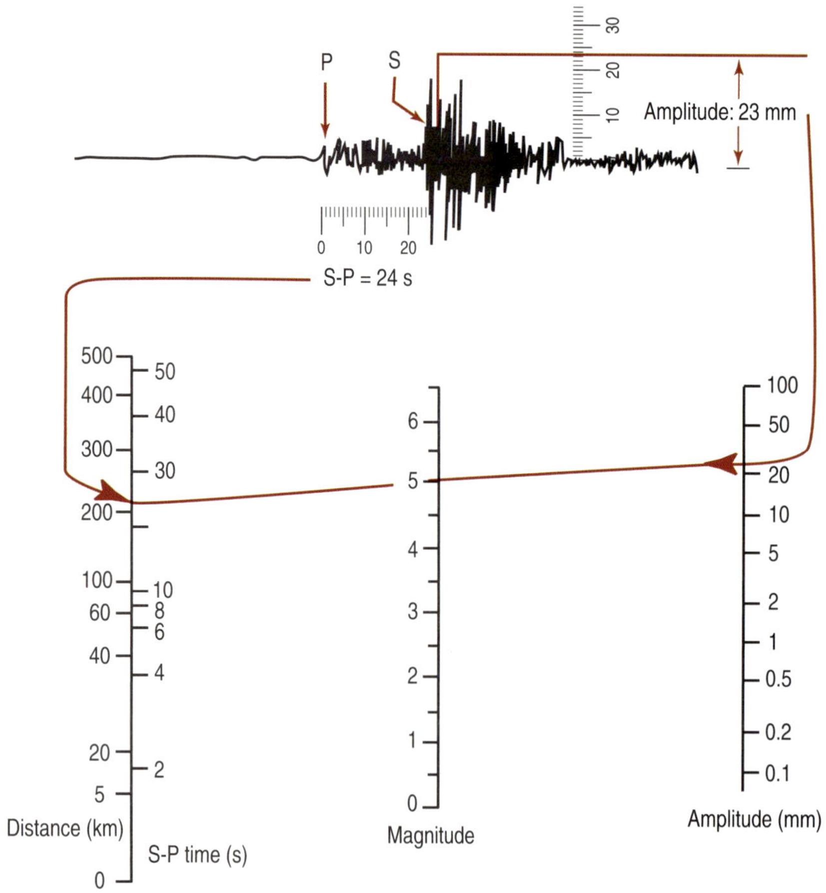

Use the diagram to determine

(a) the magnitude of an earthquake that has an S-P time of 4 seconds and an amplitude of 0.2 millimetres

(b) the S-P time of a magnitude 4 earthquake that has an amplitude of 5 millimetres

(c) the distance of a detector from an earthquake epicentre if the S-P time was 20 seconds.

Figure 2.40: Determining the Richter scale magnitude

11. Explain why the Himalayas are still increasing in height.
12. How is magma different from lava?
13. What is a pyroclastic flow?
14. Give example of locations where the following plate movements are occurring:
    (a) divergence
    (b) subduction
    (c) translation.
15. Table 2.5 shows the temperature at different depths of a hole drilled into the Earth's crust.

Table 2.5: Temperature at different depths

Depth (km)	Temperature (°C)
0	15
1	44
2	73
3	102
4	130
5	158
6	187
7	215
8	242

Use the information in the table to draw a graph and answer the following from the graph:

(a) Predict the temperature at a depth of 4.5 kilometres.

(b) At what depth is the temperature 200°C?

(c) Could you use this graph to estimate the temperature at the centre of the Earth (depth of 6475 kilometres)? Explain your answer.

(d) Why do you think that temperature increases as you go deeper into the Earth?

16. What type of crust — oceanic or continental — would the Pacific plate be made from? Explain.

17. Why does an earthquake have a single Richter scale magnitude regardless of where it is experienced, but differing Mercalli intensities in different regions?

18. Why does Australia experience earthquakes when it is not near a plate boundary?

19. What evidence is there to support the theory of continental drift?

20. Susan was woken from her sleep by a loud crashing noise. At first, she thought that a truck had hit the side of the house because she lived near a busy road but, on getting out of bed, she found that the windows in the lounge room were broken and the whole house was shaking. As she watched, the light fittings swung wildly and her favourite picture fell from the wall, along with many of the books from her bookshelf. Just as she was wondering what was going on, the shaking stopped and everything was silent. Susan realised that she had just experienced an earthquake.

(a) Use Susan's account to estimate the Mercalli intensity of the earthquake she observed.

(b) The next day, Susan learned that many of the buildings in the next town had been badly damaged by the same earthquake. Her own house seemed fine. What possible explanations are there for the difference in extent of damage in the two locations?

(c) Suggest ways in which Susan could 'earthquake-proof' her house to minimise damage.

(d) While watching a Discovery Channel program, Susan saw an earthquake program that recommended staying under a door frame or inside a small room such as the toilet in the event of an earthquake. Why would these suggestions be made?

3

Colonising other worlds

Earth is our home but we must face the possibility that we may one day have to leave it. Overpopulation, the search for new resources, the spirit of exploration, even the threat of ecological disaster, could all be the spurs that drive us to a new place in the space. However, the nearest places to us such as the Moon and Mars are inhabitable, while the promising planets found around other suns would take thousands of years to get to with our current level of technology. So where will we go from Earth?

Space stations

So far, the closest we have come to colonising space has been a succession of space stations that have been home to a handful of scientists at a time. Unlike a spacecraft, a space station has no propulsion system apart from small jets that allow it to make minor changes to its orbit, and it has no capacity to re-land on Earth. Also, while spacecraft are made to support a crew for short durations of up to 10 days or so, a space station is a living environment that can be an astronaut's home for much longer — weeks, months and even years. These orbiting research stations are generally placed in orbit a distance of 400 kilometres or so from the surface of the Earth, and pass overhead many times a day.

History of space stations

The first space station, the Russian *Salyut 1,* was launched on 19 April 1971. While it was successfully placed into orbital position, its first crew was unable to board due to the difficulties of docking their spacecraft with the poorly designed port of the station. Its second crew was able to board and stayed there for 23 days. Unfortunately, this crew died on re-entry when a pressure valve in their *Soyuz* spacecraft opened prematurely. *Salyut 1* re-entered Earth's atmosphere on 11 October 1971 after having been in orbit for less than six months.

The first truly successful space station, *Skylab,* was launched two years later in 1973 by the United States of America (see figure 3.1). *Skylab* was primarily a science and engineering laboratory that focused on observations and experiments involving astronomical phenomena and imaging of the Earth's surface. In its six years of service, a total of three crews of three men stayed on board for assignments of 28, 59 and 84 days. Eventually it was decided to scrap the vehicle and it re-entered Earth's atmosphere on 11 July 1979. While the majority of the station burned up on re-entry, some substantial pieces of the station landed near the town of Balladonia in Western Australia. These pieces have been placed in the town museum.

Figure 3.1:
Skylab

Figure 3.2: *Mir*

Learning from many of the design problems they had experienced previously and from the progress of *Skylab,* Russia launched the first module of its space station, *Mir,* on 20 February 1986. Over the next 10 years, a further eight modules were attached to the core unit until it was complete. Scientists from many countries including Russia, the US, Canada, Germany, Japan and Australia stayed on *Mir* in three-month shifts. In March 1995, medical specialist Valery Polyakov set the present record for the longest continuous stay in space — 438 days! After the station's orbit stabilisers started to fail and continuing finance for the station maintenance dried up, the decision was made to decommission *Mir.* In 2001 it was guided into the atmosphere, where most of the station burned up and the rest fell into the Pacific Ocean.

Internet activity

To complete a copy of the following table, go to the website for this book and click on the Mir modules weblink (see Weblinks, page 309). The weblink is for an interactive activity showing how and when the individual modules were added to *Mir.*

Module	Addition date	Mass (tons)	Purpose
Core			
Soyuz			
Progress			
Kvant			
Kvant-2			
Kristall			
Spektr			
Priroda			
Docking compartment			

The only space station currently in orbit is the International Space Station, or ISS (see figures 3.3 and 3.4). Like *Mir,* the ISS is being built in pieces, with the first section placed in orbit in 1998. Over time, different modules have been added to the existing structure and will probably continue to be added for a while yet. The ISS was put together as a joint project between the world's biggest space agencies: National Aeronautics and Space Administration (NASA), Russian Federal Space Agency, Japan Aerospace Exploration Agency, Canadian Space Agency and the European Space Agency.

Figure 3.3:
The International Space Station, or ISS

Figure 3.4:
Astronauts sharing a meal inside the ISS

Since November 2000, the ISS has been manned continuously by crews of two or three people, all of whom have been members of either the United States or Russian space programs. Each crew stays for six months, although visitors have come for shorter time periods. These are generally scientists, but so far three 'space tourists' have gone to the ISS. The first of these was Dennis Tito, who paid 20 million dollars (US) for the trip! While this is a lot of money, there seems to be no shortage of applicants willing to pay for a trip to space. It is even possible that, over time, money from space tourism will lead to the commercial development of space stations as an alternative living environment.

Eye on the media

Tourism takes off

The dream has been around since Hugo Gernsback founded *Amazing Stories,* the first 'scientifiction' magazine, in 1926: space travel for the price of a ticket, and rocket ships plying the spaceways as frequently as ocean liners. It's taken 80 years, but the dream is on its way to coming true. Two companies — Virgin Galactic, owned by British airline magnate Richard Branson, and an Oklahoma start-up called Rocketplane Ltd — are set to build tourist spaceships in 2006. Hawking rides for $200 000 a pop, they're not exactly bringing space to the masses. But for the first time, large numbers of people will be able to witness the daytime sky turn black, see the Earth curve beneath them, and feel themselves weightless.

Who will rule the spaceways? Virgin Galactic seems poised for success, with spaceships built by Burt Rutan's Scaled Composites, developer of the world's first, and so far only, privately-built manned spacecraft. Construction on a fleet of five eight-seat spaceships will begin this year, although the first tourist flights won't take off until 2008, after a rigorous flight-test program. As with Rutan's original *SpaceShipOne,* each *SpaceShipTwo* will ride to a launch altitude of about 50 000 feet attached to the belly of a carrier plane, where it will be released to fire a rocket engine fuelled by nitrous oxide and synthetic rubber for the boost to space.

The Rocketplane venture is decidedly more speculative. Its mad-scientist design calls for hot-rodding the fuselage of a used Lear business jet with a liquid oxygen-and-kerosene-powered rocket engine. 'We can get a Lear jet fuselage for $15 000,' says executive vice president David Urie. 'It's just from an airplane that's had some form of accident that did not in any way involve the fuselage.' If that doesn't sound very reassuring, consider that for 30 years Urie served as an engineer and program manager for aerospace giant Lockheed's famed Skunk Works division; he ought to know what he's doing. Rocketplane plans to complete its first spaceship by this fall [autumn] and launch the first tourist flights in 2007.

Their three-seat rocket plane will take off from Oklahoma under jet power, blast into space from 20 000 feet, and land by restarting its jet engines after re-entry.

At first, flights with both spacelines will amount to little more than quick joyrides. Without the extreme speed needed to reach orbit, the early tourist spaceships will allow only a few minutes of weightlessness and out-of-this-world views. But revenues from these suborbital craft will fund orbital ships now on the drawing board. And economies of scale should drive suborbital ticket prices down to the cost of an econo car within the next few years — opening up space travel to many more of the world's starry-eyed adventurers.

Source: Michael Belfiore, *Popular Science,* January 2006, p. 57.

Clarification questions

1. Which two companies are starting to sell tickets for spaceflights?
2. How long will these flights last?
3. Describe the spacecraft that will be used for these joyrides.

Design a brochure for Paradise Station, a space station built by an estate company for residential living.

The necessities of life

Wherever we go, there are some things that we simply cannot do without. Like all life forms that have arisen on Earth, we need water, food and breathable air. Any environment that we intend living in must allow us to have access to these basics.

Water

Earlier spacecraft missions to the ISS brought water up with them. This was very expensive to do at first, because water has a mass of 1 kilogram per litre and each of the soft water bags that were transported up held about 35 litres. If there were 10 of these, the transported water alone contributed 350 kilograms of extra mass, thereby using more fuel and leaving less room for other things. As there is now sufficient water on the station, water recycling has commenced. All water — including waste water, urine, water from hand washing or shaving, and air moisture — will be reclaimed, filtered and reused. By recycling water, much less will need to be transported from Earth. Without the recycling, an extra 4000 litres of water would need to be brought up each year for each crew member on board.

Investigation

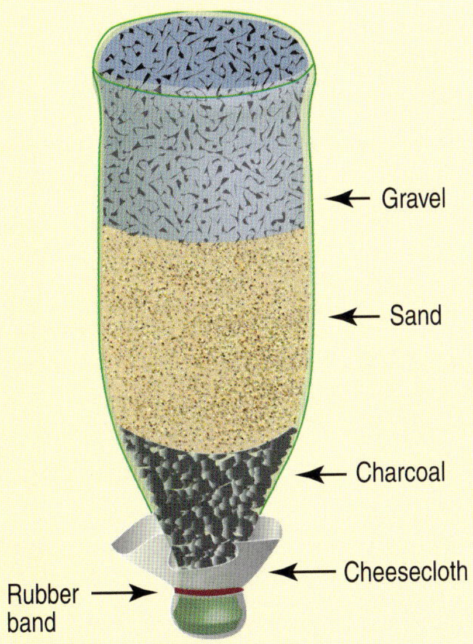

Figure 3.5: A basic water filter

Recycling water

Aim: To demonstrate the principles of purifying and recycling water on a space station

Materials: 2-litre plastic softdrink bottle; activated charcoal; aquarium gravel; sand; a piece of cheesecloth; water; soil; yellow food colouring; 250 mL conical flask; salt; vinegar; rubber bands; pH testing strips; scissors; retort stand; 250 mL beaker

Method:

1. Prepare your impure water sample by pouring 100 millilitres of water into the conical flask and adding to it a teaspoon each of soil, salt and vinegar, as well as three drops of yellow food colouring. Stir the mixture thoroughly.
2. Use the scissors to cut the rounded bottom off the softdrink bottle.
3. Cover the mouth of the bottle with two or three layers of cheesecloth. Secure these in place with a rubber band.
4. Use the retort stand to hold the bottle mouth down over the empty beaker.
5. Pour the charcoal into the bottle until it is about 8 centimetres deep. On top of this, pour a 10-centimetre layer of sand and then, finally, an 8-centimetre layer of gravel (see figure 3.5).
6. Measure the pH of your water sample using the pH strips and note the sample's appearance (colour, floating particles etc.).
7. Slowly pour the sample water into the open end of the bottle. Watch what happens as it passes through each layer.
8. When all of the water has been filtered, test its pH and note any changes to its appearance.

Discussion questions

1. What changes were made to the water as a result of the filtering process?
2. Do you think that this filtered water would be suitable for drinking? Explain your answer.
3. The water to be recycled on the space station also has a bacterial content. How could your filter be modified to remove these?
4. What do you think would be the effect if:
   (a) the layers were put in a different order?
   (b) one of the layers was removed altogether?

Oxygen

Water is needed for drinking, washing and so on, but it is also used as a source of breathable oxygen. As you will remember from Junior Science, each molecule of water is made up of two atoms of hydrogen that are bonded to one oxygen atom. When electricity is run through water in a process called **electrolysis**, the bonds between the hydrogen and oxygen atoms are broken, causing hydrogen gas and oxygen gas to be created. By using the electricity that is produced by large solar panels on the outside of the ISS, oxygen is created from water.

Electrolysis is the process by which electric current is passed through a liquid to cause a chemical change to occur.

It should be noted, however, that the atmosphere of a space station or a spacecraft is not purely oxygen; on Earth, oxygen only makes up 20 per cent of the atmosphere, with most of the rest being nitrogen gas. Pure oxygen atmospheres were abandoned as a result of the *Apollo 1* tragedy in 1967. During a routine pre-launch test, sparks from a wire bundle set the pure oxygen atmosphere alight. The fire spread rapidly, and the three astronauts inside died before they were able to get the hatch open.

Investigation

Producing oxygen

Aim: To demonstrate how oxygen gas can be produced by the electrolysis of water

Materials: Hoffman voltameter; water; 12-volt DC power supply; dilute sulfuric acid, two test tubes, two tapers

Method:

1. Fill the cylinders of the voltameter with enough water to cover the electrodes. Add two drops of dilute acid to the water. This is necessary because electrolysis works only if there are enough free ions to allow current to flow. It will not work with pure distilled water.
2. Invert the test tubes and place them over the electrodes (see figure 3.6).
3. Connect the voltameter to the power supply and turn it on. Bubbles should be produced in both of the voltameter tubes. If there are no bubbles, turn off the power and add some more dilute acid.
4. Allow the voltameter to run 15–20 minutes. During this time, you should see the water level getting lower.
5. Remove the test tubes, ensuring that they are kept inverted so that the gas inside does not escape. Label them to show which electrode (positive or negative) they came from.
6. Light a taper and hold it above the test tube marked 'negative'. Quickly turn the test tube right way up and bring the taper to the mouth of the test tube. A characteristic 'squeaky pop' should be heard, identifying the gas inside as hydrogen.
7. Light a taper and then blow it out so that it is only just glowing. Take the 'positive' electrode test tube and turn it right way up while bringing the glowing taper to its mouth. The taper should flare into flame, identifying the gas inside as oxygen.

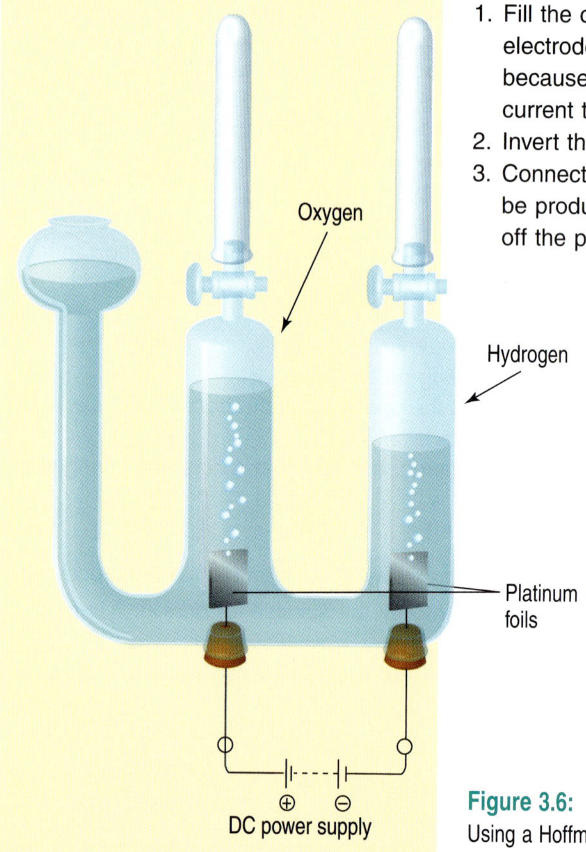

Figure 3.6:
Using a Hoffman voltameter to liberate oxygen from water

Discussion questions

1. Were your taper tests for the two gases successful? If not, suggest reasons why you could not get the results mentioned.
2. Why does oxygen gas make the glowing taper suddenly burn?
3. Estimate the volume of water that has been lost from the voltameter during the 15-minute period that it was on. If we assume that the volume of oxygen gas in the test-tube was 10 millilitres, estimate how long a voltameter would take to produce one cubic metre of oxygen gas. How much water would be consumed to do this?

Food

At this stage, food still has to be transported from Earth to the ISS. However, it is hoped that vegetables can be grown there hydroponically in the future so that some fresh food is available.

Eye on the media

Pies in the sky

'Tea. Earl Grey. Hot.' If only astronauts on the International Space Station could summon their creature comforts as easily as Jean-Luc Picard on *Star Trek's* starship, *Enterprise.* After six months on the ISS, watching the sun rise 16 times a day, astronauts have eaten more than 500 tedious meals of dehydrated, long-life food. Who wouldn't give their right arm for a pizza bubbling with browned mozzarella or a fresh cream bun?

The wacky 'food replicator' in *Star Trek* was pure make-believe. It stored 3D scans of the molecular structure of drinks and meals, then miraculously assembled individual molecules into these foods in seconds. That technology is way beyond our horizons, but a NASA-funded study is about to look into a simpler option — a compact cooking machine that will create a larderful of familiar foods from a limited range of space-friendly ingredients.

The idea is that you could request a sherry trifle, for instance, from a computer. That request would be translated into mathematical instructions controlling a food processor. A network of pipes and chambers would mix, heat and cool the ingredients to make a dessert that looks and tastes like a real trifle — even if it is made from unconventional ingredients.

Back on Earth, such machines could also invent foods to tickle the tastebuds by 'evolving' recipes. 'That would be a completely new approach to food design,' says Eric Bonabeau, chief scientist at Icosystem, a strategy research firm in Cambridge, Massachusetts. 'I think this project could have great potential not just for space, but for the food industry as a whole.'

Space food has come a long way since John Glenn squeezed apple sauce out of a tube during his pioneering three-orbit Project Mercury flight in 1962. But there is still plenty of room for improvement. Neither the space shuttle nor the ISS has a fridge or freezer, so the vast majority of food is not fresh. 'The foods have to be able to last at ambient temperatures for fairly long lengths of time,' says Vickie Kloeris, NASA's manager for shuttle and ISS food at the Johnson Space Centre in Houston, Texas. 'For the station, everything needs to have a 12-month shelf life.'

Drinks are powdered, while most of the food is either freeze-dried or the equivalent of canned food that can be sealed in lightweight pouches and heated in an onboard microwave oven. Astronauts have about 300 items to choose from, and the menus repeat every 10 days. That is dreary enough on a six-month mission to the ISS, never mind future manned trips to Mars, which could take years.

'On a Mars mission, the food could be 3 to 5 years old by the time the crew eats it,' says Kloeris. 'We have a few products which last that long, but not many. We couldn't yet provide sufficient nutrition and variety.'

NASA is researching ways of extending the shelf life of food for a Mars mission, and ways of growing fresh food in space. But Bonabeau suggested something more ambitious to the agency's Institute for Advanced Concepts (NIAC). Why not turn fiction into reality, and design a food replicator?

Source: Extract from Hazel Muir, *New Scientist*, 20 August 2005, p. 30.

Clarification questions

1. How is the food for space travellers currently packaged?
2. What is a 'food replicator' as described in this article?
3. What are the main problems with the food currently being used for space missions?
4. How has space food changed since the early days of the space program?

Which way is down again?

Freefall is the floating effect that objects experience when both they and their environment are falling at the same rate.

When a space station or spacecraft is in orbit, the people inside are in a state of **freefall** (see figure 3.7). Freefall is the floating effect that objects experience when both they and their environment are falling at the same rate. Even 400 kilometres from the Earth's surface, gravity is still at work. Although gravity is a little weaker at that distance than it is on the Earth's surface, it still pulls the station and everything inside it towards itself, and so they fall. Because the curved surface of the planet drops away beneath the station as it falls, the station never actually reaches the surface of the Earth itself. This is how an object remains in orbit — by falling towards the planet and missing!

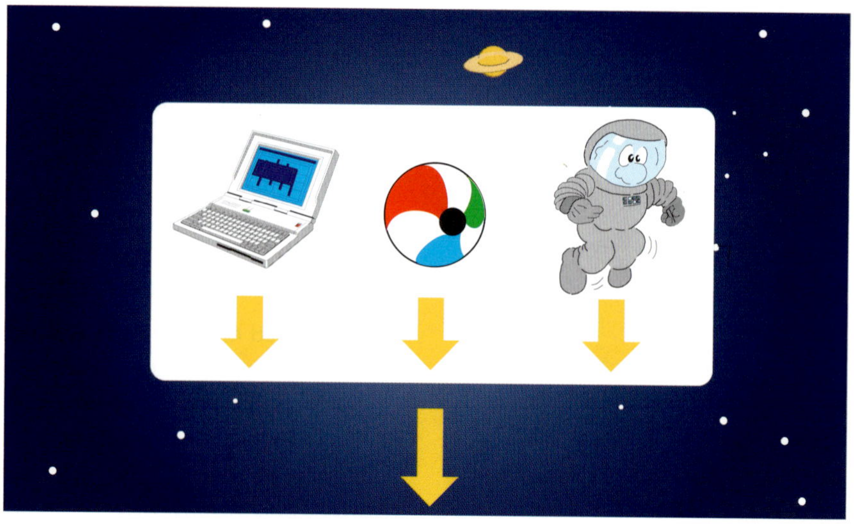

Figure 3.7: Objects in orbit experience freefall.

When objects are freefalling at the same rate as their environment, they do not experience a single orientating force of gravity that indicates 'down'. It is as if gravity has been turned off. Unfortunately, because we have evolved on Earth, our planet's gravity has been a key factor in the way our bodies evolved and the way our body systems function. So while the idea of floating around weightlessly in space on a space station or spacecraft may seem like a lot of fun, the absence of Earth-like gravity can have a number of not-so-fun effects on our bodies.

- *Digestion.* On Earth, gravity helps the muscles in the digestive tract to push material downwards. In space, the muscles have to do all the work on their own — something they were not designed for and cannot do very well.
- *Bones.* Keeping the skeletal structure erect against gravity normally keeps the bones strong. Without gravity, the body leaches calcium from the bones, so they lose their strength. Osteoporosis is found in many astronauts returning home after extended stays in space, and kidney stones are common too.
- *Muscles.* On Earth, our long skeletal muscles and our core muscles are used all the time, even if we are just walking or standing, so they are always under load. Without gravity, these muscles do very little work and so they atrophy, getting weaker, smaller and shorter. As these long muscles shorten, the astronaut develops a crouched posture called 'spaceman's crouch' (see figure 3.8). On returning to Earth after a six-month shift on the ISS, some astronauts have taken months to recover their strength again. Even after short stays of a week or so, they find walking very difficult.

Figure 3.8:
The 'spaceman's crouch'

- *Cardiac system.* The heart muscle weakens because it does not have to work so hard in freefall to pump blood against the force of gravity. As a result, the heart muscles, like the others, atrophy.
- *Anemia.* For reasons that are little understood, returning space explorers suffer red blood-cell loss, which causes weakness and breathlessness.
- *Nausea.* Our middle ears are part of our balance system and are made up of three semicircular canals containing tiny calcium 'rocks'. The orientation of these rocks depends upon gravity. In the absence of gravity, the balance centres cannot operate, causing a constant feeling of seasickness. A malfunctioning digestive system makes matters worse.

Some of the problems can be addressed with a daily exercise program using specialised equipment to keep the body strong, together with medication and a low-fibre diet, but these measures have had varying levels of success.

Creating artificial gravity

If humans experience such physical effects from a low-gravity environment over periods of less than a year, then there would have to be a way of creating artificial gravity on a space station if people are to live on them for extended periods of time. The most likely method of producing a gravitational effect to be employed involves rotating the station. Why would this work? Try the next activity to get an idea.

Material: A 330 mL clear plastic softdrink bottle with a lid; water; food colouring; 1-metre piece of strong string

Pour water into the bottle to a depth of about 3 centimetres and add some food colouring so the water level is easier to see. Screw the lid on and tie the string around the neck. Make sure the knot and the loop are very tight so that the bottle does not come off the string while it is moving (see figure 3.9).

Swing the bottle on the string in a horizontal circle above your head and get someone to observe the water level. Change your speed and see what effects these speed changes have on the water level.

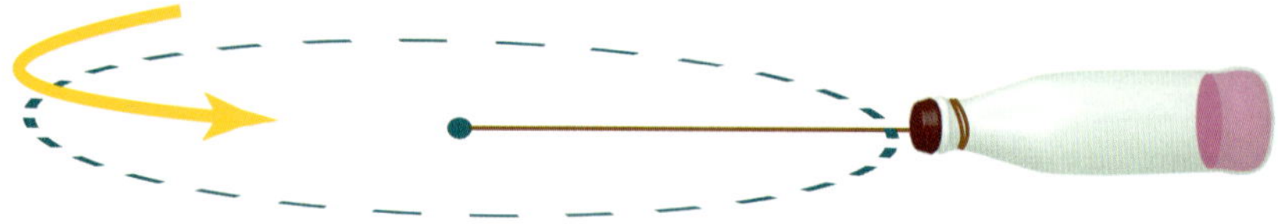

Figure 3.9:
If the bottle has enough speed, the water level stays parallel to the bottom of the bottle.

Centripetal force is the inward-acting force that acts on an object moving in a circle.

Normal force is a force that acts perpendicularly to a surface as a result of, and in the opposite direction to, the net force applied to the surface.

The water level stays parallel to the bottom of the bottle in the activity almost as if gravity is acting sideways — and, in a way, it is! When the bottle is swung in a circle, a force called **centripetal force** acts to keep the bottle moving in a circular path. This force acts towards the centre of the circle. At the same time, an equal-sized force called the **normal force** is exerted on the water, pushing it in the opposite direction — against the bottom of the bottle (see figure 3.10).

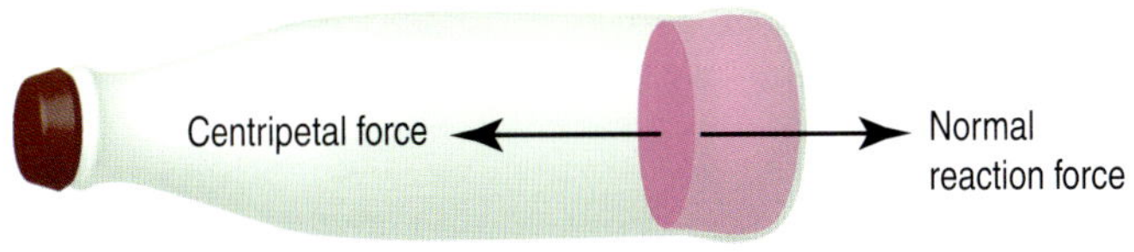

Figure 3.10:

The centripetal force is balanced by the normal reaction force, which presses the water to the end of the bottle.

The same principle can be applied to a circular space station that is spinning about its centre. In this case, the normal reaction force gives the illusion of gravity, with the outer wall of the station being perceived as 'down' while the inner wall closest to the centre of rotation seems to be 'up' (see figure 3.11). To keep the station turning, thrusters would need to be attached to the outer rim of the station and fired when needed.

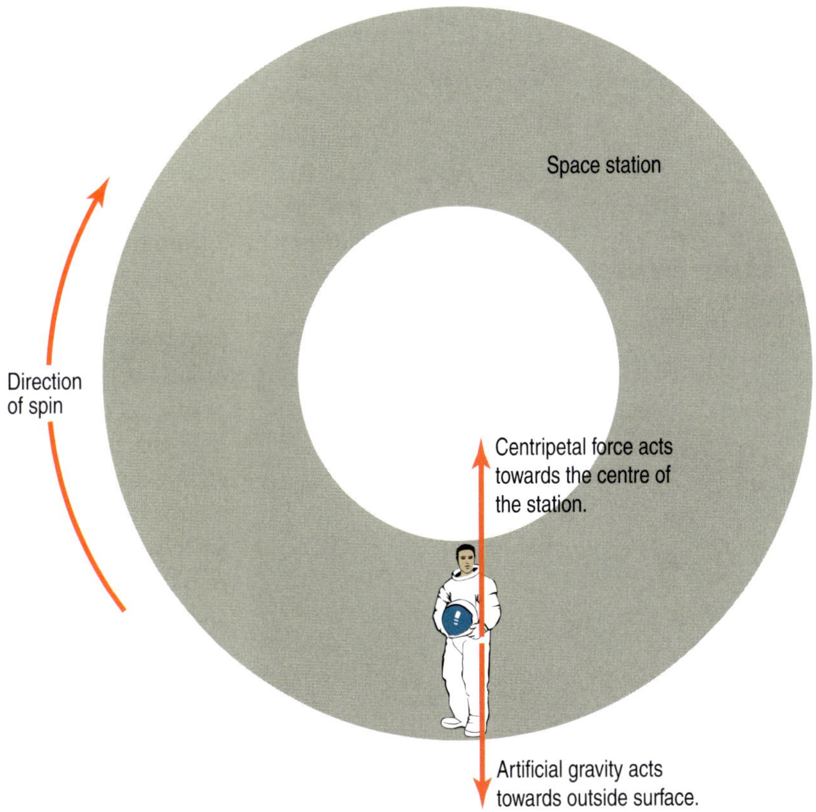

Figure 3.11:

Producing artificial gravity in a wheel-shaped space station

The gravitational acceleration (g) that is produced by this method depends upon the speed at which the rim is travelling (v) and its distance from the centre of rotation (r):

$$g = \frac{v^2}{r}$$

The faster the station spins, the larger the artificial gravity produced. To produce an Earth-type gravity of 9.8 m/s^2, a small space station with a radius of 200 metres would need to rotate so that the speed of the circumference was 44.3 metres per second or nearly 160 kilometres per hour.

$$9.8 = \frac{v^2}{200}$$

$$v^2 = 9.8 \times 200$$

$$v^2 = 1960$$

$$v = \sqrt{1960}$$

$$v = 44.3 \text{ m/s}$$

Destination: the moon

On the 20 July 1969, astronauts Neil Armstrong and Buzz Aldrin stepped out of the *Eagle* module onto the moon's surface, having been delivered there from Earth by the *Apollo 11* spacecraft (commanded by Michael Collins) now high in orbit above them. They had taken three days to get there in a spacecraft guided by computers that were less powerful than an average laptop today. Over the next three-and-a-half years, another five missions and 10 astronauts landed on the moon's ashy surface. The last of these missions, *Apollo 17,* left to return to Earth in December 1972, and humanity has not returned to the moon since. While it seemed that we had learned as much about the moon as we could at the time, new interest has arisen in the moon as the location for a base station. This could serve not only as a launch point for future missions to Mars, but could also be a new home for many people.

Mares are large lava plains found on the surface of the moon.

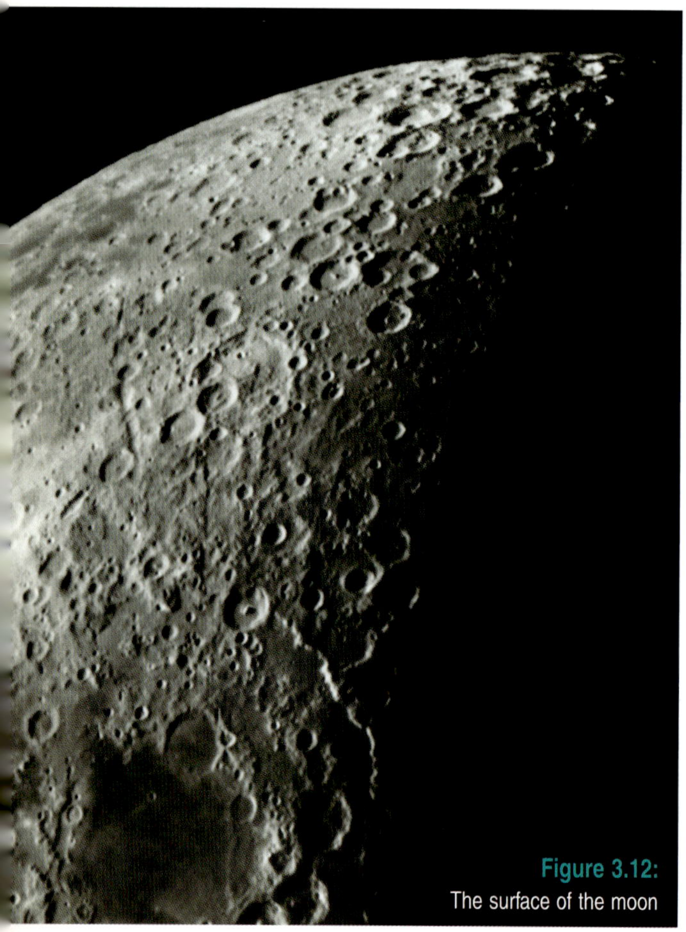

Figure 3.12: The surface of the moon

Conditions on the moon

While pictures of the moon's surface are striking, they cannot hide the fact that it is a dead world (see figure 3.12). Covered in fine white ash, it has a surface pitted with craters formed by asteroid impacts from early in the history of the solar system. Interspersed in these jagged regions are smooth lava plains, the result of asteroids punching through the young moon's crust and allowing magma from beneath its surface to flow out and cool. These lava plains are called **mares** (from the Latin word for 'sea'). These flat regions provided excellent landing sites for the *Apollo* mission vehicles.

The moon's small size means that it has much lower gravity than Earth, and this in turn means that it cannot retain an atmosphere. Without an atmosphere to protect the surface from cosmic rays, the surface is a hotbed of sterilising radiation. Much of the weight of the *Apollo* astronauts' 90-kilogram spacesuits was made up of metal shielding to protect them from this radiation.

The lack of atmosphere as well as the absence of surface water causes the moon's temperature range to be extreme — more than 100°C during the day and down to −173°C at night. Because the time the moon takes to rotate around its axis is about the same as the time it takes to orbit the Earth, the moon always presents the same side to us. People therefore often forget that the moon has a day and a night just as Earth does, although each day and each night there lasts for two weeks! When night falls, the temperature drops by about 150°C in an hour.

A dome away from home

The moon has no air, is bathed in radiation and has sterile soil and extreme temperatures. Obviously, tents are out of the question! If people are to live in this inhospitable environment, they will need a sealed, completely airtight structure. This would allow a breathable atmosphere to be retained and temperature to be regulated, and it would also shield the occupants from the solar radiation and the small meteorites (called **micrometeorites**) that occasionally strike the moon's surface.

A **micrometeorite** is a dust-sized piece of space rock.

Most science-fiction movies show this moon habitat as having a dome shape. In reality, a dome is probably the best shape for the structure because it would allow the forces exerted by the pressure of the artificial atmosphere inside to be evenly distributed (see figure 3.13). This minimises stress on the structure.

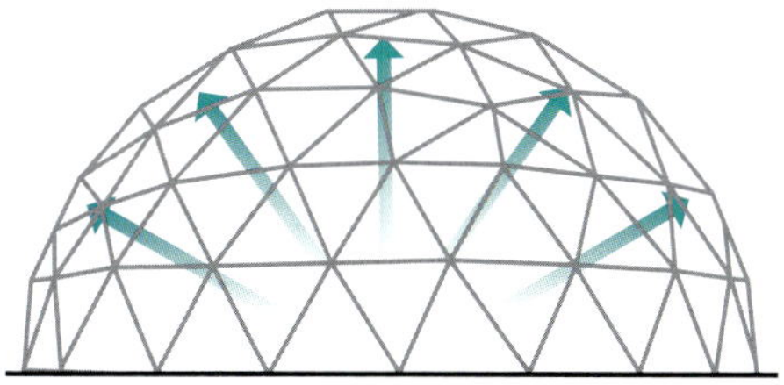

Figure 3.13:
A dome shape allows the forces exerted by the artificial atmosphere to be distributed evenly over the structure.

The structure will probably be made of some strong, opaque material such as modified concrete reinforced with steel. Normal concrete, such as that used on Earth, could have difficulty 'curing' in the conditions on the moon. Incorrectly cured concrete is weak and brittle. Also, concrete components are heavy and bulky, making them difficult to transport. However, it has been suggested that the high silicate content of the moon's surface itself could be used as a component of the concrete, so fewer materials would need to be transported to the moon.

Geodesic domes made of small interlocking triangular steel frames supporting thick glass panes would also be feasible (see figure 3.14). These have the advantage of being transportable from Earth in pieces for assembly on the site.

Geodesic domes are domes made of interlocking steel struts that are arranged in small triangles enclosing a panel (usually glass). They were first designed by the architect Richard Buckminster Fuller.

Figure 3.14:
A geodesic dome

Although the moon has a harsh environment, it has many factors in its favour:

- It has plenty of direct sunlight, which can be collected by solar panels to generate electricity for the base.

- Lunar rocks are made up of compounds that contain oxygen — equal to about half the mass of the rocks. Given an energy source, this oxygen can be liberated fairly easily and used for the artificial atmosphere inside the dome.
- While there is no liquid water on the sunny surface of the moon where the temperature is far higher than the boiling point of water at low pressure, the lunar survey craft *Prospector* discovered water ice near the poles. Water ice is also believed to exist in the deep craters that are in permanent shadow.
- Although gravity on the moon is one-sixth that on Earth, it still allows an orienting force that acts upon the body. This is healthier for the human body than freefall. Lower gravity also means that building structures is easier.
- The journey to the moon takes three days, making it close enough for rescue missions to be possible and emergency supplies to be shipped if needed.
- As the moon is 380 000 kilometres away, radio signals (which travel at 300 000 kilometres per second) take only one second to travel the distance. This makes communication between the Earth and the moon almost immediate.

Lessons from Biosphere 2

Perfectly sealed environments have been built on Earth, allowing us to experiment with moon-style living. The most notable, and certainly the most ambitious, of these environments was Biosphere 2.

Biosphere 2 was a sealed steel-and-glass structure built near the town of Oracle in Arizona, United States, by private research firm Space Biospheres Venture (see figure 3.15). It covered an area of 1.28 hectares and was 23 metres at its highest point. As part of an experiment to determine whether a colony could support itself in an environment that was completely independent of Earth, eight researchers entered Biosphere 2 on 26 September 1991. Apart from the initial oxygen in the structure and food crops that had been planted to get them started, no further oxygen, food or water was supplied. The habitat contained animals and agricultural plants, with food for the animals being grown in the dome. The plants also allowed carbon dioxide to be converted into oxygen for breathing, and all water and waste was recycled.

Figure 3.15: Biosphere 2

After two years sealed away from the outside world, the researchers emerged in 1993. The results of the study were encouraging, but not without problems:

- The researchers had lost an average of 16 per cent of their body weight due to crop yields being smaller than predicted.
- Periodic low oxygen levels led to symptoms similar to altitude sickness — fatigue, nausea, sleep problems and breathlessness.
- The structure was found to be not entirely airtight, leaking 10 per cent of its air each year.

Nonetheless, the results of the Biosphere 2 study indicated that a sustainable habitat on the moon might well be possible.

Mission to Mars

Mars has most recently been named as the next target for the United States space program, and talk of colonising the red planet has become louder. How ironic it would be if we ended up being the Martians!

The Martian environment

Supporters of a Martian colony often point out the many similarities between Mars and Earth. Certainly, they are more similar than Earth and the moon:

- A Mars day is 24 hours and 37 minutes long, only slightly longer than one Earth day.
- It has a temperature range that is not as extreme as that of the moon. Mars has a maximum temperature of 30°C during the day, dropping to –75°C at night. The poles are much colder (as are ours), with temperatures getting as low as –150°C.
- Like Earth, the rotational axis of Mars is tilted, causing the planet to have seasons.
- Mars has a gravitational strength one-third that of Earth's.

There are, however, some distinct differences that must be dealt with:

- The atmosphere on Mars is very thin, mostly made of carbon dioxide gas. Air pressure is very low.
- Enormous dust storms can cover the planet and last for weeks.

The biggest problem is the great distance to Mars. Exploration spacecraft such as *Pathfinder* (launched December 1996) and *Odyssey* (launched April 2001) took seven months to reach Mars' orbit. These were small, light, craft without any crew. They were relatively easy to launch and manoeuvre, but the larger vehicles carrying humans would take nearly two years to reach a new home on Mars.

A biosphere could be built as easily on Mars as on the Moon, but the much greater distance to Mars requires more supplies, more fuel and much larger spacecraft. Furthermore, the crew would be essentially isolated from any assistance from Earth.

Figure 3.16:
The surface of Mars as captured by a camera from an exploration craft without any crew

Making a world for ourselves

Many scientists have been struck by how similar the Martian atmosphere is to that of early Earth. Billions of years ago, our atmosphere was also composed predominantly of carbon dioxide and nitrogen gases until the metabolic processes of photosynthesising bacteria produced enough oxygen for animal life to eventually develop. This similarity has led some scientists to speculate that perhaps the same process that turned our atmosphere from unbreathable carbon dioxide into breathable atmosphere could be kick-started on Mars. This would increase the thickness of the atmosphere, allowing energy from the sun to be trapped and heat the planet's atmosphere. Eventually, Mars would be suitable for human habitation. The process by which a planet is manipulated so that it becomes Earth-like is referred to as **terraforming**.

Terraforming is the process by which a planet is artificially adapted to have an Earth-like climate and atmosphere.

The first step in a terraforming process would be to thicken and warm the atmosphere of the planet. The best way to do this is to induce a greenhouse effect — an effect that we are trying to avoid on Earth! This could be done by melting the frozen carbon dioxide at the poles of the planet. A number of methods have been proposed to commence this process, including:

- Placing large mirrors in orbit around the planet (see figure 3.17). These mirrors (which would need to be 200 kilometres in diameter) would reflect light from the sun and focus it onto the poles, melting them and liberating carbon dioxide gas. This would increase the density of the atmosphere.

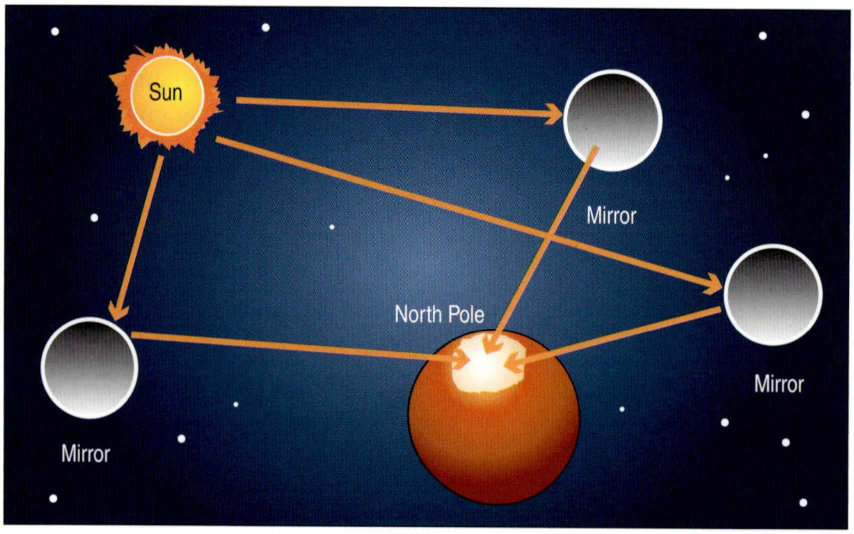

Figure 3.17: Using giant mirrors to melt the polar ice

- Placing greenhouse gas-producing factories on the planet's surface. These would pour carbon dioxide, methane, ammonia or water vapour into the Martian atmosphere, causing energy from the sun to be trapped against the planet's surface, warming it up.
- Bringing asteroids laden with frozen ammonia from the outer solar system into the inner solar system and then crashing them into the surface of Mars. The impact would cause ammonia gas to be released into the atmosphere.

These methods would result in a thicker, warmer atmosphere on Mars. In such an environment, humans would be more protected from solar radiation and could live on the Mars surface without pressure suits, although they would still need an oxygen supply.

(a) (b)

Figure 3.18:

Mars (a) as it is now (b) as it could be after terraforming

It is proposed that, once a thicker atmosphere has been established, photosynthesising organisms including bacteria and plants could be introduced en masse to turn carbon dioxide into oxygen, just as they did on the ancient Earth. It has even been theorised that oxygen factories could be operated that would do the same thing. Once oxygenated, a water cycle could be established. Eventually, Mars could become another Earth.

Regardless of how this terraforming is actually achieved, there is general agreement that the process will take thousands (if not millions) of years using the expected technology of this century.

In search of Earth, version 2.0

Given the technologies required to terraform an entire planet for our needs, perhaps it would be better to try to find a planet that already has an Earth-like environment to which we could travel. We know that none of the other planets in our solar system could be considered even close to habitable, but what about planets in other solar systems?

Our search to find another Earth-like planet orbiting another star requires that the star should:

- have a planetary system
- be similar to our sun
- have planets at a suitable distance from the star to sustain a habitable environment.

Finding exoplanets

Exoplanets or **extra-solar planets** are planets in orbit around stars outside our own solar system.

In our galaxy alone, there are about 400 billion stars of which our own Sun is only one. Statistically, astronomers have assumed that there would be planets around many of these other suns. These planets are referred to as **extra-solar planets** or **exoplanets**.

A **light-year** is the distance that light travels in a year. It is equivalent to 9.461×10^{15} metres.

Astronomers' suspicions were confirmed when a planet was discovered to be orbiting a star designated as HD70642. This planet is very large — it has a mass twice that of Jupiter — and it orbits its sun once every six years. At a distance from us of about 90 **light-years**, however, it could hardly be said to be close by!

How was a planet at such a great distance from us discovered? Unlike the planets in our own solar system, which were discovered as the result of being directly observed through telescopes, an exoplanet is found by the effect it has on the star around which it orbits — it causes the star to wobble. This wobble is caused by the gravitational attraction between the star and the planet. The bigger the planet, the bigger the wobble that is generated in its star.

Investigation

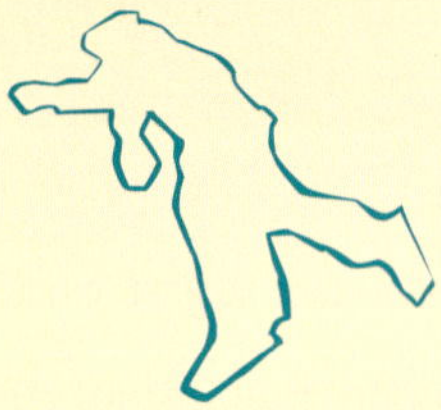

Throwing a wobbly?

Aim: To make a model that demonstrates the wobble produced in the position of a star as a result of the presence of planets in orbit

Materials: 450 g of clay or Plasticine; 1-cm diameter wooden dowel (30–40 cm long); nail; heavy base retort stand with a hollow pipe; plastic wrap; two rubber bands

Method:

1. Split the clay into a 50-gram and a 400-gram lump. Roll each of these lumps to form a sphere. The larger lump will represent a star while the smaller one will represent a planet.
2. Push the spheres firmly onto the opposite ends of the wooden dowel until the ends of the rod are located roughly at the centres of the spheres. *Note:* If it is a warm day, the clay may become soft and start to pull away from the rod. In this case, wrap each lump in plastic wrap after the rod is inserted, twist the edges of the plastic around the rod and secure tightly in place with a rubber band.
3. Find the centre of mass for the rod. This is the point where the rod can be supported on the edge of your finger and is able to stay horizontal. This point will be located between the middle of the rod and the large mass. Mark this position on the rod and drive a nail through it at this point. Be careful not to split the rod.
4. Insert the nail into the top of the hollow pipe of the retort stand. The rod should be able to rotate freely (see figure 3.19).
5. Push the larger mass so that the rod spins slowly, and carefully watch what happens to the larger mass as it rotates around the centre of mass. This is best done if you place your eyes level with the rod. Repeat this several times at different speeds — not too fast though!

Figure 3.19: A model of a wobbling star

Discussion questions

1. What do you think the effect on the 'wobble' would be if the two masses were the same?
2. Would it matter if the two masses were not shaped into spheres? Explain your answer.
3. Would the 400-gram 'sun' show more wobble if the 'planet' was 200 grams or if it was 50 grams?
4. What do you think the limitations of this model are compared to reality?

Only a large planet causes an observable wobble in its sun, so planets the size of Earth would be overlooked. Our own sun wobbles slightly due to the effects of the four large gas giants in our solar system. However, this effect is not as great as it would be if all of the gas giants moved in such a way as to all be on the same side of the sun at the same time. Instead, they are distributed around the sun so that the individual wobble effects nearly cancel each other out.

The observable wobble caused in stars by the presence of large planets is not visible to the eye. Instead, astronomers use a phenomenon called Doppler shift to determine the extent of a star's movement.

Doppler shift

The **Doppler effect** is the apparent change in frequency observed when there is relative movement between a wave source and an observer.

Frequency is a measure of how many waves pass a particular point in one second.

Before considering how Doppler shift works with light, it might be a good idea to start with the Doppler shift of sound first. You will have noticed how, when a police car comes towards you with its siren on, the siren is very high pitched. However, as it draws level with you and moves away again, the pitch gets lower. The same thing can be heard with the engine noise of racing cars as they move around the track. This change in pitch is caused by the **Doppler effect**.

Sound energy and light energy travel in the form of waves. The number of waves that the energy source produces in a second is called the source's **frequency**. The unit for frequency is the Hertz (Hz), which represents one vibration or wave per second. So, if a siren on a police car produces soundwaves at a rate of 1000 waves per second, we say that it has a frequency of 1000 Hz.

When a wave source is stationary, then all observers located around the source — in front of it, behind it or beside it — will observe the waves approaching them with the same frequency. However, if the wave source is in motion, the waves crowd together in front of the wave source and spread out behind it. An observer in front of the wave source encounters more of the crowded together waves in a second than they would if the source was stationary. As a result, they experience a greater number of waves per second and so perceive the waves as having a higher frequency. On the other hand, an observer behind the wave source encounters less waves per second and so experiences a lower frequency of waves (see figure 3.20).

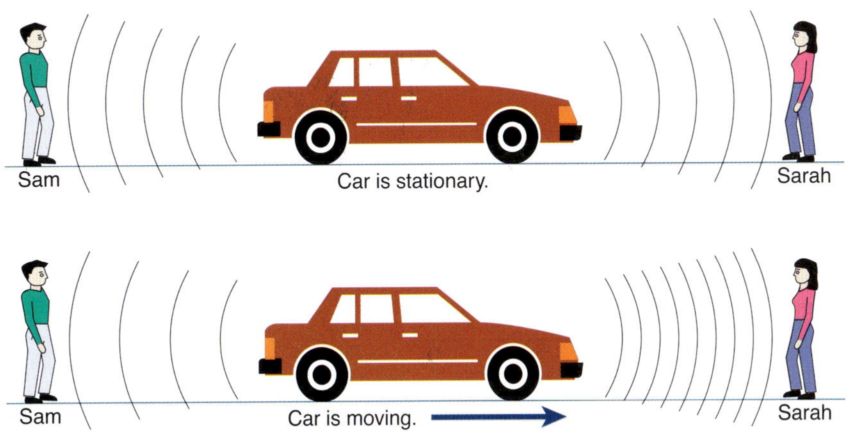

Figure 3.20:
Change in frequency heard when a sound source moves

With soundwaves, increased frequency and decreased frequency can be observed as changes in pitch. Increased frequency means a higher pitch while decreased frequency indicates a lower pitch.

With light sources, a change in frequency corresponds to a change in colour. Visible light is a part of the electromagnetic spectrum that also includes microwaves, ultraviolet, infra-red and radio waves. Our eyes have developed so that they are able to detect light that has a frequency of between 4.3×10^{14} Hz (red light) and 7.3×10^{14} Hz (violet light). All the colours that we see have frequencies between these limits; this is called the **visible spectrum**.

Each star gives off light in wide range of frequencies, some of which are in the visible spectrum. When a star is moving towards us, the frequency of the light that it emits is observed as being higher than it really is. This higher frequency corresponds with the perception of the light it produces as being further into the blue end of the spectrum. We say that the light has been **blue-shifted**. When a star is moving away from us, the frequency gets lower and the colour moves towards the red end of the spectrum; this is said to be **red-shifted**.

Astrophysicists use devices called **spectrometers** to measure the frequencies of light that a star produces. These frequencies depend upon the star's composition and its temperature. When a star has a large planet orbiting it, the star wobbles slightly forward and backward as we saw in the investigation earlier. As the star moves towards us in its wobble, the frequency of light that it emits is blue-shifted; when the star wobbles away from us, its light frequencies are red-shifted (see figure 3.21) . By using spectrometers to measure the range of frequency change, astrophysicists can determine the mass of the planet and its separation from the star.

The **visible spectrum** is composed of light that has frequencies between 4.3×10^{12} Hz and 7.3×10^{12} Hz.

Blue-shifted describes the change in light frequency of stars towards the blue end of the spectrum when they are moving towards us.

Red-shifted describes the change in light frequency of stars towards the red end of the spectrum when they are moving away from us.

An **astrophysicist** is a scientist who uses physics to explain the forces acting on and between objects in space.

A **spectrometer** is a device that measures the frequencies of light produced by a luminous object such as a star.

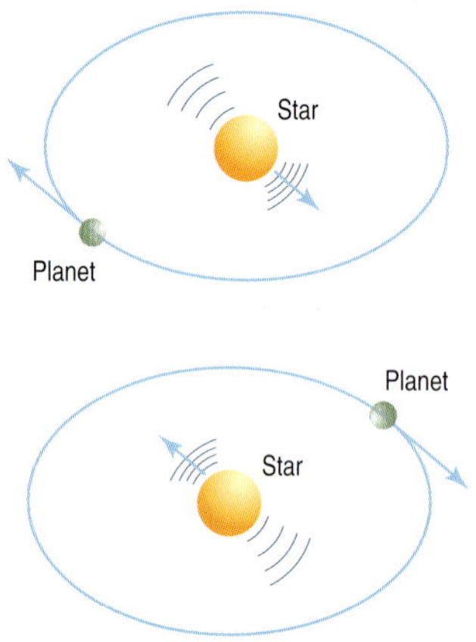

Figure 3.21: How planets are found using the Doppler effect

Suns like our own

Identifying suitable stars that may be similar to our own sun can be achieved by considering their spectral class. This method of classifying stars based upon the frequencies of light that they give off was developed by Annie Jump Cannon, and has been used by astronomers and astrophysicists since 1910. As mentioned in the previous section, the spectrum of light frequencies that a star emits and the colour that it appears to have depends upon its surface temperature and its composition. Stars that are similar in composition and temperature are placed in the same spectral class (see table 3.1). As a star ages, it gets cooler, its colour changes and its **spectral class** gets lower.

As you will see from table 3.1, the temperature of the stars is not given in Celsius. Instead, scientists use a temperature scale called the **absolute temperature scale**. This is based on the fact that the lowest temperature that is possible in theory (called **absolute zero**) is –273.15°C. The basic unit of the absolute temperature scale is the Kelvin (K). Absolute zero is defined as 0 K. As a 1-K temperature difference and a 1°C difference are the same size, we can relate the two temperature systems as shown:

Temperature (Kelvin) = Temperature (°C) + 273.15

As a result, 0°C would be equal to 273.15 K, and 20°C would equal 293.15 K.

The **spectral class** of a star is determined by its surface temperature, its colour and its composition.

The **absolute temperature scale** is devised so that absolute zero is equal to –273.15°C, and a difference of one degree in the absolute temperature scale is equivalent to a one-degree difference in the Celsius (Centigrade) scale.

Absolute zero is the lowest temperature that is theoretically possible. It is equal to –273.15°C or 0 K.

Table 3.1: Spectral classes of stars

Spectral class	Surface temperature (K)	Colour	Examples
O	28 000–50 000	Blue	10 Lacertae
B	10 000–28 000	Blue	Rigel
A	7 500–10 000	Blue-white	Sirius
F	6 000–7 500	White	Procyon
G	5 000–6 000	Yellow	Sun
K	3 500–5 000	Orange	Arcturus
M	2 500–3 500	Red	Betelgeuse

These spectral classes can be further divided into 10 categories numbered 0 to 9, representing where in the temperature range of the spectral class a particular star is located. In this system, a K5 star is located midway in temperature between K0 and M0. Our own sun is a G2 star — two-tenths of the way between G0 and K0. Over the next few billion years, our sun will become a K-class star and then an M-class star. Stars that are considered to be similar enough to our sun to have habitable planets will range from about F5 to K5. Such stars will, like our sun, have a life expectancy of a few billion years, be cool enough to allow water to exist as a liquid on planets that are the appropriate distance from them (further from F-class stars than Earth is to our sun, and closer to the K-class stars), and emit ultraviolet light strong enough to create an ozone layer on a planet but not so much that it renders all of its planets sterile. Astronomers estimate that, in our galaxy, about 5 per cent of stars are similar enough in spectral class to our own to be habitable by humans.

Figure 3.22: A portion of the night sky including the constellations of Orion (the Hunter), Canis Majoris (the Big Dog) and Canis Minoris (the Little Dog). Four of the brightest stars in this region are shown along with their distances from us.

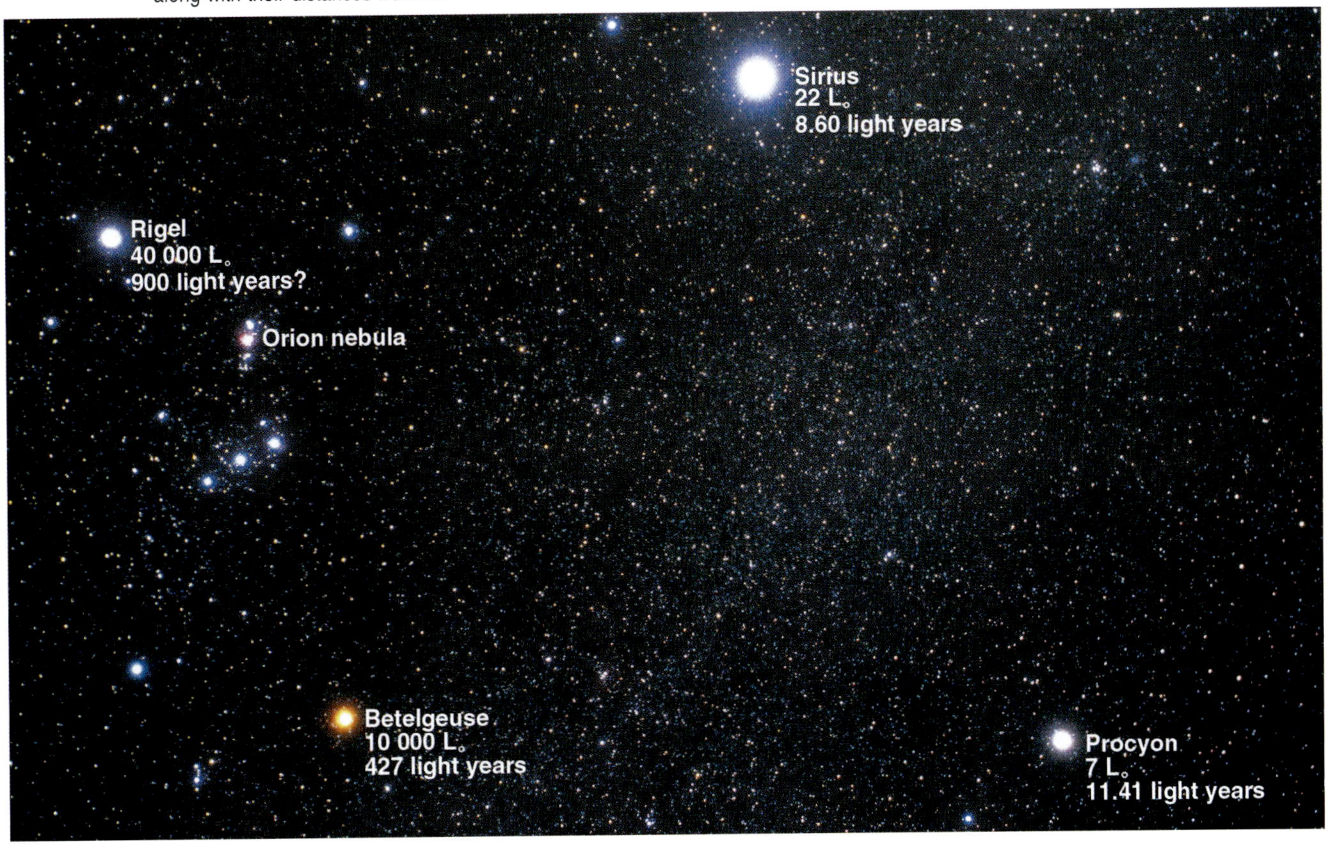

The habitable zone

Over 200 exoplanets have been found so far. However, none of these is likely to be inhabitable because the very things that make them easy to find — their large size and highly elliptical orbits — also make them unable to develop an Earth-like climate or atmosphere.

'So what use is finding them?' you may ask. The answer lies in the fact that it is unlikely that a star that has developed a planetary system has only a single planet in it. While the giant planets that we find may not themselves be suitable, they are evidence that the star developed under conditions that allowed planets to form. Just as our own solar system contains large gas giants as well as our own planet, it is possible that these distant stars have small Earth-like planets as well as large ones.

A **circumstellar habitable zone (CHZ)** is the region around a star where a planet would need to be located in order to have liquid water on its surface. Also known as a habitable zone.

These Earth-like planets would be located in what is known as a **circumstellar habitable zone (CHZ)**, a concept first put forward by Philip Morrison and Giuseppe Cocconi in 1959. They described the habitable zone as being the region around a star where a planet would need to be located in order to have liquid water on its surface, assuming that the planet had a similar atmosphere to our own. Obviously, Earth is located in the habitable zone of our solar system, but Mars is located at the edge of this region (see figure 3.23). While there is presently no liquid water on the surface of Mars, there is evidence that it existed there in the past and it is believed to exist there presently in the form of ice. It is believed that climatic and seismic changes on Mars led it to lose its liquid water.

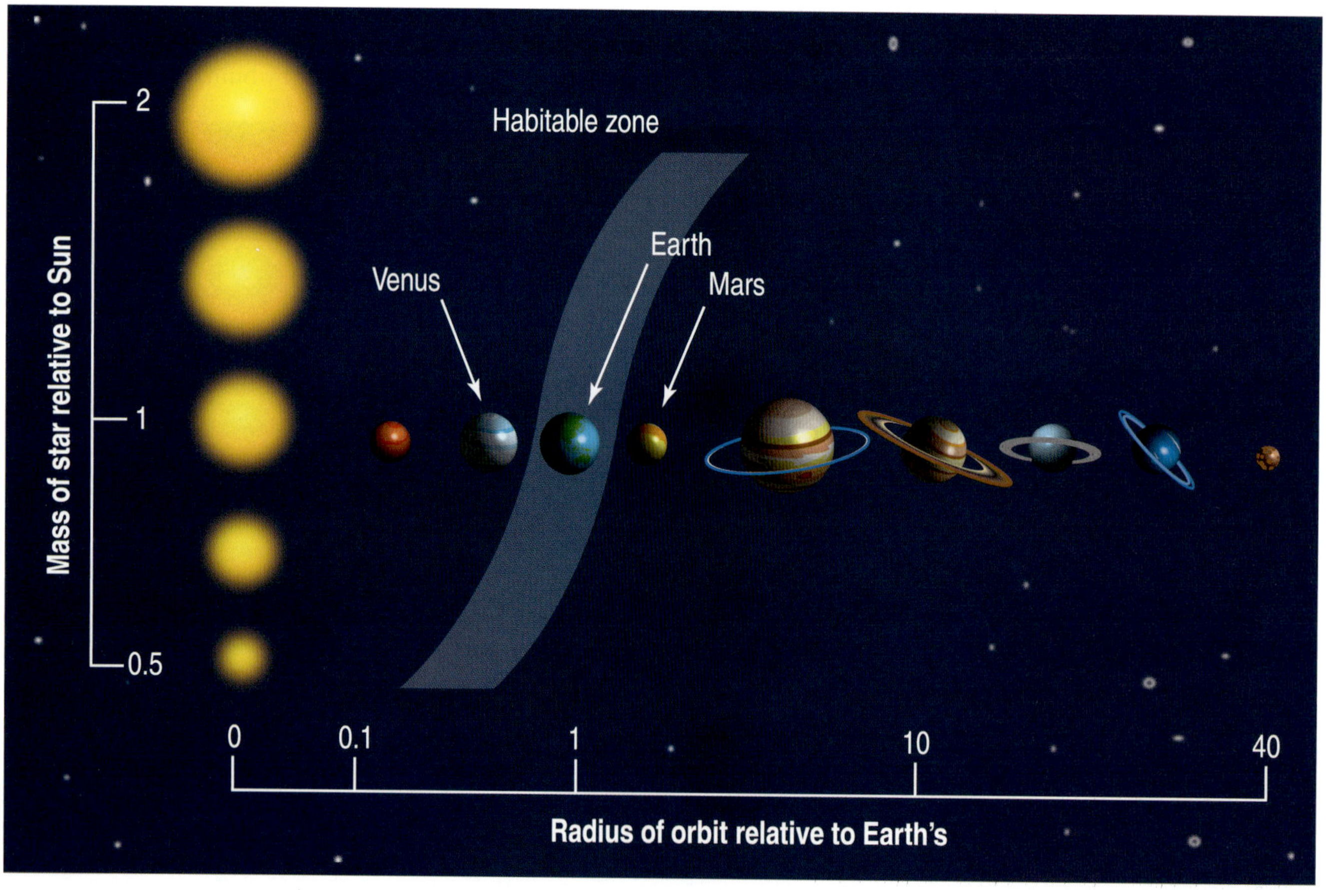

Figure 3.23: The location of the habitable zone around a star depends upon the temperature of the star.

The habitable zone around a star will be found neither too close (like Mercury), where the high surface temperature of the planet causes the water to be vaporised, or too far (like Pluto), where the temperature is so cold that even gaseous atmospheres freeze. Morrison and Cocconi theorised that the CHZ would be closer to cooler stars than our own, and further away from stars that are hotter.

A planet with a habitable zone will also need to be located far from any gas giants in the star system. The sheer size of gas giants means that they have a strong gravitational pull that acts on the planetismals that form rocky planets (like our own) when the system is being formed. Planetismals that are too close to a gas giant do not form Earth-like planets. Our asteroid belt, prevented from accreting by the nearness of Jupiter, is proof of this.

Eye on the media

Earth-like planet may be the first of many

Planet hunters have detected what seems to be the smallest extrasolar planet so far, orbiting a red dwarf 22 000 light-years away. Because red dwarfs are the most common type of star in the Milky Way, this might mean that Earth-like planets are abundant in our galaxy. In any case, it bodes well for 'gravitational micro-lensing', the technique used to find this exoplanet.

Most planet-hunting techniques pick up massive planets in tight orbits around their host stars. About 170 exoplanets have been found around sun-like stars and until now the smallest had weighed in at seven Earth masses. Now Jean-Philippe Beaulieu at the Institute of Astrophysics in Paris, France, and his team have found a planet that seems to be just 5.5 times the mass of Earth. To pick out the star they exploited the phenomenon called gravitational micro-lensing.

When one star passes in front of another as seen from Earth, light from the background star is bent and magnified, or 'lensed', by the gravity of the foreground star. If the star in the front is playing host to a planet, the planet's gravity can boost the light of the background star for a few hours.

The micro-lensing event that revealed the new planet was one of about 1000 picked up each year by OGLE, an international collaboration which monitors 170 million stars in the Milky Way's central bulge. Astronomers in Perth, Australia, who follow up some of these events for another consortium called PLANET, found the exoplanet's telltale signal on 9 August 2005.

But there is still some uncertainty as to the planet's mass — it could be anything from 2.8 to 11 Earth masses. That is because the micro-lensing measures only the ratio of the mass of the host star to the planet's mass, and researchers then have to use models to estimate the most likely mass of the star and the planet (*Nature*, vol. 439, p. 437).

The planet appears to orbit its star at about 2.6 times the distance of the Earth to the sun and is as frigid as Neptune and Pluto, with a surface temperature of about –220°C. It is significant that the planet has been found around a red dwarf because such stars make up about 70 per cent of the stars in the Milky Way, so there could be many more such planets out there.

Other techniques have shown that Jupiter-sized planets are rare around red dwarfs. 'This suggests that lower-mass planets are a lot more common than Jupiters around low-mass stars,' says team member David Bennett at the University of Notre Dame in Indiana.

The find is also an important milestone for micro-lensing. 'They've proven they can detect these low-mass planets,' says Sara Seager of the Carnegie Institution of Washington in Washington Dc. 'It's just a matter of time before they get more.'

Source: Maggie McKee, *New Scientist,* 28 January 2006, p. 12.

Clarification questions

1. What evidence is presented here for an abundance of Earth-like planets in our galaxy?
2. Describe the technique used by Jean-Philippe Beaulieu and his team to find these smaller exoplanets.
3. Why is there such a range of estimates for the new planet's mass?
4. How 'earth-like' is the planet they have found?
5. How many of the stars in our galaxy are red dwarf stars?
6. How far away is the new planet's star?

How would we get there?

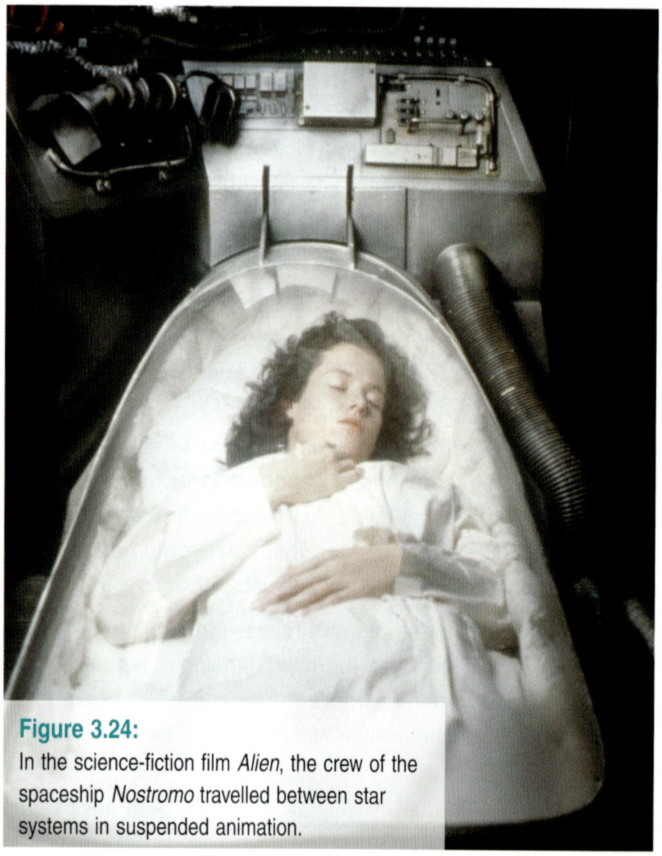

Figure 3.24: In the science-fiction film *Alien*, the crew of the spaceship *Nostromo* travelled between star systems in suspended animation.

Suppose that we do find our new Earth. The next problem is how to travel to the planet. After all, the distances involved are enormous. Our nearest stellar neighbour, Alpha Centauri, is 4.2 light-years away. As one light-year is equal to around 9.46×10^{12} kilometres and our present level of technology is such that our spacecraft can travel at about 30 kilometres per second, then the trip just to Alpha Centauri would take nearly 10 000 years! Obviously, we would need to consider carefully the design of the ships that would take us to our new homes. The design alternatives fall into two main categories:

- slow-speed spacecraft that allow a crew to survive a journey that may take tens of thousands of years
- spacecraft that are able to travel at or near the speed of light.

Sleeper ships

One solution that has been proposed is the design of a **sleeper ship**, which is a type of manned spaceship in which the crew is held in suspended animation. In this state, the body's metabolism is slowed down dramatically — breathing, consumption of nutrients and even aging virtually cease. The main advantage of this is that fewer consumable resources such as food, water and oxygen would be needed until arrival at the destination.

A **sleeper ship** is a type of manned spaceship in which the crew is held in suspended animation until arriving at the destination.

Eye on the media

'This is your captain speaking. Please prepare to enter stasis . . .'

The advantages are huge. Putting astronauts in a state of suspended animation during a trip to Mars, for instance, would greatly reduce the demand for vital oxygen as well as food, water and power — and no doubt help preserve their mental wellbeing. Unlike NASA, the European Space Agency is actively investigating ways of inducing a state of hibernation, including gradually reducing oxygen levels, attempting to reset the body's core temperature, and trying to identify the natural triggers in animals that hibernate. One substance showing promise is a small protein called DADLE, which slows down the metabolism of cells in the lab.

The finding that hydrogen sulfide gas can apparently shut down metabolism in a non-hibernating animal boosts the idea that mammals have a common hibernating ancestor, says Mark Ayre of the ESA's Advanced Concepts Team in Noordwijk, the Netherlands. We might still have vestiges of that ability, he says. Some genes that are active in hibernators are also found in people and could perhaps be switched on.

'If inducing and maintaining torpor in astronauts would require nothing more than exposure to atmospheric hydrogen sulfide, it would be almost too simple', Ayre says. 'No doubt other mechanisms would have to be used for maintaining physiological conditioning and so on . . . but the reduction in metabolic rate and core body temperature is very impressive.'

Those other mechanisms might include ways of maintaining bone and muscle mass, and dealing with waste products. Animals like ground squirrels also rev up their metabolism every two or three weeks during hibernation, perhaps to allow their immune system to deal with any invaders.

But as far as the effect's discoverers are concerned, the idea of adapting it for space travel can wait. They are focusing on the possible medical applications.

Andy Coghlin, *New Scientist,* 30 April 2005, p. 9.

Clarification questions

1. What evidence is there that humans can be placed into hibernation?
2. What do you think the words *stasis* and *torpor* mean as they are used in this article?
3. Why is the gas hydrogen sulfide mentioned in this article? How could it be useful in solving the stasis problem?
4. What is DADLE?

Bussard ramjet

The Bussard ramjet was proposed by physicist Robert Bussard in 1960 as a possible means of interstellar travel that overcame the needs of most other spacecraft designs to carry their own fuel. The ramjet would, instead, use the hydrogen gas it collected from space, as it travelled, as a fuel source.

The Bussard ramjet design has four components (see figure 3.25):

- a powerful laser that is used to ionise the hydrogen atoms ahead of the spacecraft
- a cone-shaped metal mesh 'scoop' that uses electromagnetic fields to channel the hydrogen ions into a fusion reactor

- a fusion reactor that uses a process called nuclear fusion, in which hydrogen ions are combined to form larger atoms, releasing huge amounts of energy
- a fusion engine that channels this energy into thrust, pushing the spacecraft at high speeds.

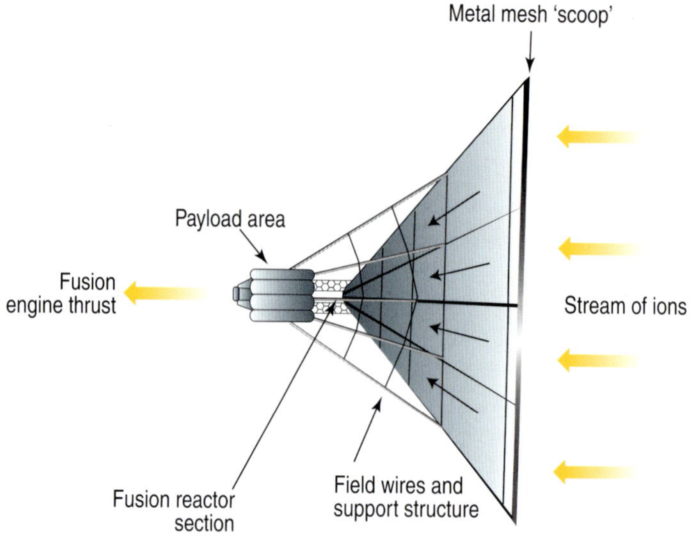

Figure 3.25: A Bussard ramjet

While the design for the Bussard ramjet is quite elegant and would probably work, it has a number of problems that would need to be overcome:

- The density of hydrogen gas between stars is not uniform. In some regions, such as in nebulae, there are about 100 hydrogen atoms per cubic centimetre, while in other regions there can be as few as one atom per cubic centimetre. As a result, there will not be a steady fuel supply and the efficiency of travel will be affected.
- The ramjet will work only when it is able to travel at high enough speeds (roughly 6 per cent of the speed of light) to collect enough hydrogen atoms to maintain its acceleration. This means that it would need another method of propulsion to get it going fast enough for the ramjet to be effective.
- Large amounts of power are needed to run the ionising laser and the magnetic field of the mesh.

A **ram-augmented interstellar rocket (RAIR)** is a modified form of a Bussard ramjet that uses hydrogen atoms to allow the fusion of lithium or boron nuclei to release energy, rather than using just the fusion of the hydrogen atoms themselves as a fuel source.

A modification of the Bussard ramjet, called the **ram-augmented interstellar rocket (RAIR)**, uses the hydrogen ions to cause fusion reactions with lithium or boron atoms. This method has the advantage of needing fewer hydrogen atoms, and releasing a larger amount of energy through the fusion of lithium or boron than could be achieved by the fusion of hydrogen alone.

Antimatter drives

Most of you who read or watch science fiction will no doubt have heard of these (see figure 3.26). An antimatter drive is an engine that runs on the energy released when matter and antimatter combine in equal amounts. When the particles of the two types meet, they completely annihilate each other, releasing 10 billion times more energy than that provided by the hydrogen–oxygen fuel used at present. Spaceships fuelled by such an

Figure 3.26: Starship *Enterprise* — powered by antimatter engines in a fictional future

engine would be able to travel at near-light speeds but, of course, there are problems with this design that must be overcome before it is a reality:

- Antimatter is not common on Earth; it has to be created in particle accelerators. The world's largest particle accelerators — Fermilab (in the United States) and CERN (in Switzerland) — only produce about 1/10 000 of a milligram of antimatter a year. Powering a spacecraft, even a small one, would need a kilogram of antimatter for a voyage to the stars.
- Antimatter cannot be stored the same way as other fuels because it would react on contact with the wall of its container. One solution to this is to store the antimatter in a large 'magnetic bottle', where a strong magnetic field surrounds and contains the antimatter particles. This would require a substantial amount of energy.

Light-speed oddities

Once we design engines that are capable of travelling at speeds near that of light, the journey time to our stellar neighbours becomes more feasible. However, physicists believe that some strange things happen when objects approach light speed. Their beliefs are based upon the calculations of a Dutch mathematician, Hendrik Lorentz, which predicted the effects of high velocities on the dimensions of space and time. Albert Einstein used these calculations (called the **Lorentz transformations**) as the basis for his famous **Special Theory of Relativity** included in his 1905 paper 'On the Electrodynamics of Moving Objects'. These two sets of writings predicted that an object travelling towards the speed of light will experience the following:

The **Lorentz transformations** are a set of calculations by the mathematician Hendrik Lorentz that predict the effects of travel at speeds approaching the speed of light.

The **Special Theory of Relativity** was part of a paper written by Albert Einstein called 'On the Electrodynamics of Moving Objects', which discussed the effect of an object's speed on the space around it.

- *Time dilation*. An observer travelling at the speed of light will appear to a stationary observer to be living in a slower timeframe.
- *Spaghettification*. Objects moving near the speed of light appear to get longer in their direction of motion and very thin in dimensions perpendicular to their direction of motion. In other words, they get stretched out like a piece of spaghetti.
- *Mass increase*. The mass of the object gets larger as it approaches light speed.

At the speeds that we are currently capable of achieving, these effects are barely noticeable. Yet experiments have been conducted showing that an atomic clock travelling in a jet at high speeds will tick fractionally more slowly than an identical clock left on the ground. As to what will actually happen to us if we build an engine capable of delivering us to other stars at light speeds — we'll just have to wait and find out for ourselves!

Research activity

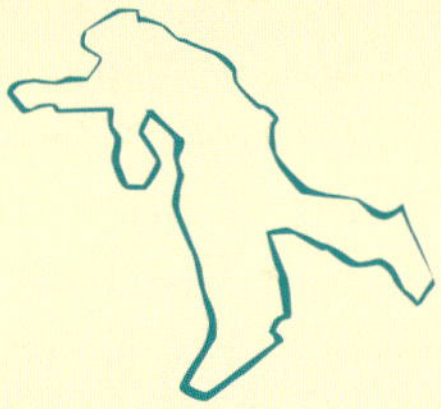

One method of interstellar travel that has been proposed is to use wormholes (also called Einstein–Rosen bridges) to take shortcuts across space. Use the library or Internet to find the following:

1. What are wormholes?
2. What conditions and hazards would spacecraft travelling through wormholes be subjected to?
3. What special characteristics should the spacecraft have to make the journey safely?

Use the Internet to find the lyrics to the song '39' by the band Queen from their 1975 album *A Night at the Opera*.

Discuss the song by considering the following questions, giving evidence from the song lyrics to support your answer:

1. When is the song set?
2. What was the mission of the volunteers?
3. What has happened to the singer while he was away?
4. Why do you think the singer asks us to pity him for his life still ahead?
5. What do you think is the significance of the phrase 'letters in the sand'?
6. The ship departed in '39 and returned in '39. How much time has passed?

End of chapter questions

1. Describe some of the negative effects that low-gravity conditions can have on the human body.
2. How can exoplanets be found when they cannot be seen through a telescope?
3. (a) Explain why a police siren changes pitch as it approaches you and then passes behind you.
   (b) How can this effect be used to determine movement in a star?
4. What considerations would have to be taken into account when designing a space station?
5. How is a space station different from a spacecraft?
6. A long, cylindrical space station is proposed for construction, as shown in figure 3.27. Habitation occurs on the inside surface, which rotates around a central axis that is driven by thruster rockets attached to the outside rim of the station. If the station has a diameter of one kilometre:
   (a) what artificial gravity is generated at the inner surface if the rim travels at 80 metres per second (288 kilometres per hour)?
   (b) how long will the station take to make one complete revolution?
   (c) how fast would the station need to rotate if the gravity on the inner surface is to be 5 m/s^2?

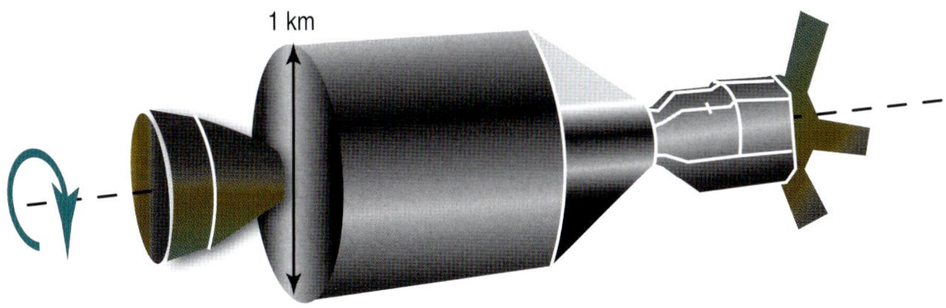

Figure 3.27: A cylindrical space station

7. Explain how oxygen can be made from water.
8. The most recent craters on the surface of the moon are as distinct as those that were formed billions of years ago. Why is there so little difference between their appearance despite the age difference?
9. What are the (a) advantages and (b) disadvantages in colonising the moon?
10. Describe the methods that might be used in order to terraform Mars.
11. How are the spectral classes of stars determined?
12. Explain how the Bussard ramjet works.
13. What effects are believed to occur to objects that travel at speeds near the speed of light?
14. Why are we unable to travel to other stars with our present spacecraft designs?
15. Radio waves from the Earth have been travelling out into space ever since Marconi first used them to send a Morse code message in 1898. What is the greatest distance in space that these first waves could have travelled in:

(a) light-years?
(b) kilometres?

16. The light from a star is often analysed by its wavelength instead of its frequency. Long wavelengths correspond to low frequencies and short wavelengths correspond to high frequencies. The spectrum of colours emitted by excited atoms of hydrogen on Earth has a wavelength of 6562.85 angstroms (where one angstrom is equal to 10^{-10} metres). Atoms of hydrogen in the star Vega emit light with a wavelength of 6562.55 angstroms. Is Vega moving towards or away from the Earth?
17. What spectral class and colour would stars with these temperatures have?
    (a) 6400 K
    (b) 9000 K
    (c) 3000 K
18. Figure 3.28 shows the different evolutionary pathways that stars may follow depending upon their mass compared to that of our sun (represented by the symbol). Use the diagram to answer the following:
    (a) What path will be followed by our sun? (We are presently on the Main Sequence.)
    (b) What is the mass of the smallest star that can go supernova?
    (c) What pathway must a star follow if it is to become a black hole?
    (d) Is it possible for a star to become a white dwarf and then form a neutron star? Explain your answer.

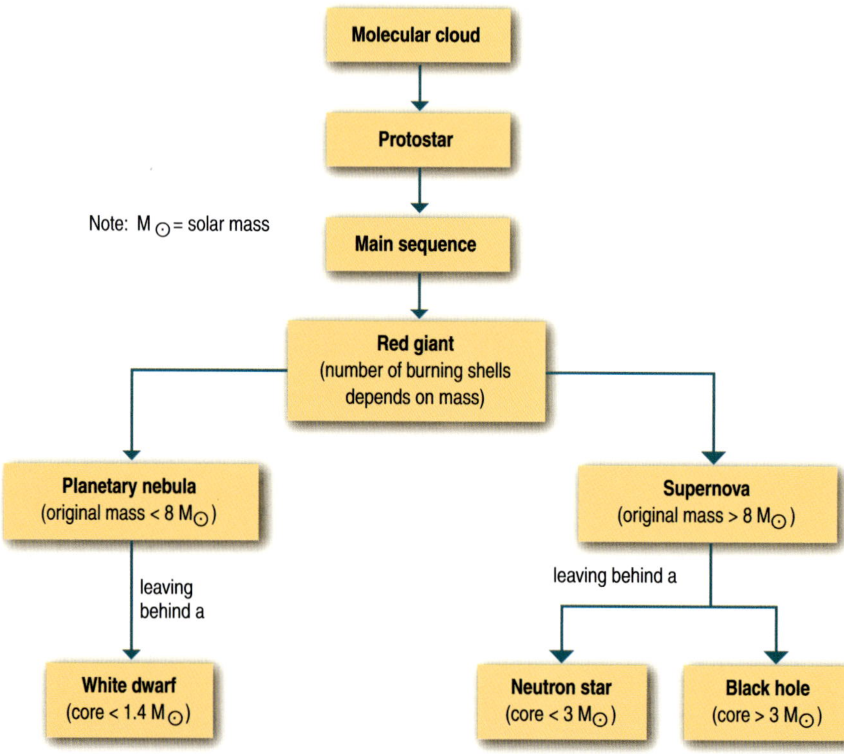

Figure 3.28:
Stellar evolution

19. The bunks in the ISS are fitted to the walls rather than the floor. Why is this comfortable for the crew?
20. Figure 3.29 shows how Earth's atmosphere filters out much of the radiation from space.
    (a) Sketch the diagram as it would appear for the radiation reaching the surface of the moon.
    (b) Estimate the altitude of the ozone layer.
    (c) Why do long waves not reach the surface of the Earth?

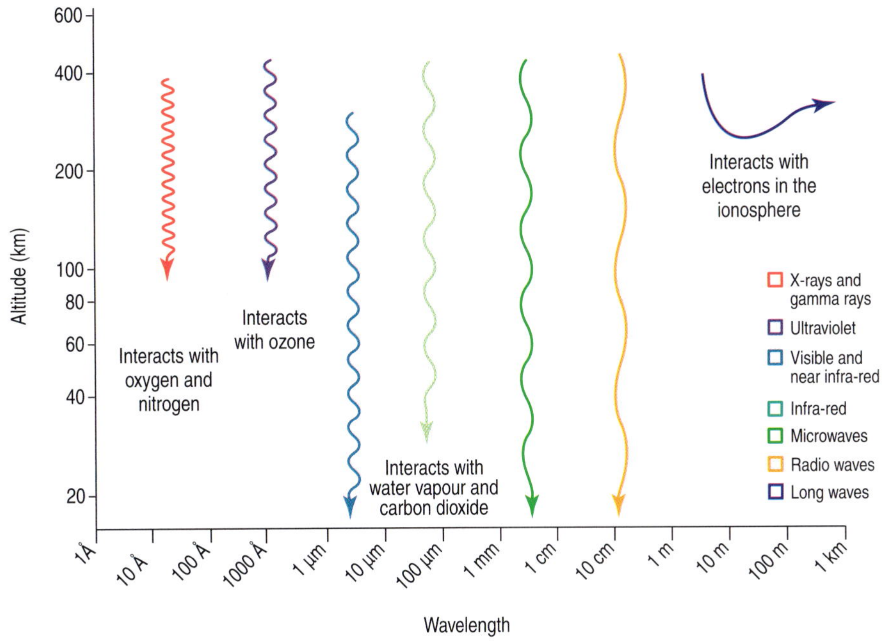

Figure 3.29:
The atmosphere filters out much of the electromagnetic radiation from space.

4

Nuclear energy — Australia's energy future?

At the beginning of the 21st century, the majority of Australia's energy is being supplied by fossil fuels. Like the rest of the world, we face an energy crisis where there will simply not be enough fossil fuels to supply the ever-increasing demand for energy to power our homes, our vehicles and our industries. So what do we do when the fossil fuels run out? Could nuclear energy be the answer?

The energy crisis

Energy comes in a variety of different forms, which you will probably remember from earlier science studies: kinetic, sound, electrical, chemical potential, elastic potential, gravitational potential, nuclear, and light energy. So many aspects of our modern way of living rely on our ability to convert energy from one form or another. Our cars move because the stored chemical energy of the petrol is converted into kinetic energy. When we listen to music, we rely on the chemical potential energy of the batteries being converted into electrical energy, which eventually becomes sound energy. The electricity in our homes is converted into lighting so we can see, heat so we can cook, and kinetic energy so our clothes can be spun dry.

Just as energy comes in a variety of forms, it is also available from a variety of sources. Generally, we can describe these energy sources as either renewable or non-renewable energy sources.

Figure 4.1:
The formation of fossil fuels

Non-renewable energy sources are finite in amount and cannot be replaced once exhausted.

Renewable energy sources can be replaced endlessly.

Non-renewable energy sources are those that are available in a finite amount and which, once exhausted, cannot be replaced. These energy sources are generally the fossil fuels — oil, natural gas and coal. They are called fossil fuels because they were formed from the ancient remains of living organisms whose remains were, over millions of years, compressed between layers of rock under enormous pressure. Having taken so long to be formed, these fuels cannot be re-created on a human timescale and, when depleted, there will be no more (see figure 4.1).

Renewable energy sources, on the other hand, are considered to be in endless supply. These include the use of solar energy, wind power, the heat of the Earth, tides and wave motion. These elemental forces cannot be depleted so long as the sun shines, the wind blows and the oceans move.

In many countries, fossil fuels are supplemented by nuclear energy. In fact, 7 per cent of the world's energy comes from nuclear power stations (see figure 4.2), with many countries planning to build even more reactors to make up for the dwindling fossil-fuel reserves in the future. Is nuclear energy also Australia's energy future?

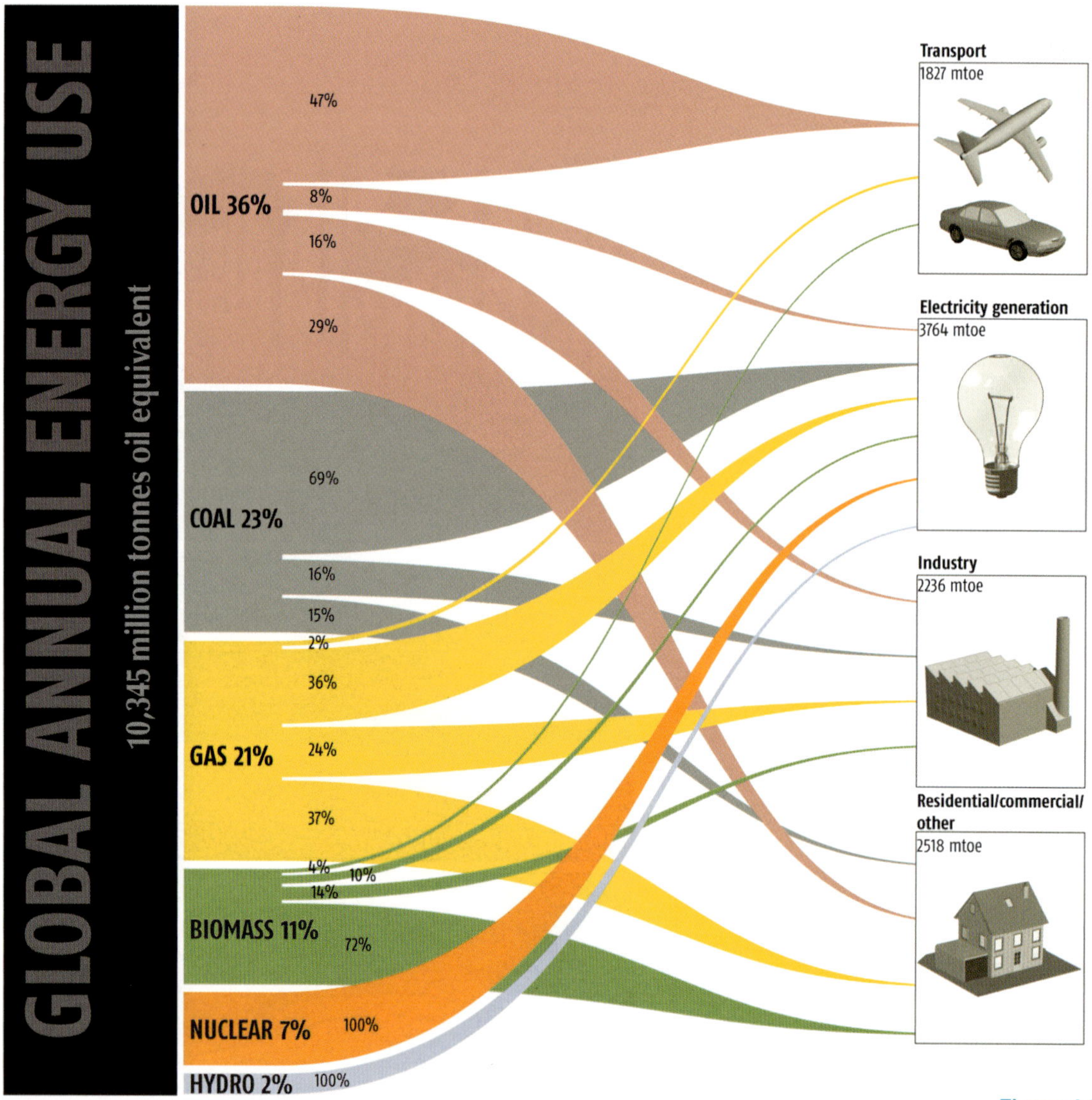

Figure 4.2: Present global energy use (mtoe—million tonnes oil equivalent)

Energy from atoms

While many people have strong feelings about whether we should use nuclear power or not, there is often a great deal of misunderstanding around the issue. In order to consider the possible benefits and dangers that could come with Australia's use of nuclear power, we first need to have a clearer picture of what nuclear power is and how it works.

Structure of the atom

The **periodic table** is a chart devised by the Russian chemist Dimitri Mendeleev that displays the known elements in rows and columns according to their physical and chemical properties.

A **nucleus** is the centre of an atom.

Protons are positively charged particles found in the nuclei of atoms.

Neutrons are particles with no charge found in the nuclei of atoms. They are similar in size to protons.

Nucleons are particles found in the nucleus of an atom. Protons and neutrons are both nucleons.

Electrons are negatively charged particles orbiting the nuclei of atoms.

All matter in the universe is made up of atoms. At present, we know of around 110 different types of atoms. You will find the names of these atoms in the **periodic table**. Atoms are all made of the same basic building blocks — protons, neutrons and electrons — but they differ in how many of each of these components they have.

All atoms have a central core called a **nucleus**, which is made up of positively charged particles called **protons** and similarly-sized particles that have no charge at all called **neutrons**. Because the protons and neutrons are found in the nucleus, they may sometimes be referred to as **nucleons**. Moving very quickly in a series of spherical orbits around the outside of the nucleus are the negatively-charged **electrons**. An electron is very small; about 1870 electrons make up the same mass as a proton. The structure of a lithium atom is shown in figure 4.3.

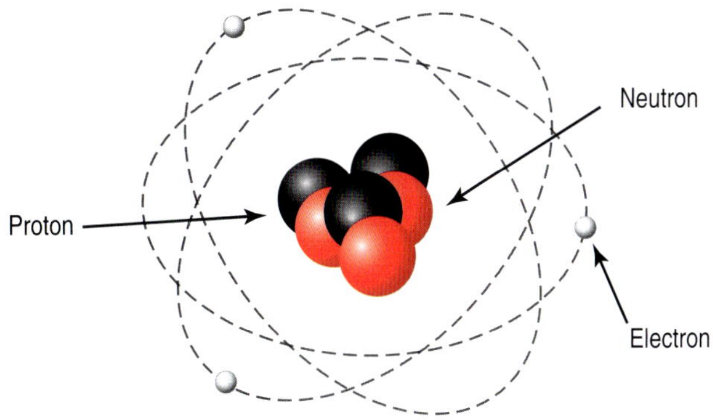

Figure 4.3:
Structure of a lithium atom

Ions are atoms that have an overall charge due to either the loss or gain of electrons.

Atoms under normal circumstances are electrically neutral; that is, they have the same number of positive charges as negative charges. This means that the neutral atom always has the same number of protons as electrons. As you can see, the lithium atom in figure 4.3 has three protons and three electrons. Atoms can lose or gain electrons in different circumstances, causing the atoms to become particles called **ions**. A positive ion is an atom that has lost electrons, and a negative ion is an atom that has gained electrons. Regardless of whether an atom loses or gains electrons, it will still have the same characteristics as an ion as it did when it was an atom. A hydrogen atom that loses an electron to become a positive hydrogen ion is still a particle of hydrogen.

However, the essential characteristics of an atom depend upon its number of protons. Removing or adding protons causes the atom to change its nature. For example, helium has two protons in its nucleus. If we were to add a proton to the nucleus of helium, it would become lithium. If we were to take a proton from the helium, it would become a hydrogen atom. Of course, the removal or addition of protons is no easy thing, as we will find a little further on!

Neutrons and isotopes

If you look at the lithium atom in figure 4.3 again, you will see that it has four neutrons in its nucleus. This is the most *common* form of lithium, but it is not the *only* form. While all lithium atoms have three protons and three electrons, and most have four neutrons, some will be found with three neutrons and others with five or even six neutrons. We refer to these different forms of lithium as lithium *isotopes*.

The **atomic number** is the number of protons in the nucleus of an atom.

The **atomic mass** of an atom is equivalent to the number of nucleons (protons and neutrons) in its nucleus.

To distinguish between different atoms and isotopes, scientists use a special form of notation that involves the atomic number and the atomic mass of the atom. The **atomic number** (A) is equal to the number of protons in the nucleus of that atom. Lithium's atomic number, for example, is 3; while the atomic number of hydrogen is 1. The **atomic mass** (Z) is equal to the number of nucleons in the nucleus of the atom; that is, it is equal to the number of protons plus the number of neutrons.

For example: the lithium atom in figure 4.3 has three protons; therefore its atomic number is 3. As it has seven nucleons (three protons and four neutrons), this isotope of lithium has an atomic mass of 7.

If we wanted to represent an isotope as a symbol, we would first write the isotope's element symbol. In front of the symbol we would write the atomic mass as a superscript and the atomic number as a subscript. In this way, our lithium isotope would be written as:

$^{7}_{3}\text{Li}$

If we had a different isotope of lithium that had five neutrons, say, then it would still have an atomic number of 3 but it would now have an atomic mass of 8 (three protons and five neutrons). We would write *this* isotope of lithium as:

$^{8}_{3}\text{Li}$

Sometimes scientists may simply refer to isotopes in terms of their element and their atomic mass. In this case, they would write our two lithium isotopes as lithium-7 and lithium-8.

The symbol, atomic number and atomic mass of every element can be found in the element's box on the periodic table, usually as shown in figure 4.4.

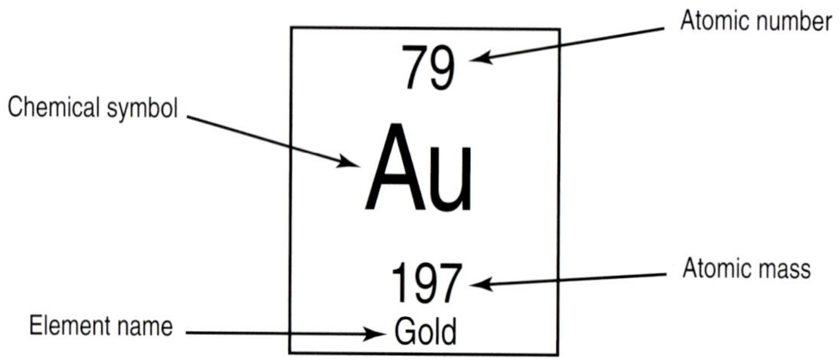

Figure 4.4:
How to read the periodic table

Neutrons and stability

The **strong nuclear force (SNF)** is the force that holds the nucleons together in the nucleus of an atom.

As you may remember from earlier studies, like charges repel each other. The nucleus contains protons that are positively charged; the bigger the atom, the more protons it has. So how is it that all of these positive charges do not all push each other away, causing the atom to disintegrate?

The answer to this lies with one of the fundamental universal forces called the **strong nuclear force (SNF)**. This force, which acts between all of the nucleons, is an attractive force that is stronger than the repulsive forces acting between the protons, provided that the protons are fairly close to each other. If the protons are separated by too great a distance, then the repulsive force overcomes the strong nuclear force and the nucleus will break up. Similarly, if the protons are too close together then, once again, the repulsive forces will overcome the strong nuclear force and the nucleus will break up.

The separation of the protons in the nucleus depends on the number of neutrons present. Too few neutrons means that the protons will be too close together, while too many leads to the protons being too far apart. In either case, the nucleus will break up. In such cases, the atom is said to be unstable. For the smaller atoms, stability is achieved when there is roughly the same number of neutrons as protons. As the atoms get bigger and heavier, the proportion of neutrons to protons needs to be bigger for stability.

Radioisotopes are isotopes that are unstable and likely to decay, releasing radiation.

For most elements, there is an isotope that is stable. Carbon-12, for example, is a stable isotope of carbon and is its most common form. Its heavier isotopes, carbon-14 and carbon 15, are unstable. Isotopes that are unstable are called **radioisotopes**.

Radioactive decay

Radiation is made up of the particles released when an unstable nucleus decays or a nucleus is forced to undergo a nuclear reaction.

A **daughter nucleus** is the nucleus formed when an unstable nucleus undergoes decay

When the nuclei of radioisotopes start to break up, they break up into smaller nuclei and release energy and particles in the form of **radiation**. This deterioration is called *radioactive decay*. For example, when the unstable isotope polonium-210 breaks up, it releases energy and radiation, and forms a smaller, more stable atom: lead-206 (see figure 4.5). As lead-206 was formed from the polonium-210, it is said to be a **daughter nucleus** of polonium-210.

A **decay chain** or **decay series** is the sequence of decay that an isotope undergoes until it forms a stable nucleus.

Sometimes an unstable isotope will decay to form a new isotope that is also unstable. In this case, decay occurs again. Thus uranium-238 decays 14 times, releasing energy and radiation at every stage, before it finally forms the stable isotope lead-206. When decay occurs in a series like this, it is referred to as a **decay chain** or a **decay series**.

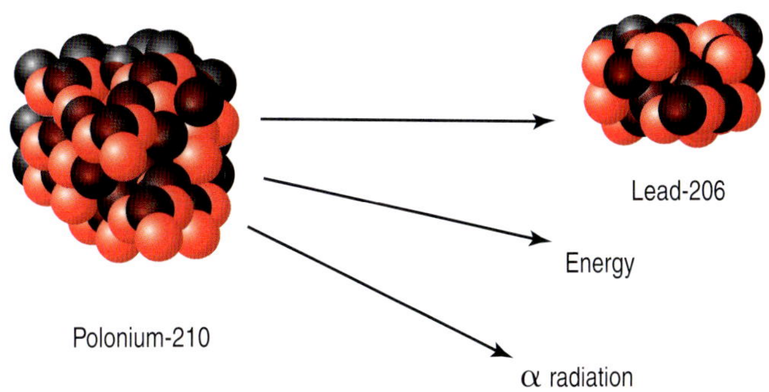

Figure 4.5: The decay of polonium-210

Nuclear radiation

In describing the decay of radioisotopes, we mentioned that radiation is released as a larger, unstable isotope breaks up into smaller nuclei. There are three types of radiation: alpha (α), beta (β) and gamma (γ).

- *Alpha radiation* is made up of alpha particles (α particles). Each α particle comprises two neutrons and two protons — essentially the same as the nucleus of a helium atom. These particles are the largest of the radiation particles and move relatively slowly, at about 20 000 kilometres per second.
- *Beta radiation* is made up of beta particles (β particles). A β particle is a fast-moving electron. Smaller than the α particles, β particles travel at 99 per cent of the speed of light (which moves at about 300 000 kilometres per second).
- *Gamma radiation* is different from the other two because it is made up of electromagnetic waves (as are radio waves and microwaves) rather than particles. These γ rays, as they are known, have no mass and travel at the speed of light.

The penetrating power of the various forms of radiation is shown in figure 4.6.

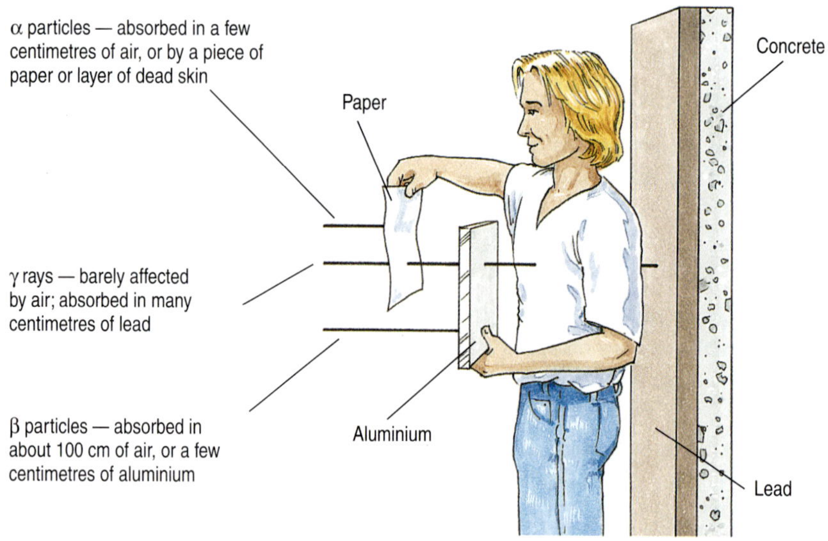

Figure 4.6:
Penetration power of alpha, beta and gamma radiation

Background radiation is radiation that is always present, coming from sources such as the sun, space, soil, building materials and medical equipment.

Many people get worried at even the *mention* of the word radiation. What most of them do not realise is that we are surrounded by radiation all the time. This is known as **background radiation**, and it comes from a number of different sources including building materials, the food we eat, X-ray machines, microwave ovens, mobile phones, the sun — even coffee!

The amount of hazard represented by a radiation source depends on the:

- amount of radiation
- type of radiation
- radioactivity of the material
- rate at which the radiation is received.

The **absorbed dose** is the amount of radiation energy that a person has absorbed per kilogram of body mass. It is measured in grays (Gy).

When a person is exposed to radiation, energy is deposited in the tissues of the body. The amount of energy that has been deposited per kilogram of body mass is known as the **absorbed dose**. This is measured in units called grays (Gy). The biological risk of exposure (the degree to which a person's

The **dose equivalent** is the biological effect of radiation that has been absorbed by living tissue, and is measured in sieverts (Sv).

health might be damaged by radiation) is measured in sieverts (Sv). This biological risk is also referred to as the **dose equivalent**. Table 4.1 shows the typical dose equivalents for a number of different radioactive sources.

Table 4.1: Some typical dose equivalents

Source	Dose equivalent (Sv)
Background radiation (Australia)	0.002
One mammogram	0.0007
One chest X-ray	0.0001
One dental X-ray	0.00004–0.00015
Exposure to cosmic rays during an aircraft flight Brisbane–Perth (return)	0.00003

Acute radiation syndrome is caused when a person is exposed to high levels of radiation over a short period of time.

The amount of radiation that a person can experience before suffering damage to their health depends on the size of the dose received and the length of the time over which they received it. High radiation doses tend to kill body cells while low doses damage the structure of the DNA within the cell. In fact, high doses can kill so many cells that entire organs are destroyed. The higher the dose, the sooner the effects of radiation poisoning will appear and the higher the probability that the exposure will be fatal. **Acute radiation syndrome** is caused when a person is exposed to a high dose of radiation over a short period of time — minutes or hours. This can result in extreme nausea, diarrhoea, internal bleeding, bone marrow depletion and organ failure. The effects of acute radiation doses are shown in table 4.2.

Table 4.2: Acute effects of radiation

Radiation dose (Sv)	Effects
0.75	• Causes vomiting in 10% of people
1	• Short-term effects such as nausea and diarrhoea • Development of cancer after many years in around 5% of people
3–5	• Damage to bone marrow, resulting in infection and haemorrhage • Can lead to death within two months in about 50% of people if medical treatment is not available
>10	• Death within 10 days due to fluid and electrolyte imbalance • Bone marrow and gastrointestinal damage and infection
>40	• Death within 48 hours due to damage to the vascular systems, resulting in accumulation of fluid in the brain.

The plant workers and emergency teams at Chernobyl were exposed to between 0.8 and 16 Sv in a matter of hours, and 20 per cent of them died within two months of the accident. Most people will generally experience less than 0.1 Sv over the course of their lives. Those who work with nuclear sources for a living — such as radiographers, uranium miners and nuclear engineers — would receive more than this. Industry standards are applied to these occupations to minimise their risk to prolonged exposure effects. For example, people who work with radiation sources (radiographers, nuclear engineers, nuclear power-plant workers) are limited to an exposure of 0.1 Sv over a five-year period (0.02 Sv/year average), while uranium workers are not permitted to be exposed to more than 0.013 Sv in a year.

Energy from chain reactions

Nuclear energy is the energy emitted by a nucleus as it becomes more stable.

The energy that is released when an unstable isotope decays into another (known as **nuclear energy**) is the key to nuclear power. However, the rate at which radioisotopes normally decay is too slow to use as an energy source on a large scale. Instead, scientists employ a process called nuclear fission in the harnessing of nuclear energy.

Nuclear fission is the process by which large nuclei break up to form smaller nuclei.

Nuclear fission is the process in which a large unstable nucleus splits, forming two (or more) smaller, more stable nuclei and releasing neutrons and energy. The only naturally occurring isotope that undergoes sustainable fission is uranium-235. Bombarding an atom of uranium-235 with slow neutrons (neutrons travelling at about 2 kilometres per second) forces it to split, forming the isotopes barium-92 and krypton-141 and releasing three neutrons and 20 million times more energy than could possibly be obtained from burning an atom of carbon (from which coal is made).

There are several things that can happen to these released neutrons. In the case of uranium-235, the neutrons can:

- be captured by other uranium nuclei in a non-fission reaction, usually leading to beta decay
- be captured by other materials in the sample of uranium ore or by materials surrounding the sample
- escape from the sample without being captured (most likely because capture requires a head-on collision between the neutron and a nucleus)
- be captured by uranium nuclei and cause a fission reaction.

Critical mass is the smallest spherical mass that will sustain an uncontrolled chain reaction.

An **uncontrolled chain reaction** occurs when all of the neutrons released during fission cause the fission of further nuclei, allowing all of the fissionable material present to be involved in the chain reaction.

If the mass of the uranium sample is too small, most of the neutrons will escape or be absorbed by other non-fissionable materials (materials that cannot be made to undergo fission) before they can continue the chain. If the sample is above a certain mass (referred to as the **critical mass**), then it is possible that every neutron released will cause another fission reaction in what is known as an **uncontrolled chain reaction**. This is shown in figure 4.7. (In an atomic bomb, the critical mass is obtained by bringing together two smaller masses of the material just on detonation.)

The size of the critical mass depends on the proportion of fissionable material to non-fissionable material present in the sample. The critical mass for a sphere-shaped sample of pure uranium-235 is about 50 kilograms. In its natural state in the Earth's crust, only

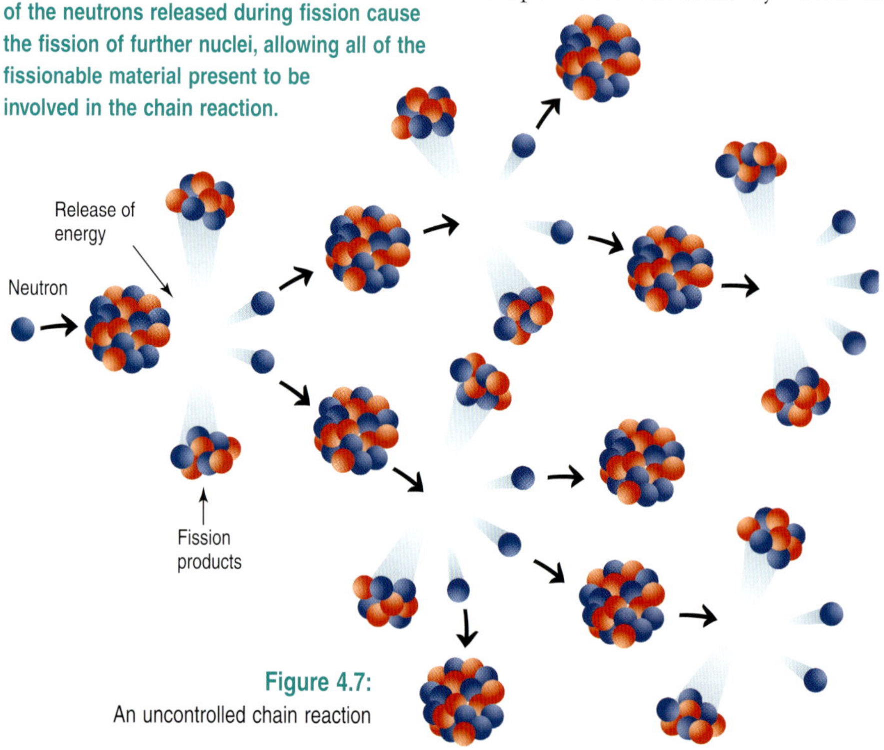

Figure 4.7: An uncontrolled chain reaction

0.7 per cent of uranium ore is made up of fissionable uranium-235, while the remainder is non-fissionable (at least under these circumstances) uranium-238. Due to the low proportion of fissionable uranium, uranium ore does not undergo an uncontrolled chain reaction.

In a **controlled chain reaction**, not all of the neutrons released cause the fission of further nuclei.

Nuclear reactors use a **controlled chain reaction** to produce nuclear energy. In this situation, only some of the released neutrons will go on to strike other uranium atoms and cause them to undergo fission (see figure 4.8). The rest are absorbed by non-fissionable material that prevents them interacting any further. By manipulating the amount of non-fissionable material present, the rate at which the chain reaction occurs can be controlled. Ideally, each fission triggers only a single further fission rather than three.

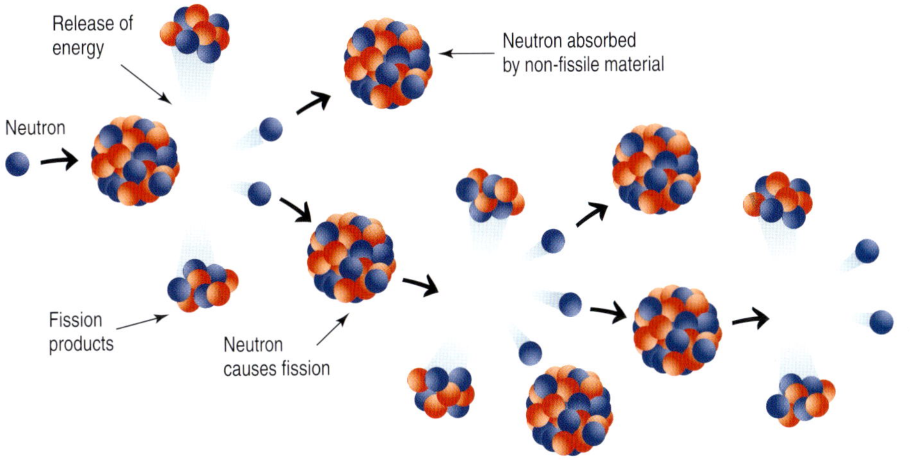

Figure 4.8: A controlled chain reaction

Figure 4.9: Placing dominos

Materials: A box of dominoes

1. Place a single domino upright on its end on a level surface.
2. Place two more dominos close together in front of the first.
3. Continue to place dominoes as shown in figure 4.9, making sure that each domino is half of a domino's length from the one behind it.
4. Tip the first domino over and watch what happens.

Nuclear reactors

A **nuclear reactor** is a device in which controlled fission is used to provide energy.

A **nuclear reactor** is a device in which controlled fission is used to produce energy and new substances. In this controlled fission, one neutron on average from each fission leads to the fission of one other nucleus.

Nuclear power plants are similar in parts to coal-burning power plants, as shown in figure 4.10. All power plants require the movement of enormous turbines that generate electricity by rotating coils in a magnetic field. The way the turbines are turned makes the power stations different. In a coal-burning station, coal is burned to heat water into steam, which then turns the turbines. In a nuclear power station, the energy produced by the fission of uranium provides the heat that turns the water into steam, which goes on to turn the turbines and generate electricity.

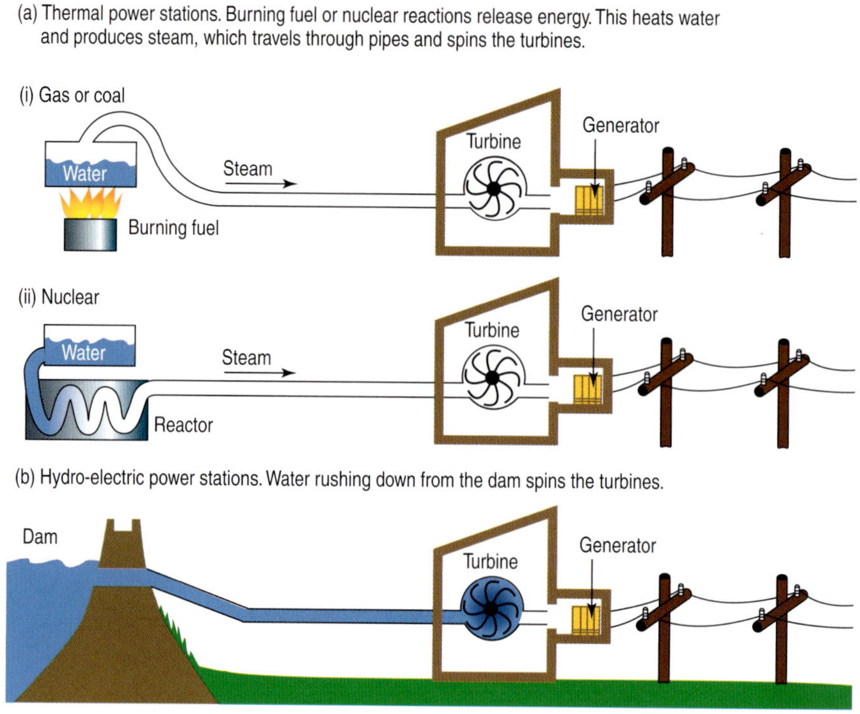

Figure 4.10:
Various methods can be used to turn the turbines that allow electricity to be produced.

Many countries use nuclear reactors to produce electricity; 77 per cent of France's electricity comes from nuclear reactors. In Australia, there are presently only two small nuclear reactors, located at Lucas Heights in Sydney. These reactors are used only for nuclear research and for producing the radioactive materials needed for medical use.

The design of a nuclear reactor

There are many different types of nuclear reactors, but the most common style is like that shown in figure 4.11.

Both uranium-235 and uranium-238 can be made to undergo fission and release energy. However, uranium-238 undergoes nuclear fission only when it interacts with fast neutrons, and it is difficult to keep up the supply of fast neutrons because they tend to lose their energy rapidly in collisions with other atoms. This makes it very problematic to use in nuclear reactors.

Uranium-235 readily undergoes fission with slow-moving neutrons, but the difficulty here is that uranium-235 makes up only 0.7 per cent of uranium ore, with 99.3 per cent of the ore being uranium-238. Therefore, the nuclear material used in reactors is **enriched** with uranium-235.

The neutrons are slowed down to speeds and energies comparable to those of gas molecules at normal temperatures, so they are referred to as **thermal neutrons**.

Most reactors are **thermal reactors** in which the neutrons produced by nuclear fission are slowed down after being emitted so that further fission can occur. There are also some **fast reactors**, in which neutrons producing fission have high energy. It has been found safer and easier to build thermal reactors.

Enrichment is the process by which the percentage of uranium-235 in a uranium sample is increased so that it can sustain a chain reaction. Rather than the 0.7 per cent uranium-235 found in normal uranium ore, enriched uranium may contain 1–4 per cent uranium-235.

Thermal neutrons are used in nuclear reactors to cause fission in uranium nuclei.

Thermal reactors are nuclear reactors in which the neutrons produced in each fission reaction are slowed down so that they may cause further fission of nuclei.

Fast reactors use neutrons that have high speed and energy to cause nuclear fission.

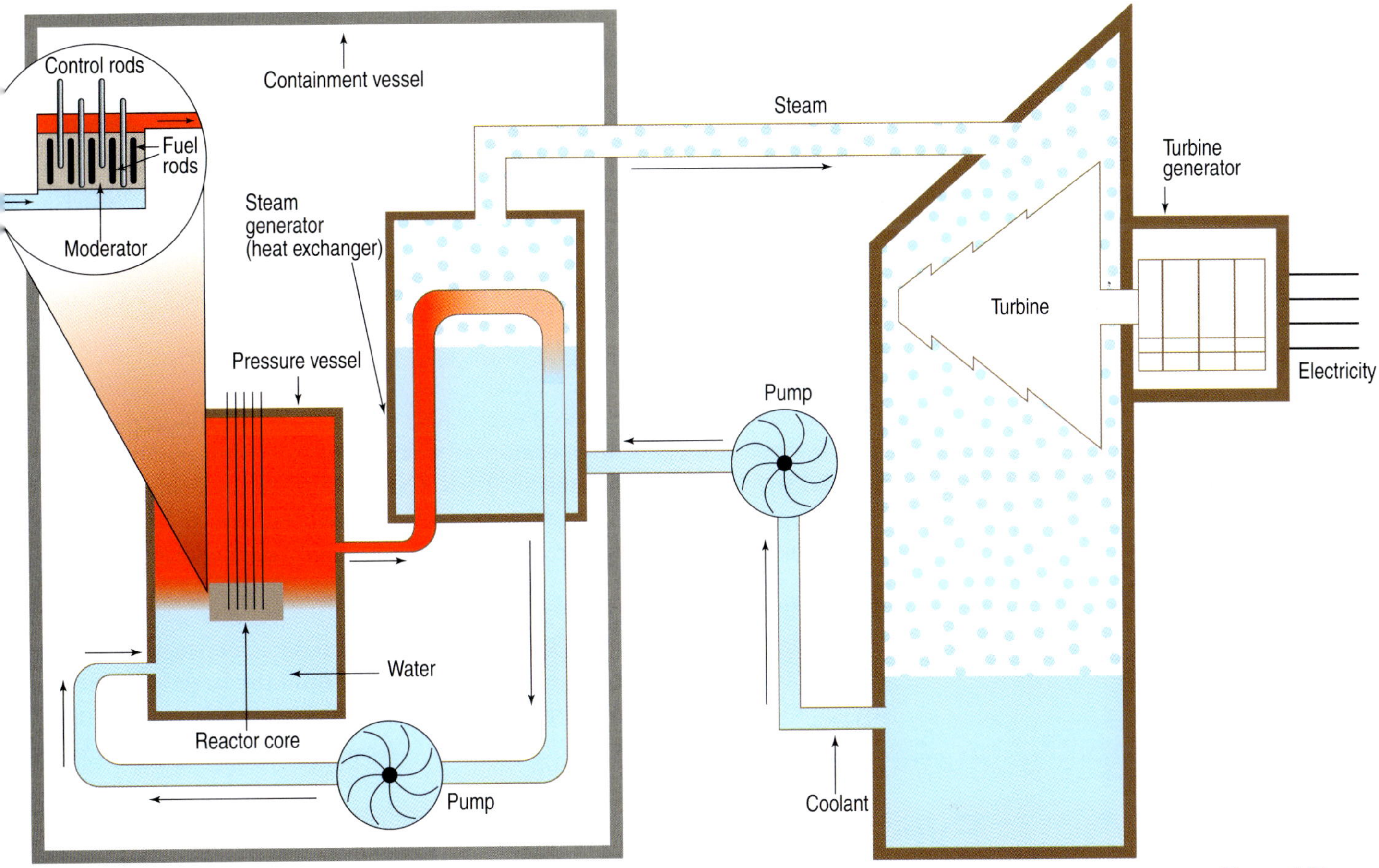

Figure 4.11: A typical nuclear-power plant

The **reactor core** is the heart of the nuclear reactor.

Components of a nuclear reactor

The main components of the reactor are found in the **reactor core**. They are the fuel rods, the control rods, the moderator and the coolant.

Fuel rods

The nuclear fuel is made up of pellets of fissionable material, usually uranium ore enriched with U-235 nuclei. The enriched fuel is placed in airtight containers or rods. Some reactors may contain as much as 200 tonnes of uranium fuel. The fission process where nuclear energy is released occurs within these rods. Over time, the uranium is used up, replaced by the smaller nuclei materials (such as barium and krypton) made in the fission process. The used rods are pulled from the core and replaced with new ones as needed. The used rods are usually stored in the bottom of the reactor pool until they can be disposed of safely.

Control rods

Control rods usually made of cadmium or boron steel are inserted into the reactor to control the rate at which fission occurs by absorbing neutrons. When fully inserted, the control rods absorb nearly all of the neutrons, causing the fission process to cease. If pulled out completely, the reaction may become uncontrollable or runaway. Most reactors usually contain two sets of control rods — one for routine control and the other set (called safety rods) for use in case of safety concerns or emergency core shutdown.

Moderator

A moderator is a material that is found around the fuel and control rods. Its purpose is to slow down neutrons released during fission. Remember, only slow neutrons can cause fission of uranium-235; faster neutrons bounce off the uranium atoms rather than splitting them. The materials used as moderators tend to have small nuclei, which slow the neutrons down without actually absorbing them. Materials such as ordinary water, heavy water (deuterium oxide), graphite and beryllium can be used. Because graphite can catch fire if there is an accident in the core, it is now rarely used in Western countries.

Coolant

The coolant is needed to absorb excess heat produced by the fission process in the reactor, keeping it operating at a fairly constant temperature. This coolant may be either a liquid or a gas. Carbon dioxide (used in reactors with graphite moderators), heavy water and liquid sodium have all been used in various reactor designs. Water is the most commonly used coolant.

Containment vessel

The reactor is generally enclosed in a thick, airtight concrete-and-steel vessel in order to protect those outside the reactor from the large quantities of gamma radiation and neutrons being given off. The vessel also reflects neutrons back into the core.

Eye on the media

Core cracks may force shutdown of UK reactors

Reactors in many UK nuclear power stations are in danger of developing cracks in their graphite cores. This could force some plants to close down earlier than expected, dealing a blow to the idea that nuclear power can become a 'green' option in the fight against global warming.

Documents obtained by *New Scientist* under the UK's Freedom of Information Act have revealed unsuspected problems with the country's aging advanced gas-cooled reactors (AGRs). Government nuclear inspectors say they have uncovered weaknesses in the safety analyses carried out by British Energy, the company that runs the reactors.

The UK's 14 AGRs provide nearly a fifth of the country's electricity. The graphite bricks that form part of their core help sustain the nuclear reaction by slowing down fast-moving neutrons. They also play a vital part in maintaining the core's structural integrity.

While it has long been known that irradiation and thermal stress would eventually cause the graphite bricks to crack, new estimates suggest these cracks could develop up to two years earlier than thought, according to British Energy. In a letter last August, the company warned the Nuclear Installations Inspectorate (NII) of 'possible errors' in the computer models used to predict the onset of cracking.

In December, British Energy found cracks in the graphite bricks at one of the two AGRs in Hartlepool, County Durham. Cracking is now suspected in six other reactors at Heysham in Lancashire, Hinkley Point in Somerset and Hunterston in Ayrshire.

Widespread cracks in these bricks could cause the core to distort, overheat and leak radiation. To ensure this does not happen, the NII has asked British Energy to conduct more inspections of the bricks. This means that reactors may have to be closed down for maintenance more frequently and for longer, at a cost in lost income of £250 000 a day for each reactor that is shut down.

Two reactors at Hinkley Point and Hunterston are scheduled for maintenance shutdowns later this year. In letters sent in February, the NII told the stations' directors that their safety analyses of the graphite cores were 'weak', and demanded answers to 47 technical questions about cracking before it will permit the reactors to restart.

British Energy has warned its shareholders that graphite cracking could kill its hopes of extending the lives of AGRs, and that there may have to be 'early closures'. The reactors, which were built in the 1970s and 1980s, were expected to operate for 35 years, and not start closing until after 2010. British Energy stresses that, at present, no reactors have been shut down because of graphite cracking. 'The implication is one of periodic monitoring and inspection within our normal program of planned maintenance shutdowns,' a spokeswoman says.

Critics of nuclear power argue that the safety and economic implications are more serious. The worst-case scenario is that cracking could cause part of the core to collapse and the reactor to be written off, says John Large, an independent nuclear engineering expert.

Pete Roche, a Greenpeace consultant, says graphite cracking highlights the unreliability of nuclear power. 'Generic problems can shut down several large nuclear stations all at the same time,' he says. 'So they are not the best solution to combat climate change.'

Source: Rob Edwards, *New Scientist,* 26 March 2005, p. 9.

Clarification questions

1. Why are cracks in the graphite cores of the UK reactors a problem?
2. What is an AGR?
3. What has caused the cracks?
4. How many reactors have recorded cracks?
5. What is the worst-case scenario for continued cracks?
6. How much of the UK's electricity is generated by AGRs?
7. What solutions have been proposed for the problem?

The nuclear debate

As the gap widens between the world's need for energy and the availability of the fossil fuels it depends on, there is great pressure to find energy alternatives. While nuclear energy has many factors in its favour, many people consider that the problems associated with its use make it a poor choice for use in this country. Let us look at some arguments from both sides of the debate.

Arguments for nuclear energy

1. Ready access to fuel

As Australia has 30 per cent of the world's uranium supply, we are capable of supplying our own nuclear fuel for any reactors built here. Presently, we export nearly all the 7000 tonnes of uranium ore that comes from mines in the Northern Territory (Ranger and Nabarlek) and South Australia (Roxby Downs).

Global warming is the term used to describe the present increase in the average temperature of the Earth's atmosphere over time.

2. No greenhouse gases

In using coal for our power stations, we release enormous amounts of greenhouse gases into the Earth's atmosphere. These gases are believed to contribute to **global warming**, acid rain and climate instability. The average coal-fired power station produces about seven million tonnes of carbon dioxide each year, as well as around 200 000 tonnes of sulfur dioxide (a major source of atmospheric pollution), and about the same amount again of other waste products that contain toxic metals such as arsenic, cadmium and mercury. Nuclear power stations, on the other hand, release only water vapour into the air.

3. Efficiency

One of nuclear power's greatest advantages is that it is extremely efficient in terms of the amount of energy derived from a small amount of uranium fuel. One tonne of uranium produces more energy than is produced by several million tons of coal or several million barrels of oil. As can be seen in table 4.3, uranium is far more efficient than any fossil fuel.

Table 4.3: Typical heat values of various fuels

Fuel source	Energy produced from 1 kg of fuel (MJ)
Brown coal	9.7
Firewood	16
Black coal	30
Natural gas	39
Crude oil	46
Uranium (light water reactor)	500 000

Note: 1 MJ = 1 million joules.

Source: www.nvnpp.vrn.ru/en/about/advantages.html.

4. Essentially renewable

Like fossil fuels, uranium is found in the Earth in the form of reserves that can be mined. Unlike fossil fuels, however, much of the fuel used in reactors can be processed and so reused. Combined with the fact that only small amounts of uranium are needed to fuel reactors and that Australia has a large supply of accessible uranium ore, this makes uranium (unlike fossil fuels) a resource that is almost endless.

Arguments against nuclear energy

1. Cost

While nuclear fuel and its efficiency make it more competitively priced than fossil fuels, the costs of building a nuclear reactor are much higher than those involved in coal-burning power stations, especially as the safety measures that must be incorporated into their design make them more expensive. When nuclear reactors were being built in the 1980s, it was believed that their initial cost would be offset against a long working life, believed then to be 100 years or more. It is now known that the lifetime of a reactor is more like 40 to 50 years, which is comparable to that of a traditional power station. It has also been argued that the money spent in building nuclear reactors would be better invested in the development of alternative energies such as biomass fuel, solar energy and wind power.

2. Safety concerns

Undoubtedly, safety is what concerns people most about nuclear reactors. This concern is difficult to dismiss in light of the widely reported mishaps at Three Mile Island in 1979 and Chernobyl in 1986 (which will be discussed in more detail in the next section). Accidents at nuclear facilities that release deadly radiation into the atmosphere could affect whole regions. Figure 4.12 shows how radiation levels in Europe, Africa and Asia were affected by the radiation released from Chernobyl.

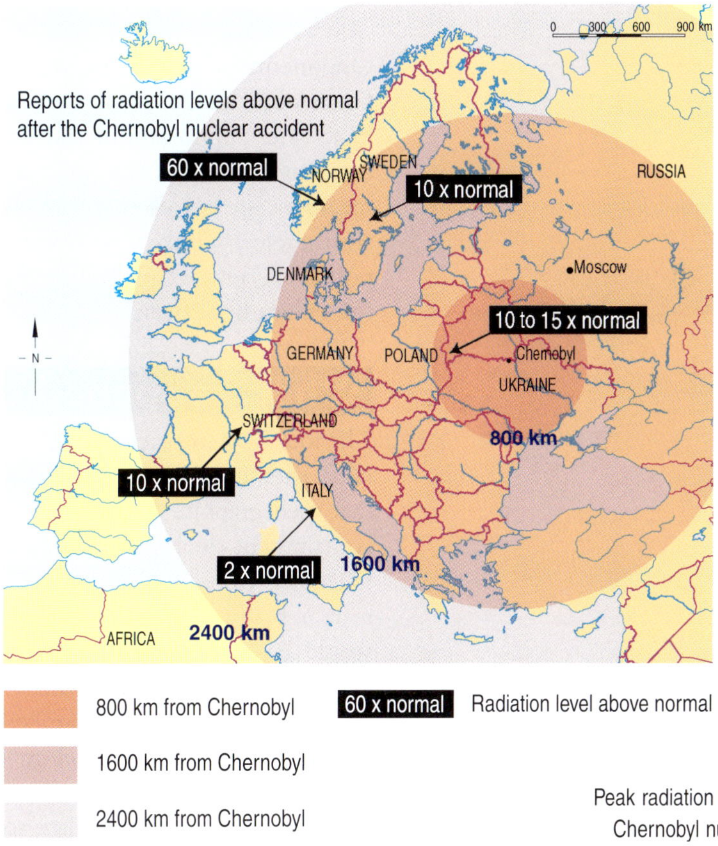

Figure 4.12: Peak radiation levels after the Chernobyl nuclear accident

3. Nuclear waste management

During the fission process, uranium-235 nuclei are broken into smaller nuclei such as barium and krypton, both of which will decay into other isotopes. The average nuclear reactor produces around 25 tonnes or so of this spent fuel each year it is in operation. This waste is highly radioactive and gives off a great deal of heat. While most of this waste can be reprocessed for reuse, it must be stored until it can be reprocessed. Further, the 700 kilograms of this waste that is not reprocessed makes up high-level radioactive waste that is potentially hazardous and needs to be isolated from the environment for a very long time. It has been pointed out by some critics that, if the Roman Empire had used nuclear reactors, we would *still* be dealing with the wastes from them today! While many methods of storing or disposing of this waste have been used — including storage in steel or concrete containers that are then buried in deep disused mineshafts or in isolated areas such as deserts none has as yet proved satisfactory due to the risk of leakage and subsequent damage to the environment.

Imagine that you have been asked to design a series of signs to be placed around a nuclear waste site somewhere in the desert, warning of its danger. You have been asked to do this in the form of a picture and/or symbols rather than using words, on the basis that the nature of the hazard needs to be understood by anyone — including those who cannot read or who have no idea what nuclear waste is. Design a sign that fulfils these criteria.

Case study: Chernobyl

Opponents of nuclear power often point to the disaster that occurred at Chernobyl on 26 April 1986 as clear evidence of the dangers inherent in its use. Supporters of nuclear energy claim just as vigorously that the incident resulted from poor design, human error and inadequate safety measures — factors that have since been addressed in other reactors and are therefore unlikely to occur again. In order to better understand these points of view, let us look at what happened, how it happened and what the effects have been.

What happened at Chernobyl?

The four nuclear reactors that made up the nuclear power plant in Chernobyl (near Kiev in what used to be the Soviet Union) were built to an old design that used graphite moderators and water as a coolant, and had no radiation containment shields around their cores. On 25 April, reactor number 4 was scheduled to be shut down for routine maintenance. This would involve the reactor being completely turned off. It was decided by engineers that this would be a good time to test recent modifications on the electrical generators that would allow, in the event of an emergency reactor shutdown, enough electricity to be supplied in order to operate cooling pumps in the core until the backup diesel generators could be activated. Remember, the nuclear core generates huge amounts of heat during the fission process so it must be supplied with coolant to prevent overheating.

During the shutdown process, the engineer in charge of the distribution of electrical power generated by the reactor refused to allow the complete shutdown of the reactor because there was a high demand for electricity in the region at the time. The decision was made to run the reactor at about half of its normal capacity so that the electrical demand could be met. The tests on the generators went ahead. As part of these tests, the emergency core-cooling system was turned off.

The demand for electricity in the region had dropped at around 11 pm, so the electrical distribution engineer allowed the reactor power to be lowered further.

Due to an operational error, the reactor power dropped too low, so technicians attempted to raise the fission rate by manually raising the control rods to compensate for the power loss. As the control rods absorb neutrons, raising them allows more neutrons to be available to cause fission in the fuel rods.

Not long after this, there was an increase in the coolant flow and a drop in the steam pressure to the electricity turbine, once again causing a drop in power. In trying to raise the fission rate, the technicians withdrew nearly all of the control rods. This caused the reactor to become rapidly unstable.

The electricity turbine normally provided power to run the coolant pumps in the reactor core. As the turbine slowed down, the coolant pumps also slowed, causing the coolant to turn into steam. While an attempt was made to fully insert all of the control rods (absorbing all of the neutrons in the core and stopping the fission reaction), a reaction with the graphite tips of the control rods suddenly caused an uncontrollable power surge in the reactor. The time was 1 am on 26 April. In four seconds, the power rocketed up to 100 times its normal value and the reactor core reached 5000°C (about the same temperature as the surface of the sun).

The sudden increase in temperature caused some of the fuel rods to rupture. The fuel particles then reacted with the water, resulting in a steam explosion that destroyed the reactor core. A second explosion added to the destruction two minutes later. The graphite core burned and, because there was no containment shield around the core, 4 per cent of the radioactive fuel was vaporised into the atmosphere. As you can see in figure 4.13, the nuclear core of reactor 4 was completely exposed and pouring out radiation and neutrons.

Figure 4.13: The exposed core of reactor 4 in the Chernobyl nuclear power plant

Eye on the media

Interview: Cheating Chernobyl

Alexander Yuvchenko was on duty at Chernobyl's reactor number 4 the night it exploded on 26 April 1986. He is one of the few working there that night to have survived. He suffered serious burns and went through many operations to save his life, and he is still ill from the radiation. He recently broke his silence for a documentary to be shown on the Discovery Channel. Here he speaks to Michael Bond about what happened that night.

How did you end up working at Chernobyl?

I chose it. It was one of the best stations in the Soviet Union, it was a good town to live in, and I had been there for practical work as part of my studies. And it was a good wage. Being a nuclear engineer was a prestigious career — in those days. Nowadays people in Russia prefer to be businessmen and lawyers.

What were you doing the night the reactor exploded?

I was on the night shift. When I turned up I found out that the safety test that had been planned for the day had been put off until the evening. The reactor had already been powered down and so we would just be overseeing its cooling, which is a very easy job. I was thinking that I wouldn't have much to do that night.

What were you doing when you heard the explosion?

I was in my office, talking to a colleague who had come in to ask for some paint, and reading some documents.

What happened?

The first thing I heard wasn't an explosion, it was a thud, a shaking. Then two or three seconds later came the explosion. The doors of my office were blown out. It was like when an old building is demolished, with clouds of dust, but combined with lots of steam. It was a very damp, dusty, powerful movement of air. There was a lot of shaking, a lot of things were falling. The lights went off. Our first thought was to find somewhere we could safely hide. We headed towards the transport corridor, where there was a small passage with a low ceiling. We were standing there and everything was falling around us.

What did you think it was?

When I heard the thud I thought it was something very heavy that had fallen. After that I didn't know. I thought that maybe war had begun.

Did you imagine that it might be the reactor?

I couldn't imagine it was something to do with the reactor. Before it happened there were no vibrations, no sounds, nothing to indicate there was something wrong. We were trained for various emergency situations. We were engineers, and we were trained in what the reactors could or could not do and what could go wrong. We were prepared for fire and other things, but we were not trained for this. We all thought the safety measures were reliable, that if you pressed the emergency stop button to lower the control rods into the reactor — which is what my friend Leonid Toptunov in the control room did that night — that it would stop the power as it was supposed to. But it didn't. People make mistakes, but we thought the safety measures would compensate for that. We believed what we were told in the work manual.

What did you do after the explosion?

I went back to my office and tried to ring the control room for reactor number 4 to find out what had happened, but there was no line. Suddenly the phone from control room number 3 rang. I got a command to bring stretchers. I grabbed the stretchers and ran. Outside the control room I met a friend who had been close to the centre of the explosion. I didn't recognise him. His clothes were black and his face was disfigured because he had been covered in scalding water. I only recognised him by his voice. He told me to go to the site of the explosion because there were others injured. This friend was being tended by others, so I got a torch and ran to find the other operator who had been near the huge coolant tanks.

What did you find?

I got to where I expected to find this person but I couldn't find anything, there was a huge mess. I found him on the other side, he had managed to crawl away. It was the same picture: he was wet, dirty, with serious scalding burns. He was standing up but was in an extremely shocked state, shaking. He told me I had to go to where the main blast happened, which was where my friend Valera Khodemchuk was. This guy couldn't see it, but there was nothing there in that direction, it was just empty space.

What happened then?

At this point I saw Yuri Tregub, who had been sent from control room number 4 by Anayoly Deatlov, Chernobyl's deputy chief engineer, to manually turn on the emergency high pressure coolant water to flood the area. Realising that he wouldn't be able to do this on his own, I told my friend where to go to get help and I went with Tregub to turn on the water.

Did you succeed?

We weren't able to get to the taps. The coolant tanks were in a hall close to the reactor. There were two doors in. We couldn't get in the first because the walls had collapsed, so we went down a couple of floors to the other door. We were in water up to our knees. We couldn't open the door but we could see a little through it, and all we could see were ruins. The huge water containers had been blown apart. There was just a wall and a door left. We were looking into open space.

Literally?

To get a clearer idea of what had happened we walked outside. What we saw was terrifying. Everything that could be destroyed had been. The entire water-coolant system was gone. The right-hand side of the reactor hall had been completely destroyed, and on the left the pipes were just hanging. That was when I realised that Khodemchuk was definitely dead. The place where I was told he'd been standing was in ruins. The huge turbines were still standing, but everything around them was rubble. He must have been buried under that. From where I stood I could see a huge beam of projected light flooding up into infinity from the reactor. It was like a laser light, caused by the ionisation of the air. It was light-bluish, and it was very beautiful. I watched it for several seconds. If I'd stood there for just a few minutes I would probably have died on the spot because of gamma rays and neutrons and everything else that was spewing out. But Tregub yanked me around the corner to get me out [of] the way. He was older and more experienced.

What did you do then?

We started to make our way to control room number 4, but on the way we met three workers who had been ordered by Deatlov to go to the reactor hall and lower the control rods manually. Tregub ran back to the control room to report what we'd seen, and I went with these three to help them. I told them that the order they had been given was senseless because there was no reactor hall anymore and it was highly unlikely there were any control rods. But they said I had only seen it from the lower level and they had to see it for themselves from the upper level.

Did you realise how dangerous that was?

Yes, we did.

What happened when you got back to the reactor hall?

We climbed up to a ledge but there was very little room. Because I had come up the stairs last I stayed behind propping open the door. They took the torch from me and went in. I stood there listening to their reaction to what they saw, which looked like a volcano crater. They said there was nothing they could do, they had to get out.

What happened to those three?

All three of them died very soon afterwards. That wall and the door basically saved my life. I received quite a high dose propping open the door. We had done everything we could. That was the worst feeling: that there was nothing else we could do.

At what point did you start to feel ill?

About 3 am, one-and-a-half hours after the explosion.

How did you feel?

I began to feel sick. I knew one of the first symptoms of radiation illness was vomiting, but I was thinking, have I eaten something? I was trying to keep the worst thoughts at bay. Half an hour after the explosion I had met a man with a dosimeter, he was fully covered so I don't know who it was, and I asked him what the reading was. He showed me the counter, which was off the scale. That was a frightening moment. It was impossible to say how much radiation we were taking in, but I knew it was a large dose. I was taken to the local hospital at about 5 am because I was too weak to walk. I was taken to Moscow that evening.

Did you think you would die?

The most frightening thing was to lie there and hear how one after another the others were dying. I was thinking, when will it be my turn? I'm not a religious believer and I don't know any prayers, but I did pray every evening that I would wake up the next morning.

Source: Michael Bond, *New Scientist,* 21 August 2004, pp. 44–7.

Clarification questions

1. What was Alexander Yuvchenko's job at Chernobyl?
2. After the explosion, Yuvchenko encountered a number of people who suffered scalding (burns from hot water and steam). What was the most likely source of these burns?
3. What caused the bluish light that poured out from the reactor's core?
4. How soon after the explosion did Yuvchenko start to experience the symptoms of radiation sickness?

Causes of the Chernobyl accident

It is claimed that a number of factors contributed to the destruction of reactor 4 at Chernobyl:

1. Disregard of safety concerns

In the subsequent decades after Chernobyl, it was revealed that the government organisations responsible for the nuclear reactors were aware of construction weaknesses in the Chernobyl reactor. However, it was decided that these weaknesses were too expensive to fix and that they were unlikely to cause a problem during the working life of the reactor.

2. Communications breakdown

Lack of adequate communication between the technicians running the test on the generators and the technicians overseeing the electricity distribution caused conflicting actions that led to the destabilisation of the core.

3. Reactor design flaw

The Chernobyl reactor was made to an older design that was known to be unstable at low power and to have uncontrolled power surges. Later versions of this design have incorporated design features that prevented the instability from occurring.

4. Violation of safety protocols

While running the test on the generators, station technicians violated the safety procedures that were a normal part of the reactor's management. The critical violations were the disabling of the reactor's emergency-cooling system and using only a third of the required minimum number of control rods needed to keep the reactor running safely.

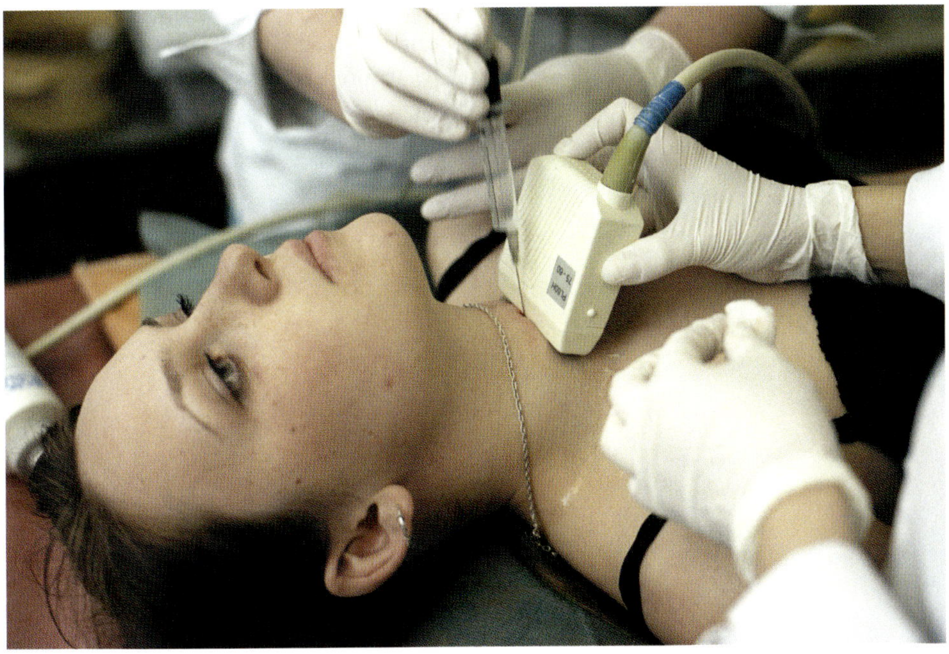

Figure 4.14:
The rate of thyroid cancer in children around Chernobyl has increased 10 times since the accident. This girl undergoes a procedure where the thyroid is punctured to control the cancer.

Effects of the Chernobyl accident

Health effects

- While only two people were killed in the original explosion, three others died during the night and 50 emergency workers died from acute radiation poisoning.
- A United Nations study estimates that 4000 people will die from long-term cancers as a result of the accident.
- Since the accident, the rate of thyroid cancer in children (15 years and under) has been 10 times higher in the region around Chernobyl.
- Some 600 000 people were contaminated by radiation with unknown long-term effects.

- There have been significant increases in psychological health disorders such as anxiety and depression, as well as disorders resulting from mental stress caused by trauma from the accident or from the subsequent evacuation and displacement.
- Promiscuous behaviour arising from feelings of mortality and loss of hope for the future has increased.

Social effects

- Following the accident, 116 000 people had to be evacuated, and between 1990 and 1995 an additional 210 000 people were resettled. This created a need to build a new town (Slavutich) for the former personnel of the Chernobyl power plant.
- Villages had to be decontaminated and major work had to be carried out on utilities such as water and gas. The destruction of reactor 4 and the halting of construction of new reactors created an electricity shortage.
- Population in the areas around Chernobyl are declining due to a falling birth rate and the movement of the workforce away from the plant, which was the major employer in the region. The loss of population has led to increased social isolation and decreased availability of services to those few who have remained.
- Mistrust of officials and government who, it is felt by many, misled the people with regard to their safety or actively spread misinformation.

Economic effects

- The cost to the Soviet government for cleaning the site, evacuation, relocation and healthcare is estimated at $12.8 billion.
- The region's industrial and agricultural exports have decreased due to the mistrust of products coming from contaminated areas. This has led to reduced incomes or unemployment for the people who remain.
- Investment in affected areas has been reduced due to a perception of an 'unsafe lifestyle'.
- The migration of the workforce from contaminated areas to uncontaminated areas has resulted in a shortage of professionals such as doctors and teachers.

End of chapter questions

1. How do renewable energy sources differ from non-renewable energy sources? Give examples of each.
2. Why are coal, natural gas and oil referred to as 'fossil fuels?'
3. What do the following terms mean?
   (a) Enrichment
   (b) Radiation
   (c) Chain reaction
4. Explain the function of each of these components of a nuclear reactor:
   (a) coolant
   (b) control rods
   (c) moderator.
5. Why is graphite rarely used as a moderator material in Western countries?
6. Uranium-235 has an atomic mass of 235 and an atomic number of 92. How many:
   (a) protons
   (b) neutrons
   (c) nucleons

   does it have?
7. How is uranium-235 different from uranium-238?
8. Describe the three main types of radiation and their penetrating power.
9. How do control rods allow the fission rate in a reactor to be manipulated?
10. Explain why a 2-kilogram lump of uranium-235 could not explode like an atom bomb.
11. Determine how many X-rays you would need to have in a short period of time before you developed acute symptoms of radiation poisoning.
12. It has been found that crews of international passenger jets develop cancers at a slightly higher rate than people in many other professions. Why do you think this is the case?
13. How does a controlled chain reaction differ from an uncontrolled chain reaction?
14. It has been decided that a nuclear power station will be built in Queensland. Suggest possible locations for the site as well as a place to store the waste. Justify your choices.
15. It has been suggested that a good way to dispose of nuclear waste would be to load it onto a rocket and send it into the sun. Do you think that this would work? Explain your answer.
16. Why is radiation therapy used in the treatment of cancerous tumours when radiation can *cause* cancer?

5

Weapons through the ages

It seems that, for the whole of human history, we have managed to develop increasingly efficient ways of harming each other! However, in the process of developing weaponry, great advances have been made in our understanding of a number of scientific principles. What are these principles and how are they involved in making a weapon find its target?

The boomerang

The boomerang is regarded by many as the most advanced and elegantly designed weapon of early man.

History of the boomerang

While weapons similar to the boomerang have been used in many parts of the world through the ages, the earliest ones were used by Australian Aboriginal peoples for hunting. The earliest preserved specimen, found at Wyrie Swamp in South Australia, is about 10 000 years old. It is difficult to know how the boomerang was developed due to the extreme length of time that has elapsed since its invention, but many historians believe that it evolved from a non-returning throwing club that was in use previously. The main characteristic of the Australian boomerang is its ability to return to the thrower if it does not hit anything — a somewhat unique flight pattern. Boomerangs were not used by all tribes, but were fairly widespread nonetheless. They were usually crafted from wood and were sometimes decorated with ochre pigments (see figure 5.1). While primarily a hunting weapon, the boomerang was occasionally used against other people with equal efficiency.

Figure 5.1: A traditional boomerang

Design of the boomerang

The boomerang is made from a single piece of wood that is bent into a V-shape. The angle between the two arms where they meet at the elbow (the V-shape) is about 105 degrees. Many students who attempt to make their own boomerangs in woodwork class make them flat. In fact, the cross-section of the arms is the same as for an aircraft wing. This shape is called an **aerofoil**.

An **aerofoil** is any surface that extends into the air so as to create lift, such as an aircraft wing.

As you can see in figure 5.2, one edge of an aerofoil is thicker than the other. This thicker edge is called the *leading edge* and the other is the *trailing edge*. As the aerofoil moves through the air, the air particles strike the leading edge and split into two streams — one above the wing and the other underneath the wing. The airstream flowing over the top of the wing has a greater distance to travel because it must pass over the curve. This causes the upper airstream to move faster than the lower one. The greater speed of the upper airstream creates less pressure on the top than on the bottom, so the wing is pushed up by the greater air pressure on the bottom. This is called **lift**.

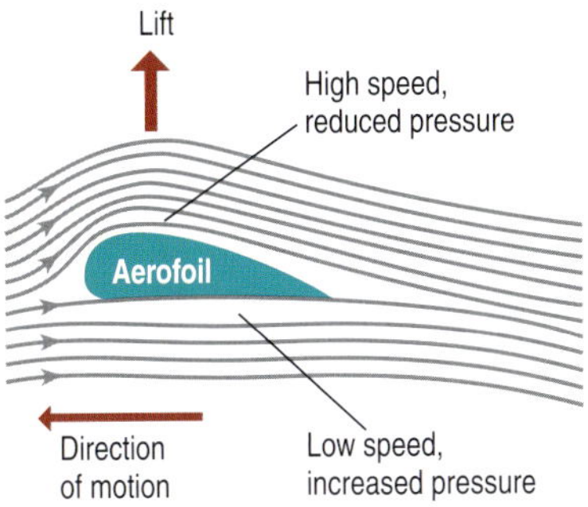

Figure 5.2: Lift acting on an aerofoil

Lift is the upwards-acting force created by a wing moving through the air.

On an aircraft, the leading edges of both wings face the front. On a boomerang, the leading edges face in opposite directions (as shown in figure 5.3), which means that, when thrown correctly, a leading edge is always moving forward in the direction of the throw as it rotates.

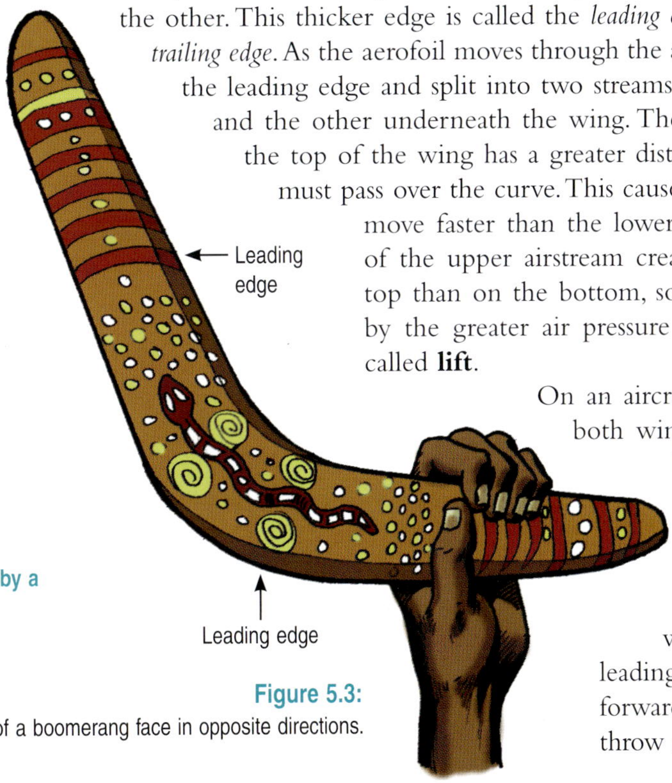

Figure 5.3: The leading edges of a boomerang face in opposite directions.

Dynamics of the boomerang

Apart from the shape, one of the key factors in the boomerang returning to the thrower lies in the throwing technique. The correct throwing technique causes a number of different forces to act on the boomerang in such a way that it moves in a circle rather than a straight line as a propeller does. The boomerang should be held slightly off the vertical, with the boomerang elbow pointing towards the thrower and the flat surface facing to the right (most boomerangs were made to be thrown with the right hand). The thrower draws their arm back and throws the boomerang forward by snapping their wrist at the last moment, causing the boomerang to spin vertically as it flies (see figure 5.4).

Figure 5.4:
Throwing a boomerang correctly

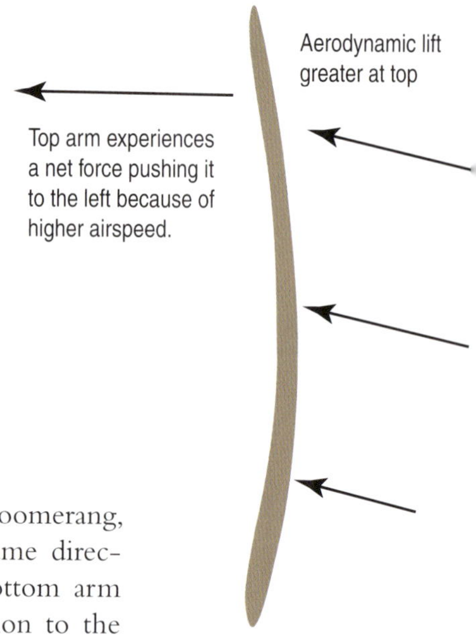

Figure 5.5:
Unequal lift acts on the top arm and the bottom arm, causing the top arm to experience a force pushing to the left.

Due to the propeller action of the boomerang, the top arm is always moving in the same direction as the throw motion, while the bottom arm is always moving in the opposite direction to the throw motion. As a result, while the two arms are both moving with the same rotational speed, the air particles travelling over the top arm will be moving faster than those over the bottom arm. So the lift acting on the top arm will be greater than that on the bottom one. The result is that, for our right-handed boomerang, the imbalance in lift forces pushes the top arm to the left (see figure 5.5).

Gyroscopic motion is demonstrated by any object spinning about its axis. Such an object resists any changes in the direction of the axis by giving rise to opposing forces.

When forces are exerted on spinning objects, they are subject to **gyroscopic motion**, causing the effect of the force to be delayed. So instead of the top arm moving to the left the moment that the force due to unequal lift acts on it, the effect occurs when the top arm has rotated forward into a horizontal position. As a result, the path of the boomerang is pulled to the left. This happens every time an arm is vertically upwards. After a few rotations, the entire boomerang is travelling in an anticlockwise arc, eventually forming a full circle back to the thrower. The rotational forces acting on the boomerang also cause it to move into a horizontal position rather than the vertical position from which it was thrown. When the boomerang returns to the thrower, it is still moving very quickly so it tends to be caught flat with one hand on either side (see figure 5.6).

Figure 5.6: The boomerang spins in the horizontal plane by the time it returns to the thrower.

Investigation

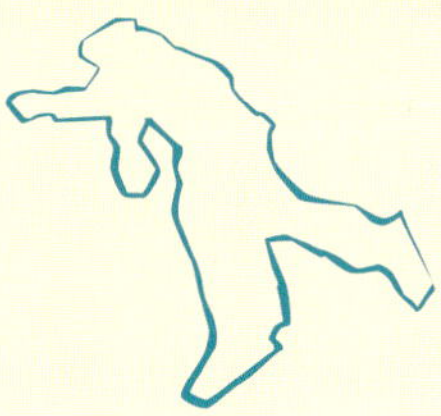

Building a model boomerang

Aim: To design and build a working model of a boomerang

Materials: 20 cm × 20 cm sheet of 2mm thick wood (pine or cedar, but not balsa or masonite); figure 5.7; carbon paper for tracing, hacksaw or electric jigsaw, felt pen; sandpaper

Method:

1. Photocopy and enlarge the boomerang shape in figure 5.7 so that it just fits on your wooden sheet.
2. Trace the shape onto the wood using carbon paper. (The carbon paper is placed with the ink side towards the wood and the shape placed on top of that.)
3. Using a hacksaw or electric jigsaw, carefully cut the shape out of the wood. (This could be done by your school's manual arts department).
4. Mark the position of the leading edges in felt pen as shown in figure 5.8.
5. Use sandpaper to give each arm of the boomerang an aerofoil shape. This means that you will be removing less wood from the top of the leading edge than from the trailing edge (see figure 5.9).

Caution: If you cut the shape out yourself, do it with care and under adult supervision!

(continued)

Figure 5.7:
The basic boomerang shape

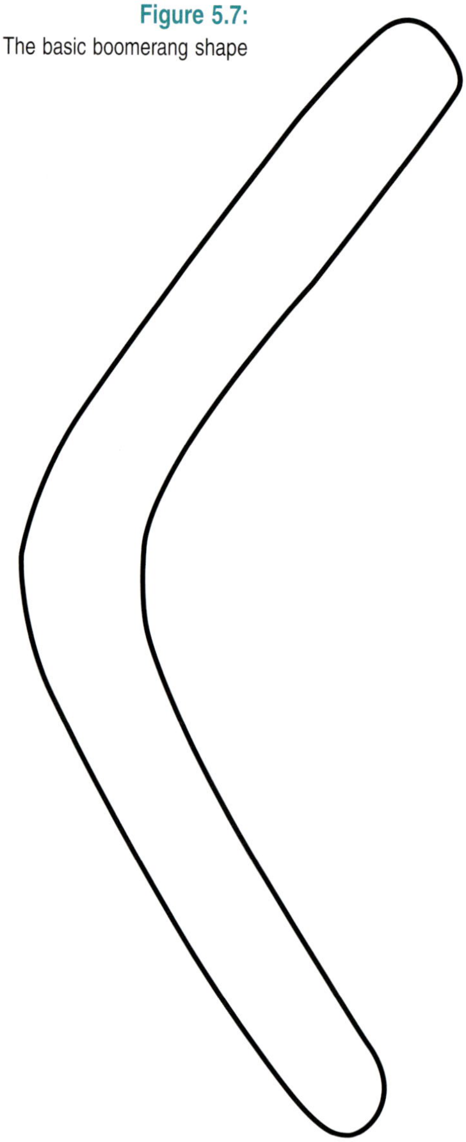

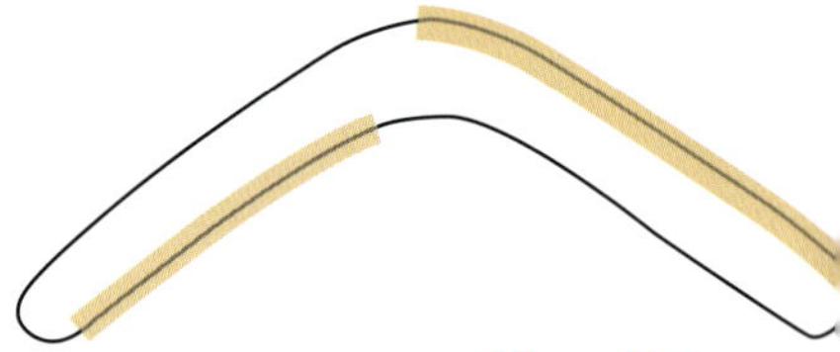

Figure 5.8:
Mark the leading edges on each arm.

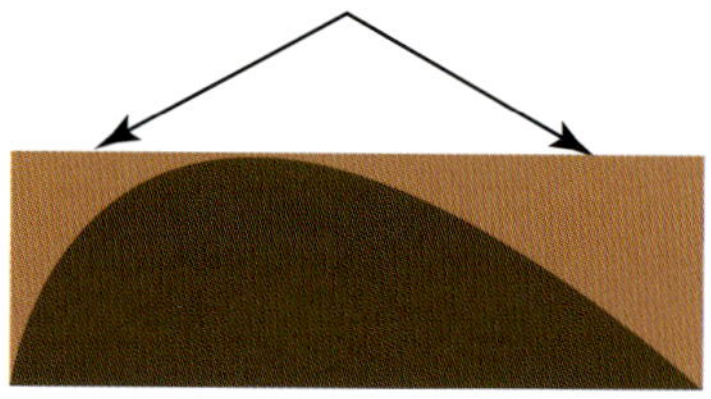

Figure 5.9:
Cross-section of arm showing the wood to be removed, giving the boomerang an aerofoil shape

6. Test your boomerang when you think you have the right profile. Make sure you do this in an open area! It might be wise to practice with a commercially bought boomerang first to perfect your throwing technique.
7. Wobbles in the flight of the boomerang tend to indicate that the arms have not been equally shaped. Fine-tune the shape of your boomerang by sandpapering it until the boomerang flies smoothly. Take your time doing this because it is better to sand away a little bit at a time than to go too fast and end up with a disaster!
8. You may choose to paint or varnish your boomerang to finish it off.

The trebuchet

The word *trebuchet* comes from the Old French word *trabucher,* meaning 'to overturn' or 'to fall over'. While the trebuchet is often classified as a type of catapult, it is probably more accurate to think of it as an enormous sling.

History of the trebuchet

Figure 5.10:
A simple traction trebuchet

The simple trebuchet is one where men pulled back the long arm, readying it for launch (see figure 5.10). It is thought to have originated in China somewhere between the 5th and 3rd centuries BC. Its use spread through Persia and Byzantium, reaching Europe around 500 AD, where men were replaced with a large fixed or pivoting counterbalance weight (see figure 5.11). The *gravity-counterweight trebuchet* has been described as the most powerful and efficient non-explosive mechanical weapon ever developed, and it was the major form of heavy siege artillery in the Middle Ages throughout Europe, the Mediterranean and Asia during the 14th, 15th and 16th centuries. With the invention of gunpowder, however, the trebuchet was replaced in its role as an artillery weapon by the cannon.

Trebuchets were enormous machines whose frames typically rose to heights of 14 metres, and their counterweights (generally wooden 'baskets' filled with rocks) weighed many tonnes. The largest trebuchets could weigh dozens of tonnes. They were not readily transportable, so they had to be built on the spot

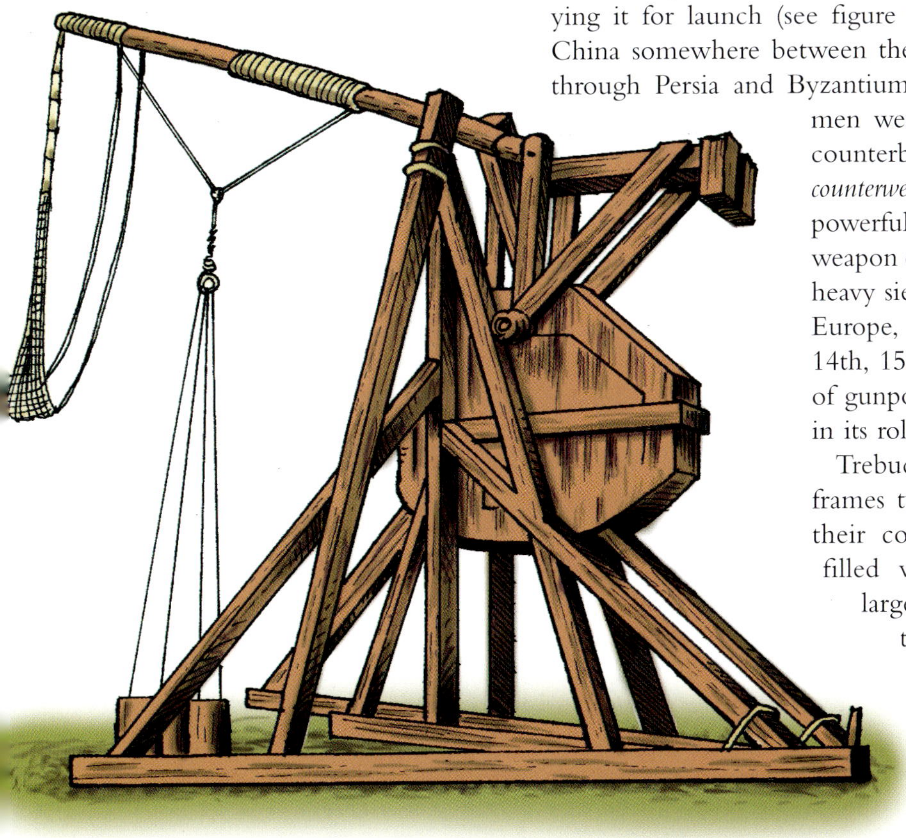

Figure 5.11:
A gravity-counterweight trebuchet

Figure 5.12:
The remains of Castle Urquhart

where they were to be used. Their chief advantage, aside from their large range, was their accuracy: they were capable of striking the same place on a wall repeatedly, once it was sighted. This repeated hammering was how the army of Edward I used the trebuchet *War Wolf* to bring down the walls of Castle Urquhart in Scotland (see figure 5.12).

Trebuchet mechanics

A **first-class lever** is one in which the fulcrum lies between the load and the effort.

The basic principle behind the trebuchet is that of a *lever* system. While there are three main classes of levers, all levers are made up of the same three basic components: the *load* (which is lifted), the *fulcrum* (or pivot) and the *effort* (which does the work). To explain the principles behind the trebuchet (which acts on the principle of a **first-class lever**), let us first look at a simple seesaw, which is also a first-class lever.

Imagine that a woman sits on one end of a seesaw. Her weight causes her end of the seesaw to touch the ground. In order for her to be lifted up, the other end of the seesaw will need to be pressed downwards towards the ground (usually this is done by someone else jumping onto the other end). The woman is the load, the seesaw pivot acts as a fulcrum, and the force that is to be exerted on the other end will be the effort (see figure 5.13). When the fulcrum is in the middle of the lever, the effort needed to lift the woman will need to exceed the woman's weight.

Figure 5.13:
To lift the woman (load), effort must be applied to the other end of the lever.

With a seesaw, the woman's end will rise quite slowly when the effort is applied. This is not going to be useful in a trebuchet because the load needs to travel quickly. So how can we make the load rise more quickly? The following investigation may give us a few clues.

Investigation

Trebuchets for beginners

Aim: To demonstrate the basic principle of the trebuchet

Materials: Ruler; eraser; 5-cent coin; a large space

Method:

1. Arrange the ruler, eraser and coin as shown in figure 5.14.
2. Press the free end of the ruler down to the desk. Did the coin move much?
3. Rearrange your equipment in its starting position. Quickly slap the free end of the ruler to the desk. Did the coin travel further this time?
4. Place your eraser near the edge of the desk and move the ruler and the coin into their positions at opposite ends of the ruler. The free end of the ruler should overhang the edge of the desk. Once again, slap the end of the ruler downwards. The free end should have a much greater range of motion. How far did the coin go this time?

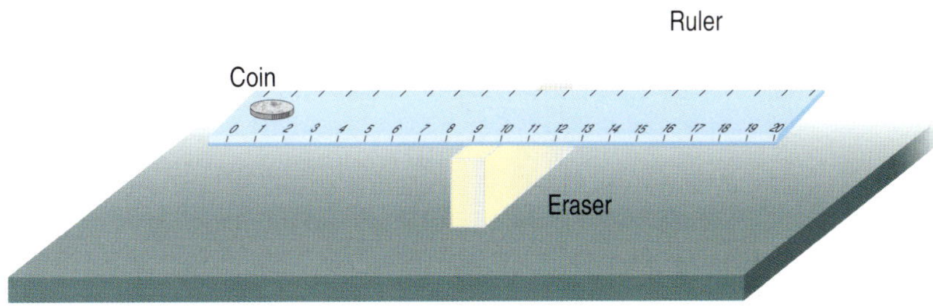

Figure 5.14:
Model of a very simple trebuchet

Structure of the trebuchet

The lever on the trebuchet is a long, thick beam that has a counterweight on one end that acts as the effort and a sling on the other end that carries the load. The pivot (in this case, an axle) is located so that the two arms of the lever are of unequal lengths, with the longer arm carrying the sling and the shorter arm the counterweight (see figure 5.15).

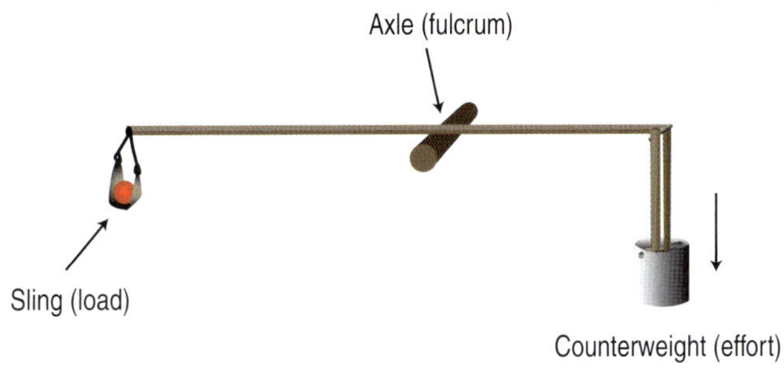

Figure 5.15:
The trebuchet arm as a lever

The things that were to be thrown by the trebuchet (stones, rocks, burning blocks — occasionally even the severed heads of the enemy) were loaded into the sling. This ammunition could weigh anywhere up to 300 kilograms. The sling was usually braided from rope or strips of leather. One end of the sling was permanently attached to the arm while the other end had a loop that slipped over a hook on the arm.

The counterweight had a much greater mass than the ammunition — up to 20 tonnes. At times, the counterweight consisted of a large metal (usually lead) weight that was fixed into a permanent position at the end of the arm. However, the trebuchet worked most effectively when the counterweight was able to fall in as straight a line directly downwards as possible. This was achieved by using as the counterweight an enormous wooden bucket that was jointed so that it swung on the short end of the beam. The bucket was then filled with sand or stones. Using a bucket had the advantage of being cheaper than using a large, fixed metal weight. Also, it was difficult to transport large metal weights to a siege site whereas sand and rocks for a bucket were easy to find on location.

Trebuchets in operation

The **range** of a projectile weapon is the furthest horizontal distance that it is able to send a missile.

The **range** of the trebuchet depended upon the mass of the counter-weight, the arm length and the length of the sling line. In operation, the long, unweighted end with the sling was pulled towards the ground and held by a trigger. The sling was then loaded. When the trigger was released, the counterweight dropped and caused the long sling arm to be lifted upwards in a circular arc. At the top of its swing, the sling loop came free of the hook, releasing the contents of the sling (see figure 5.16).

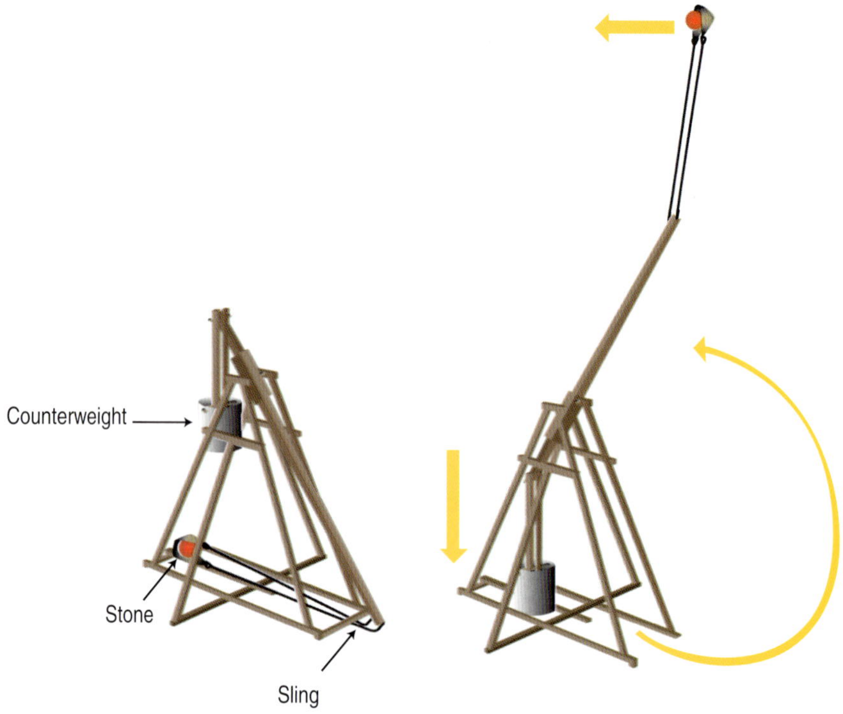

Figure 5.16: The firing motion of the gravity-counterweight trebuchet

The operation of the trebuchet was found to improve by placing the entire platform on wheels so that, after firing, the entire frame would rock backwards and forwards. The firing arm was then wound back down and the whole process would start again. As the winding time was often quite long due to the sheer size and mass of the machine, the trebuchet would often fire only three or four shots an hour.

Investigation

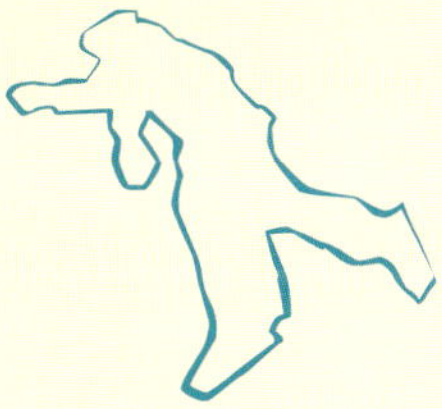

Designing and building a model trebuchet

Aim: To design and build a working model of a trebuchet from common household items that will fling a Jaffa the greatest possible distance

Materials: A packet of plastic drinking straws; several Jaffa lollies; wooden tongue depressors; string or thread; sticky tape; bolts, screws or coins to use as counterweights; large steel paperclips; scraps of cloth

Method:

1. Use the pictures of the trebuchets in the previous sections as a guide.
2. Make sure that your trebuchet has the following components:
   - Two triangular pieces to form the sides of the trebuchet frame. These will need to support the throwing arm.
   - A long throwing arm that pivots on an axle. The short end will have a counterweight attached to it. Note that the counterweight section should be able to swing freely without touching any other part of the trebuchet.
   - A sling attached with string or thread to the longer part of the throwing arm. This will hold the Jaffa and release it at the top of the arm's motion.

Discussion questions

1. What were the main difficulties that you found in making your trebuchet work?
2. How did you overcome these problems?
3. What was the furthest distance that your Jaffa reached?
4. What effect did the size of the counterweight have on your device?

The cannon

The name cannon comes from the Latin word *canna,* which refers to a hollow reed, essentially describing the hollow cylindrical shape of the device.

History of the cannon

Chinese and Arabic armies are recorded as having used fairly small tubes filled with explosive powder to fire arrows, but these weapons lacked the power that would be provided by the perfection of the cannon's key ingredient — gunpowder. The smoothbore cannon first appeared in the early 14th century in southern Europe, and there are records of cannons being used during a number of key battles at this time, including the siege of Stirling Castle in 1341. Early cannons tended to be uniformly cylindrical in shape, but they gained their more recognisable tapered form at the beginning of the 17th century. This tapered design persisted unchanged to the mid-19th century, when the cannon was superseded by more modern gun designs.

The Eastern cannons were constructed of materials that had a naturally cylindrical shape (such as tree trunks) or that could be shaped into a cylinder (such as stone) and have a cavity bored through the centre. Unfortunately, these materials had a tendency to explode, often wiping out the people who were firing the cannon. When the design reached Europe, where metal technology had been flourishing for some time, the metal cannon was created (see figure 5.17). These were at first made from a series of long iron bars that were forge-welded into a tubular shape and bound with iron hoops, much as the slats of a barrel are held together. These cannons fired balls of stone that could weigh up to 50 kilograms.

Eventually, as metallurgy progressed further and the process of casting improved, cannons were cast from a single piece in bronze (an alloy of copper and tin). The cannonballs were made of iron, stone or lead. Later, as cheaper cast iron became available, both cannons and cannonballs were made of this instead of bronze.

Figure 5.17:
Defensive cannon mounted on battlements

Structure of the cannon

The structure of the cannon was a thick metal tube that tapered externally from its base to the mouth. The explosive charge (gunpowder) was placed in the base and lit by means of gunpowder packed into a hole (the touch hole) that ran through the cannon's shell. Cylindrical knobs called *trunnions* were cast in a piece with the cannon's barrel. These were used to support the cannon on its carriage (usually made of wood) and also acted as pivots, allowing the angle at which the cannon was fired (called the *elevation*) to be altered (see figure 5.18).

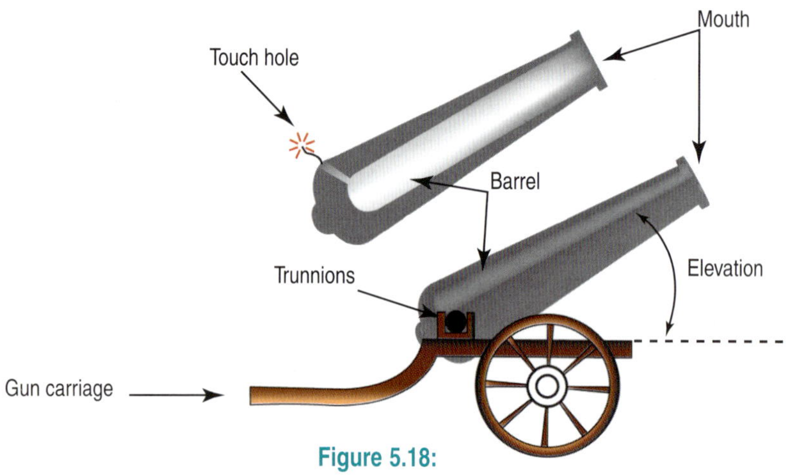

Figure 5.18:
Structure of the cannon

Gunpowder

The most important ingredient of the cannon was *gunpowder*, or black powder as it is often called. This was composed of a mixture of saltpetre (potassium nitrate), sulfur and carbon (which came from wood charcoal). The proportions in which these were mixed determined the effectiveness of the cannon, so recipes for good mixtures were highly prized. The best results seem to have been achieved when the weights of the individual powders were measured out as four parts saltpetre, one part sulfur and one part carbon. The quality of the resulting explosive depended on additional factors such as air humidity and the purity of the ingredients.

Firing the cannon

Taking into account the range of the target and the type of cannonball that was going to be fired, the appropriate amount of powder was measured out and placed in the bottom of the bore with a long-handled scoop. Then a wad of cloth (usually cotton) was rammed in firmly on top of the powder to form a solid plug at the base of the bore. The cannonball was rolled down the bore to sit on the top of the packed wadding and powder. More gunpowder was poured down into the touch hole to meet up with the powder in the base of the bore, forming a fuse. When the fuse was lit, the gunpowder ignited.

When lit, the components of the gunpowder reacted. This reaction can be written in the form of a word equation as follows:

Potassium nitrate + sulfur + carbon → Potassium sulfide + nitrogen + carbon dioxide

or as a balanced chemical equation:

$$2KNO_3 + S + 3C \rightarrow K_2S + N_2 + 3CO_2$$

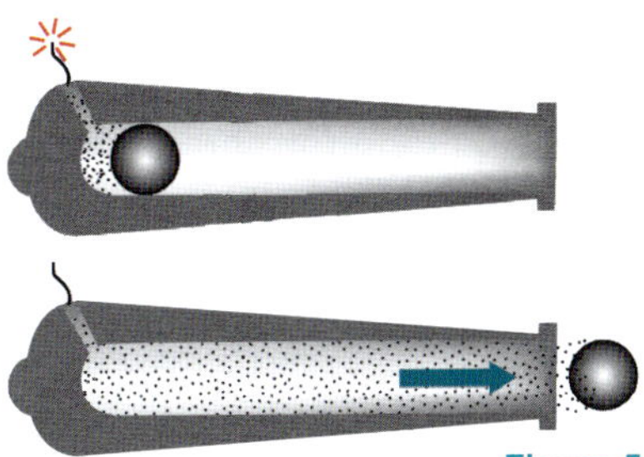

Figure 5.19: Expanding gas produced by the chemical reaction of the gunpowder ingredients caused an increase in pressure that fired the cannonball out at high speeds.

This reaction caused a large amount of nitrogen and carbon dioxide gas to be produced as well as powdered potassium sulfide, which was blown out of the cannon as a thick white smoke. The volume of these products was 4000 times greater than the volume of the gunpowder, and this increase of volume occurred in a fraction of a second, resulting in a huge increase of pressure behind the wad. The expansion of the gas pushed the wad forward like a piston that in turn pushed the cannonball out of the mouth of the cannon at 100 to 200 metres per second (see figure 5.19).

Between firings, the barrel of the cannon was cooled with a water-soaked rag bound on the end of a stick. This was to ensure that there were no stray embers or remaining heat that could ignite the gunpowder as it was being poured in.

Targeting the cannon

Where the cannonball landed depended upon the amount of gunpowder used in the cannon and the vertical angle at which the cannon was pointing. Until the 17th century, the physics of motion was not very well understood. It was believed that cannonballs flew in a straight path until they ran out of 'impetus', whereupon they fell straight down again. Targeting the cannonball correctly was a matter of trial and error, dependent more upon the experience of the cannon crews than upon mathematics. After the work of scientists such as Galileo and Newton, it was realised that the path of the cannonball was in fact a curved shape called a **parabola** (see figure 5.20). The mathematics of projectile motion allowed range calculations to be made rather than guessed.

A **parabola** is a symmetrical curved path. In geometric terms, it is formed by the intersection of a plane through a cone.

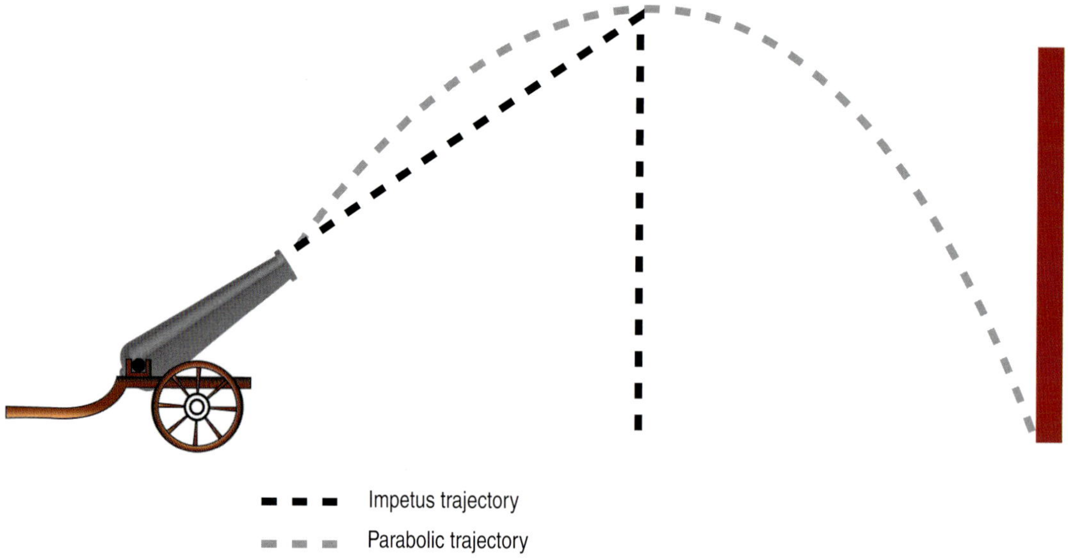

Figure 5.20:
The 'impetus' model of cannonball motion versus parabolic motion

Cannons were usually positioned behind the troop lines. When the cannon was completely horizontal, its position was referred to as point blank. As this gave only limited range, it was used only when the enemy was very close (hence 'at point blank range'). By elevating the cannon from the horizontal, the range was increased and it was possible to fire cannonballs over your troops onto the enemy beyond.

In general, the range R of the cannonball depended upon the angle of elevation θ and the speed V at which the cannonball left the cannon. The muzzle speed depended upon the amount of gunpowder loaded — the more gunpowder that ignited, the faster the cannonball left the cannon. Too much gunpowder in a cannon could lead to it bursting upon explosion, usually killing everyone around it for tens of metres!

The maximum range of the cannonball over horizontal ground can be calculated by using the equation:

$$R = \frac{V^2 \sin 2\theta}{9.8}$$

So, for a cannonball with a muzzle velocity of 300 metres per second fired at an elevation of 20 degrees above the horizontal, the range can be calculated as:

$$R = \frac{(300)^2 \times \sin(2 \times 20)}{9.8} = \frac{90000 \times \sin 40°}{9.8} = 5903 \text{ metres}$$

This is nearly six kilometres!

Similarly, the maximum height H that the cannonball will reach can be found as:

$$H = \frac{R \tan \theta}{4}$$

The maximum height that the cannonball from the previous example would reach could be calculated as follows:

$$H = \frac{5903 \times \tan 20°}{4} = 537 \text{ metres}$$

In reality, however, the range would not be quite as far and the height nowhere near as high due to air resistance, which slows the cannonball down as it moves through the air particles. In modern firing tables calculated for artillery, air resistance is factored in for a variety of conditions, but such tables could not be fully calculated until the invention of the computer in the 20th century!

Internet activity

To use a cannon simulator that allows you to see the effect of air resistance, muzzle velocity and elevation angle on targeting and range, go to the website for this book and click on the Cannon simulator weblink (see Weblinks, page 309).

1. Use the website to determine the best angle of elevation to hit the target in each of the following circumstances:
   (a) a standard cannonball fired at 100 metres per second when there is no wind
   (b) a cannonball that has a density of 0.75 fired at 43 metres per second while there is a tail wind of 10 kilometres per hour
   (c) a cannonball with a density of 0.3 that is fired with a muzzle speed of 100 metres per second and a head wind of 20 kilometres per hour.
2. Are there any cases where the shot is impossible under the imposed conditions?
3. Explain what would have to be changed in order to hit the target if the wind conditions and cannonball were kept the same.

Recoil

Momentum is the product of an object's mass and its velocity.

When a cannon is fired, it has a tendency to move backwards. This backward motion is called *recoil* (see figure 5.21) and can be explained in terms of momentum. The **momentum** of any object p is the product of its mass m and its speed v:

$$p = mv$$

Figure 5.21:
Recoil causes the cannon to move in the opposite direction to the cannonball.

The principle of the Conservation of Momentum states that the total momentum of a system of objects before an event has to be equal to the total momentum of all the objects in the system after the event. In the case of our cannon, the system is made up of the cannon and the cannonball. Before the cannon is fired, both the cannon and the cannonball are stationary. Therefore, because each of them has a speed equal to 0, they must each have a momentum equal to 0. This means that if momentum is to be conserved, the total momentum of the cannon and the cannonball after firing must still equal 0. That is:

$$0 = p_{\text{cannon}} + p_{\text{cannonball}}$$

In other words:

$$p_{\text{cannon}} = -p_{\text{cannonball}}$$

The negative sign in front of the term for the cannonball momentum in the equation above indicates that the cannon and the cannonball move in opposite directions; this describes our cannon's recoil. We will drop the negative from our equations from now on.

The recoil speed of the cannon v_{recoil} depends upon the mass of the cannon m_{cannon}, the mass of the cannonball $m_{\text{cannonball}}$ and the speed with which the cannonball leaves the cannon $v_{\text{cannonball}}$:

$$v_{\text{recoil}} = \frac{m_{\text{cannonball}} \times v_{\text{cannonball}}}{m_{\text{cannon}}}$$

For example, we can calculate the recoil speed of a 5000-kilogram cannon that fires a 100-kilogram cannonball at a speed of 250 metres per second as:

$$v_{\text{recoil}} = \frac{100 \times 250}{5000} = 5 \text{ metres per second}$$

As you can see from this equation, the bigger the cannon is, the smaller the recoil speed. However, a larger cannon is also less portable and so harder to move during battle.

Eye on the media

Slingshot firestorm — call to ban 'lethal' weapons

Deadly slingshots can be bought legally across the counter in Queensland for as little as $22.

Banned in other states, the catapults have been used to fire ball bearings and other missiles at shops and vehicles.

Victims, backed by safety bodies, want the weapons banned.

Police admit there is a risk of serious injury and State Police Minister Judy Spence has asked for community input before considering making the weapons illegal.

The Sunday Mail this week found slingshots for as little as $22 at camping and disposal stores.

Some are pre-loaded with ball bearing ammunition and boast ranges of up to 160 m and 200 km/h velocities.

The shanghais are classed as a weapon in Queensland, and while it is an offence to carry or use them in public, it is legal to own one.

Among recent incidents involving slingshots were:

- At Ferny Grove six men were charged with more than 200 offences relating to firing marbles, ball bearings and stones from slingshots in April.
- In Rockhampton [there were] more than 50 drive-by slingshot attacks on residents and businesses in January.
- In Townsville 30 vehicles were damaged in slingshot attacks in September.

Attacks were spread from Ingham to the Gold Coast.

Bowen Shire's Mayor, Michael Brunker, said an attack in September left ratepayers footing the bill for new windows on council offices.

'It's just bloody disgraceful. They should ban them,' he said. 'These types of things are fairly lethal weapons. They ricochet. There's no way they should be sold — they're bloody deadly.'

Bowen Senior Sergeant Mick Wilkie said the availability of slingshots and ammunition encouraged some offenders.

'If it hits you in the temple, or the eye, they could cause serious injury,' he said.

Queensland Injury and Surveillance Unit manager Richard Hockey estimated six injuries yearly were directly related to the weapons.

Source: Caitriona Murtagh, *The Sunday Mail,* 11 December 2005, p. 30.

Clarification questions

1. Use the equation for range to determine whether the claims for range and velocity in this article are possible.
2. At what angle would the ball bearings have to be launched for the claims to be true?
3. If the mass of a ball bearing is 25 grams, what would be its momentum when it leaves the slingshot? (Momentum is measured in Newton seconds when mass is in kilograms and velocity is in metres per second.)

Mustard gas

Mustard gas is the common name used for the chemical compound thio-dichloroethane (chemical formula: $C_4H_8Cl_2S$). Despite its name, pure mustard gas is a colourless and odourless liquid whose toxic effects were first noted by chemist Frederick Guthrie in 1860.

History of mustard gas

Mustard gas (so called because the impurities in it had a smell like mustard) was first used as a weapon during World War I near the town of Ypres (Belgium) on 12 July 1917. A variety of poisoned gases such as arsenic gas and tear gas had already been used on and by the British forces and the Germans. These gases quickly became ineffective with the invention and use of gas masks that filtered out the fumes and allowed soldiers to continue fighting (see figure 5.22). Mustard gas had a number of advantages over other gas weapons: it could penetrate gas masks and clothes, it was difficult to detect until too late, its effects were debilitating and it was easy to use.

Figure 5.22:
Gas masks were effective against many gas weapons but were useless against mustard gas.

How mustard gas works

Mustard gas works by reacting with the water found in the skin to form thiodiglycol and hydrochloric acid, which form blisters and cause cell damage. The moist areas of the body such as the face, eyes, lungs and armpits are the main areas affected. One of the nastier characteristics of this chemical is that it is not very soluble in water, so trying to wash it off made its effects worse.

Effects of mustard gas

The actual effects of exposure to mustard gas varied enormously, depending upon the dose size, length of exposure, air temperature and humidity, the degree of moistness in the person's skin and whether the exposed skin had open wounds through which the chemical could penetrate further. Mustard gas was not originally intended to be fatal; rather, it was meant to cause pain, panic or incapacitation to enemy troops. Nonetheless, many soldiers were killed by it when exposed to large doses.

The effects of mustard gas exposure included reddening and blistering of the exposed areas. If breathed in, it blistered the lining of the lungs, causing breathing difficulties or even suffocation. If the eyes were exposed to the gas, the moist outer layers of the eyes (the corneas) were destroyed, causing blindness. The skin affected by mustard gas took a long time to heal, usually resulting in severe infections.

Many years after exposure, victims of mustard gas would experience effects from the destruction of their bone marrow. Because bone marrow produces red blood cells, loss of bone marrow also led to chronic anaemia. Mustard gas also attacked the white blood cells, which are an important part of the immune system, so victims often died from what would normally be fairly low-grade infections. It was also implicated as a cause in the high rate of cancer and leukaemia in attack survivors.

Delivery methods

Mustard gas was usually delivered in artillery shells that were fired at enemy lines. Other poisonous gases such as phosgene and chlorine were difficult to deliver in this way because they had to be released in the form of a concentrated cloud; the landing of the shell tended to cause these gases to be dispersed quite rapidly, so the gases were not as effective as intended. If there was a wind blowing, the gases were diffused, further reducing their effectiveness. Mustard gas, on the other hand, did not need to be a concentrated fog to be effective, and wind served to increase the area of effect. Also, because mustard gas is heavier than air, the gas would sink towards the ground and fall downwards into the trenches where the enemy troops were sheltered. Shell delivery also had the advantage of surprise because, unlike conventional weapons, they could be delivered without any audible warning.

Nuclear weapons

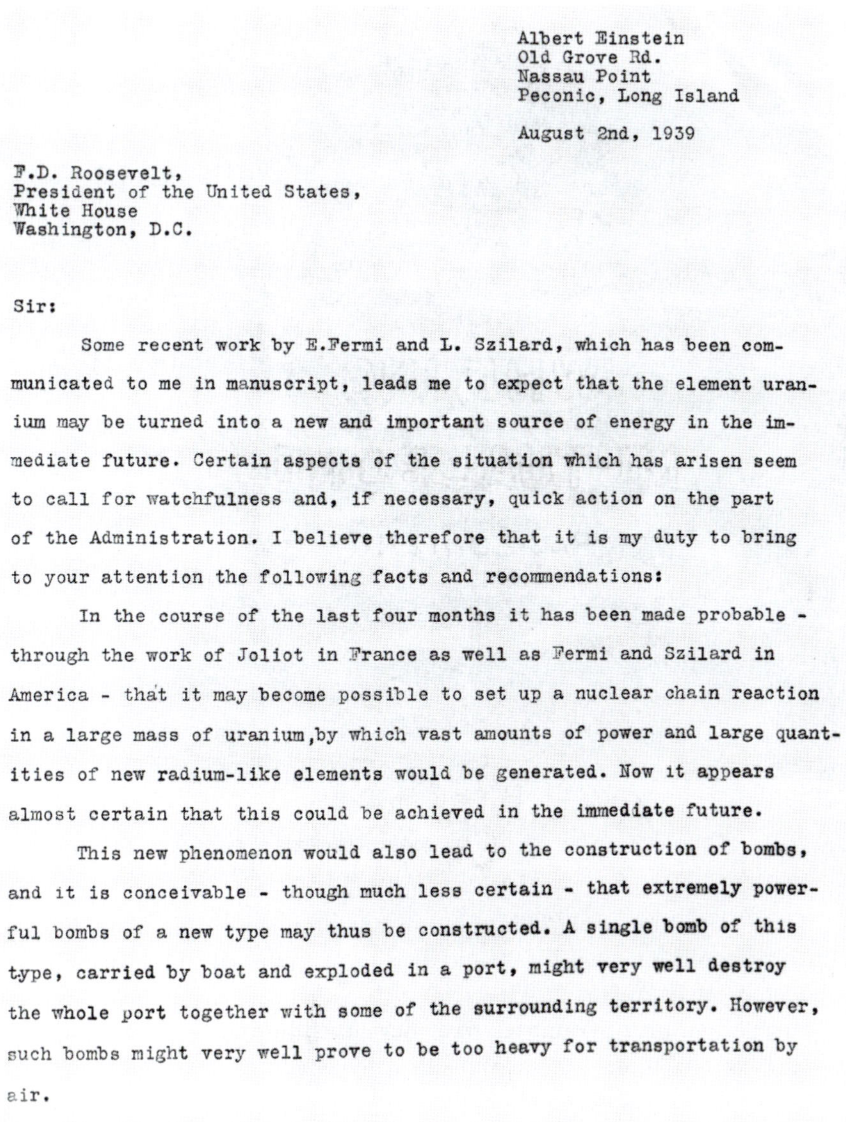

Albert Einstein
Old Grove Rd.
Nassau Point
Peconic, Long Island

August 2nd, 1939

F.D. Roosevelt,
President of the United States,
White House
Washington, D.C.

Sir:

Some recent work by E.Fermi and L. Szilard, which has been communicated to me in manuscript, leads me to expect that the element uranium may be turned into a new and important source of energy in the immediate future. Certain aspects of the situation which has arisen seem to call for watchfulness and, if necessary, quick action on the part of the Administration. I believe therefore that it is my duty to bring to your attention the following facts and recommendations:

In the course of the last four months it has been made probable - through the work of Joliot in France as well as Fermi and Szilard in America - that it may become possible to set up a nuclear chain reaction in a large mass of uranium,by which vast amounts of power and large quantities of new radium-like elements would be generated. Now it appears almost certain that this could be achieved in the immediate future.

This new phenomenon would also lead to the construction of bombs, and it is conceivable - though much less certain - that extremely powerful bombs of a new type may thus be constructed. A single bomb of this type, carried by boat and exploded in a port, might very well destroy the whole port together with some of the surrounding territory. However, such bombs might very well prove to be too heavy for transportation by air.

Figure 5.23: Einstein's letter to President Theodore Roosevelt

We use the term 'nuclear weapons' to describe weapons that rely on the nuclear processes of fission or fusion for their destructive energy. Most recently, they have also been referred to as 'weapons of mass destruction' (WMD).

History of nuclear weapons

The so-called Nuclear Age is said to have begun in 1945 with the bombing of Hiroshima, but the history of nuclear weapons really begins with a letter written by Albert Einstein on 2 August 1939 to the US president, Franklin D Roosevelt. In this letter (reproduced in figure 5.23), Einstein warned the president that scientists in Nazi Germany were purifying Uranium-235, which could be used to make a terrible weapon.

The **Manhattan Project** was the codename used in World War II for the secret US project to beat Germany in developing the first atomic bomb.

Roosevelt's response was to begin a similar project in the US in an effort to beat the Germans to it. Called the **Manhattan Project**, this venture included the work of many of the world's top physicists, among them Robert Oppenheimer, Niels Bohr, Richard Feynman, Enrico Fermi and Edward Teller. Their work culminated in the testing of the first atomic bomb in the desert at Alamogordo in New Mexico on 16 July 1945. The device was winched to the top of a tower and detonated by remote control. The test was a spectacular success, but the sheer power of the explosion made many of the observers doubt what they had achieved. Upon seeing the explosion, Oppenheimer is famously reported as having quoted from the Bhagavad-Gita (a sacred Hindu text): 'I am become Death, the destroyer of worlds'.

With the success of the initial test, it was decided to use the atomic bomb on selected Japanese cities in order to persuade the Japanese to surrender. At 8.15 am on 6 August 1945, the B52 bomber *Enola Gay* dropped an atomic bomb on the city of Hiroshima. In seconds, 66 000 people were killed and 69 000 were injured. Everything within a kilometre of the blast point was vaporised, and everything within 2 kilometres was totally destroyed (see figure 5.24).

Figure 5.24:
The immediate after-effects of the nuclear bomb explosion in Hiroshima, previously a built-up city

Three days later, a plutonium bomb was dropped on Nagasaki. It killed 39 000 people, injuring 25 000 and levelling half the city. The Japanese surrendered on 10 August 1945.

Since then, many countries have developed atomic weapons, although they have so far been used only in tests. There at least 30 000 nuclear weapons in the world today, enough to destroy the planet many times over and effectively obliterate life from its face.

Fission bombs are another name for atom bombs — devices that use an uncontrolled chain reaction to produce destructive energy and heat.

Fission bombs

Fission bombs (also called *atom* bombs or *A-bombs*) rely on uncontrolled chain reactions in unstable radioactive materials such as uranium-235 (see figure 5.25). Recall from chapter 4 that a chain reaction occurs when slow neutrons strike a uranium-235 nucleus, causing the nucleus to break into smaller nuclei and release energy, radiation and more neutrons. These neutrons in turn strike more uranium nuclei and so on. In nuclear reactors, the rate at which neutrons strike uranium nuclei is controlled. If a sample of uranium is larger than the critical mass, however, then an uncontrolled chain reaction occurs in which all of the released neutrons cause further reactions that release devastating amounts of energy (mainly in the form of heat and light) and radiation. It is also possible to use plutonium-239 as the fissionable material (see figure 5.26).

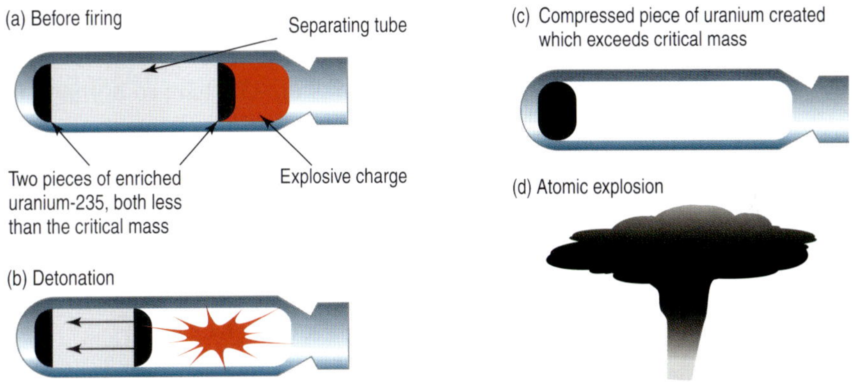

Figure 5.25: This gun-style bomb using enriched uranium-235 is similar to the bomb codenamed *Little Boy* that was dropped on Hiroshima on 6 August 1945.

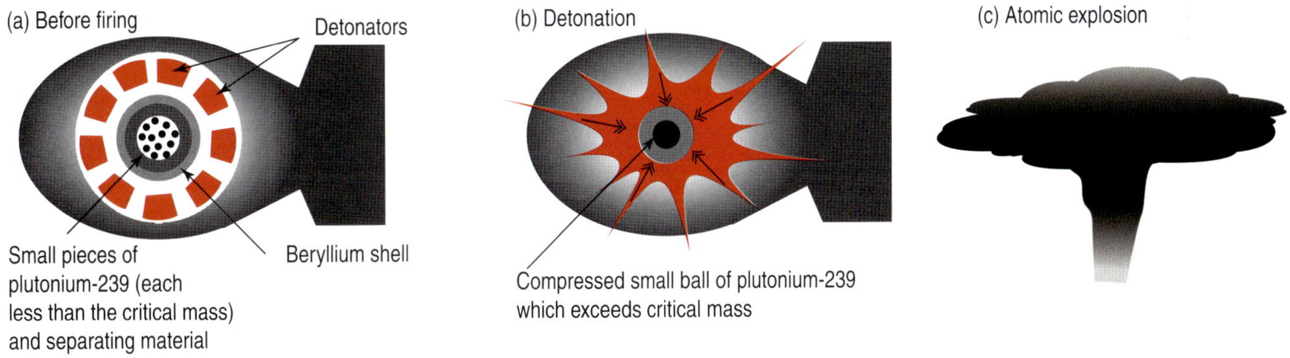

Figure 5.26: This implosion bomb using plutonium-239 as its energy source is similar to the bomb code-named *Fat Man* that was dropped on Nagasaki on 9 August 1945.

Fission bombs use lumps of weapons-grade enriched uranium ore, 97 per cent of which is uranium-235. Each lump is smaller than the critical mass. The fission bomb works by bringing two or three of these sub-critical masses together, usually using an explosive charge. When the masses are combined, the mass exceeds the critical mass and causes a nuclear explosion.

Fusion bombs

While fission relies on large nuclei being broken into smaller nuclei to release energy, fusion occurs when small nuclei such as hydrogen are forced to fuse together to form larger nuclei. Fusion reactions can take place only when the nuclei are very hot, overcoming the tendency of the positively charged hydrogen nuclei to repel each other and therefore allowing them to combine. Fusion releases more energy than fission but can occur only when the surrounding temperature is very high — about 20 000 000°C. Because of the high temperature, fusion reactions are also called *thermonuclear reactions.*

It is fusion that provides the energy that keeps the sun burning, releasing energy by combining superheated hydrogen nuclei to form larger helium nuclei in a process called the *proton–proton chain*. This occurs in three stages, with energy being released at each stage (see figure 5.27):

1. Two hydrogen nuclei join to form deuterium, which is an isotope of hydrogen that has a proton and a neutron in its nucleus.
2. Each deuterium nucleus combines with a hydrogen nucleus to form helium-3, a helium isotope that has one less neutron than normal helium.
3. Two helium-3 nuclei combine to form one helium nucleus and two more hydrogen nuclei.

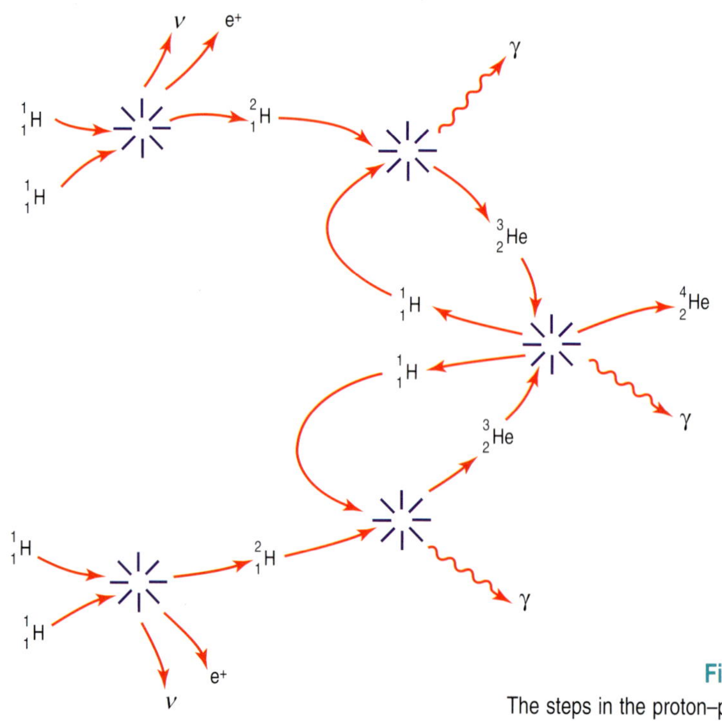

Figure 5.27: The steps in the proton–proton chain

Fusion bombs use the nuclear fusion of hydrogen nuclei to release energy.

Fusion bombs (also known as hydrogen bombs or H-bombs) detonate in two stages. The first stage involves the explosion of a small fission bomb that creates the necessary high temperatures, and the second stage occurs when the superheated hydrogen nuclei combine.

Ground zero is the location on the ground directly under the detonation point of an atomic weapon.

A **thermal flash** describes the enormous amounts of heat and radiation that rapidly spread out from the centre of an atomic blast.

Effects of nuclear weapons

When nuclear weapons are detonated, enormous amounts of heat and radiation spread out from the centre of the blast (known as **ground zero**) in what is called a **thermal flash**. This radiation forms a fireball

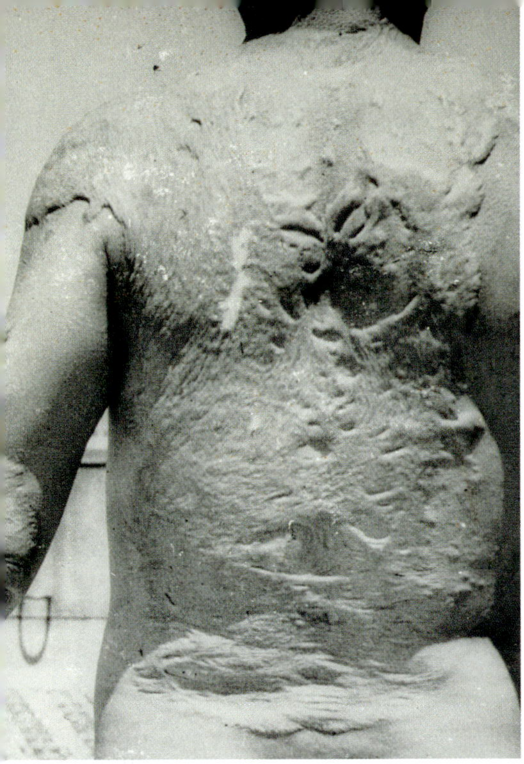

Figure 5.28: Radiation burns on the back of a Hiroshima survivor

A **shock wave** is the rapid expansion and contraction of particles in a gas or liquid caused by a sudden change in pressure or density. Shock waves travel at or above the speed of sound.

An **electromagnetic pulse** occurs when the very strong electromagnetic fields formed by an atomic blast suddenly collapse; this can cause electrical and electronic systems to burn out.

Nuclear fallout is the return to Earth of the irradiated material sent into the atmosphere by an atomic blast.

that generates the distinctive mushroom cloud associated with nuclear weapons. The fireball from the Hiroshima bomb formed a mushroom cloud 7 kilometres wide. At locations close to ground zero, most substances were melted or burned. Organic matter (including people) was vaporised. People up to 50 kilometres away received serious burns (see figure 5.28), and those who looked directly at the flash were blinded.

After the initial blast, the vaporisation of particles close to the blast causes an implosion of air from further out. When these inrushing air particles collide, they cause a high-pressure **shock wave** that spreads outwards at speeds of up to 3000 kilometres per hour. This shock wave causes the destruction of buildings, blowing them away from the centre of the blast.

The detonation of nuclear devices releases large amounts of radiation in the form of gamma rays, which cause very strong electromagnetic fields to form and then collapse. These electromagnetic fields can burn out electrical and electronic systems, including computer networks and power grids, and even disrupt the electrical systems that control cars, planes and weaponry. This burst of electromagnetic activity is called an **electromagnetic pulse**.

The most devastating effects for survivors arise from radiation exposure. As discussed in chapter 4, large amounts of radiation as experienced by those closest to the blast would be fatal, with the level of exposure dropping with increased distance from ground zero. The DNA in many of a survivor's cells can be damaged, leading to diseases such as cancer, leukaemia and immune system collapse later in life. Damaged DNA in the sex cells means that the children and even grandchildren of survivors may suffer mental or physical defects.

The radioactive nuclei formed during the nuclear reactions, as well as tonnes of irradiated dust, are blasted high into the atmosphere during detonation and the formation of the mushroom cloud. In the weeks following the nuclear explosion, these return to Earth as **nuclear fallout**. This radioactive fallout increases the background radiation for many years in the areas where they land, known as fallout zones. People in these zones are exposed to higher radiation levels with damaging effects on their health.

Napalm

Napalm is actually the name of the powder that is mixed into gasoline to form a thick, sticky substance called *jellied gasoline*. However, the term 'napalm' is often used to refer to the liquid mixture itself.

History of napalm

Napalm was first used during World War II in the invasion of Tinian in June 1944 and before the invasion of Iwo Jima. It was also used as the fuel in the bombs dropped on Dresden and in Japan by the Allies.

Napalm was invented in 1942 by a combined team from Harvard University and the US army. Earlier forms of jellied gasoline had been known, but they needed rubber as the principal ingredient — a substance that was in very short supply during World War II. This new form used a soap derived from naphthene and palmitate (hence the name 'napalm') that was then mixed with gasoline. When napalm is ignited with white phosphorous, it becomes an incendiary weapon that burns fiercely and is difficult to extinguish.

A more modern version of napalm developed by Dow Chemicals (called *napalm B*) was used during the Vietnam War. This upgraded form used chemicals such as benzene or polystyrene rather than naphthene and palmitate. While the use of napalm on civilians was outlawed by international law in 1980, it continued to be used in conflicts until the end of the 20th century. The last two canisters of napalm in the possession of the US military were disposed of in April 2001.

Figure 5.29:
A napalm air strike during the Vietnam War

How napalm works

When mixed with gasoline, napalm was used as a tactical weapon to remove vegetation cover and subdue enemies hidden in it. Napalm was usually delivered in the form of bombs dropped by aircraft. Napalm itself is very difficult to set alight, so the bombs contained white phosphorous mixed with the napalm. When the bombs hit the ground, they burst and released the phosphorous and napalm contents. Phosphorous is notoriously reactive to air, so the rupturing of the bomb caused it to ignite. This in turn ignited the napalm. The fuel-gel mixture was stringy and sticky, and readily adhered to most surfaces including human skin. It could not be washed off with water, so victims would try to scour it off their skin using sand or dirt.

Effects of napalm

Napalm kills through immolation (burning) and suffocation. Victims of napalm tend to suffer horrendous injuries because it actively sticks to the skin while it burns, and most people suffer either second- or third-degree burns (see figure 5.30). A person will probably survive if splashed with a small amount of napalm, but they will develop extremely painful second-degree burns that result in thick keloid scars and nerve damage. Ironically, the most severe burns — third degree — are not painful because the burn damages the full thickness of the skin layer and thereby destroys the pain receptors. These burns tend to be extensive, however, leading to a rapid drop in blood pressure and often death from respiratory failure.

Figure.5.30:
Nine-year-old Phan Thi Kim Phuc ran out of her village with her clothes burnt off after a South Vietnamese aircraft accidentally dropped its napalm bomb on her village of Trang Bang on 8 June 1972.

Suffocation results when the burning napalm consumes the oxygen in the immediate area of the strike and carbon monoxide replaces the oxygen. When breathed in, carbon monoxide attaches to the haemoglobin molecules in the red blood cells that would normally carry oxygen. Without a supply of oxygen, suffocation results and the body cells die. The body can tolerate air that is less than 0.4 per cent carbon monoxide, but napalm produces regions where the air is 20 per cent carbon monoxide! Those people who did not suffocate to death often fell unconscious and were then burned to death.

The bunker-buster bomb

A bunker-buster bomb is designed penetrate underground targets such as caves, and hardened targets buried deep underground such as concrete bunkers. Such targets are hard to find and destroy by other means because they are given a great deal of protection by the amount of earth above them.

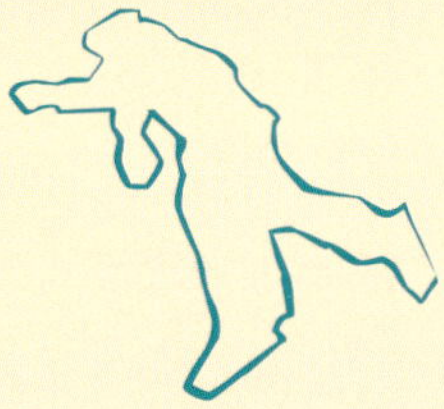

Investigation

Bunker busting

Aim: To demonstrate how a bunker-buster bomb penetrates to underground targets
Materials: Six eggs; six shoeboxes filled with sand; food colouring; a large ball bearing; metre ruler
Method:

1. Take six eggs and bury them at different depths in the shoeboxes filled with sand. Once each egg is in place, pat the sand down firmly, taking care not to break the egg. Mark the location of each egg with a drop of food colouring and write the depth of each egg on the outside of the box.
2. Drop a large ball bearing from a height of 1 metre onto each of the egg locations (see figure 5.31).
3. Uncover the eggs and note which ones have been damaged.
4. Rebury the ones that are intact at the same depth as they had been previously and repeat the bombing with a larger ball bearing.

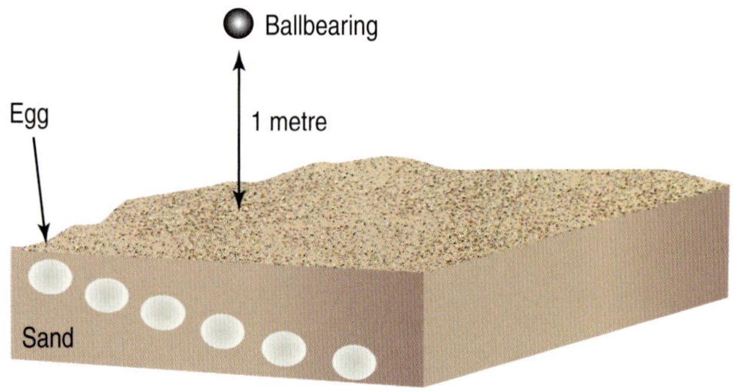

Figure 5.31: Generally, the deeper the egg is buried, the safer it is.

History of the bunker-buster bomb

The bunker-buster bomb was developed during the 1991 Gulf War when Allied Forces were frustrated by the inability of their normal bombs to penetrate the underground bunkers of the Iraqi military. These bunkers were deep underground and heavily reinforced, so they were virtually invulnerable to the Allied artillery. This frustration led the US air force to develop the GBU-28 bunker-buster bomb. Such bombs have since been used in the war on Iraq, which began in March 2003.

Structure of the GBU-28

The GBU-28 bunker-buster bomb has four main components (see figure 5.32)

- *The barrel.* This is a hardened steel tube that is 5.8 metres long and has a diameter of 37 centimetres.
- *The explosive.* The bomb carries nearly 300 kilograms of explosive, made up of a mixture of TNT and aluminium (which improves the bomb's penetrating power).
- *The laser guidance assembly.* This is attached to the front of the barrel. This section allows the bomb to home in on the target, which has

been 'painted' with a laser by either a soldier on the ground or from the bomber itself. The assembly steers the bomb by manipulating the steering fins near the front of the barrel.

- *Rear fins.* These are larger fins that are fixed in position to stabilise the bomb's flight.

The completed bomb has a mass of nearly 2000 kilograms and is dropped from B2 stealth bombers or jets such as the F-111.

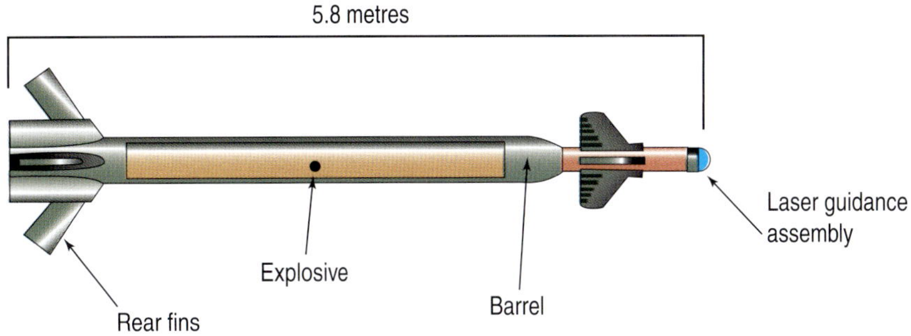

Figure 5.32: The structure of the GBU-28 bunker-buster bomb

How bunker-buster bombs work

Bombs in general work by transferring large amounts of kinetic energy to the target. You may recall that the kinetic energy KE of any object depends upon its mass m and its velocity v:

$$KE = \frac{1}{2} mv^2$$

The amount of kinetic energy in a bomb influences its ability to penetrate the ground. Increasing the mass of the bomb and its velocity also increases its kinetic energy.

You might think that to make a really effective bomb, all you need do is make the bomb very heavy and drop it from a bigger height. After all, the longer the drop, the faster the speed with which the object hits the ground. Unfortunately, there are problems with this plan.

First, the bombs need to be carried by aircraft, so they cannot be too heavy. A 10 000-kilogram bomb would do a lot of damage but a plane could not carry many.

Second, falling objects continue to speed up as they fall *until* they reach terminal velocity. When an object falls, it is pulled downwards by gravitational force and is simultaneously pushed upwards by the force of air resistance. The faster the object moves, the larger the air resistance. Eventually, if the object falls for long enough, the upward push of air resistance is equal to the downward pull of gravity and the object stops accelerating. The object then falls at a constant speed (the **terminal velocity**) until it hits the ground (see figure 5.33).

As a result, there is a limit to the velocity that the bomb can reach before it hits terminal velocity. Generally, the narrower the bomb is, the smaller the air resistance it encounters and the faster it is travelling when it reaches terminal velocity. So how does the width affect the penetration of the bomb? This question is tackled in the following investigation.

Terminal velocity is the speed at which a falling object ceases to accelerate because the upward force of air resistance is balanced by the downward force of gravity.

Figure 5.33: An object reaches terminal velocity when the gravitational force is equal to the force of air resistance.

Investigation

Effect of cross-sectional area on penetration depth

Aim: To test how the cross-sectional area of an object influences how far it will penetrate into a sand bed

Materials: Shoebox; sand; three acetate sheets (OHTs); sticky tape; metre ruler; whiteboard marker; electronic scales; sandwich bags; funnel

Method:

1. Fill the shoebox full of sand. Pat the sand down so that it is firm
2. Cut the acetate sheets in half to create six pieces of acetate. Roll them up to form tubes with diameters of 1 centimetre, 1.5 centimetres, 2 centimetres, 2.5 centimetres, 3 centimetres and 3.5 centimetres. Use sticky tape to join the seams.
3. Cut out sections of sandwich bag to cover one end of each of the tubes. Sticky tape these sections into position, making sure to wrap the tape several times around the tube.
4. Using the funnel, fill the 1-centimetre tube with sand until it has a mass of 50 grams.
5. Hold the tube upright and, with the sealed bottom of the tube 1 metre above the sand surface, drop the tube (see figure 5.34).

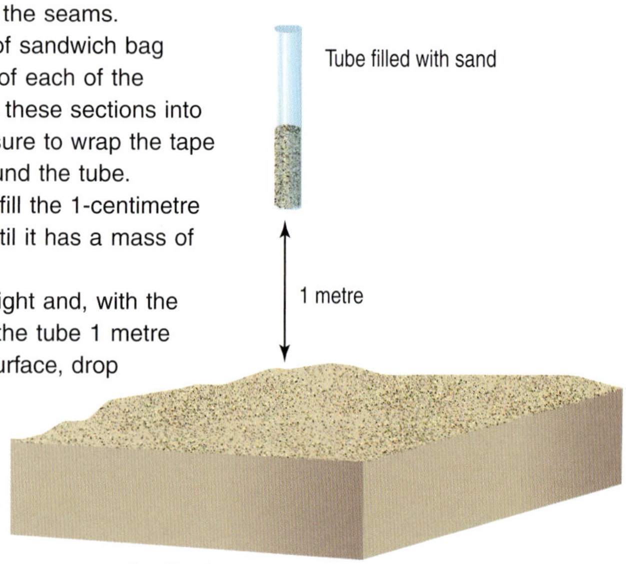

Figure 5.34: Investigating depth penetration

6. Mark the tube at the sand level with the whiteboard marker and use a ruler to measure the distance from the mark to the sealed end of the tube. This is the penetration depth of the tube.
7. Rub off the mark from the tube, level the sand surface and repeat steps 5 and 6 four more times. Enter the average penetration depth for the tube into a copy of the data table.
8. Repeat steps 4 to 7 for all of the other tubes.

Data table

Tube diameter (cm)	Penetration depth (cm)					
	Trial 1	Trial 2	Trial 3	Trial 4	Trial 5	Average
1						
1.5						
2						
2.5						
3						
3.5						

Discussion questions

1. What were the variables that were kept the same in each test (i.e. the controlled variables)?
2. Why was each trial done five times and the average found?
3. Draw a line graph of your results showing tube diameter on the horizontal axis and the average penetration depth on the vertical axis. Using your graph, determine the penetration depth of a tube that has a diameter of 2.25 centimetres.
4. What problems did you encounter in this investigation? How would you change the method to fix these problems?
5. State the general relationship between the tube diameter and the penetration depth.

As you should have found in the investigation, the wider the bomb, the harder it is for it to penetrate very far into the ground. The narrower the bomb, the easier it can move earth out of its way.

So the design of the GBU-28 bunker-buster bomb takes advantage of all these factors. It is extremely heavy, but very narrow for its weight due to the use of high-density materials. Its barrel is extremely strong so that it is able to withstand the crumpling forces that act on it as it enters the ground, which would slow its progress. As a result, the GBU-28 can penetrate bunkers that are under 30 metres of earth or 6 metres of concrete.

Eye on the media

A bomb to bust the deepest bunkers

Despite the intelligence failure that led the Bush administration to believe Iraq had weapons of mass destruction, the Pentagon is pressing ahead with the development of technologies designed to destroy WMDs. Its latest idea is a bomb that can destroy deeply buried WMD storage bunkers by cutting through earth and concrete inside a bubble of air.

Traditional 'bunker busters' are streamlined bombs that rely on sheer weight to force their way through soil, rock or concrete. But the new design has a blunt nose that forces the earth ahead of it out to the sides, creating a cavity the bomb can easily slide through, allowing it to reach much deeper buried structures than conventional bunker busters.

New Scientist has learned that the novel warhead is being developed for the Pentagon's Defence Threat Reduction Agency by Lockheed Martin Missiles and Fire Control of Dallas, Texas, in conjunction with the US navy's Surface Warfare Centre at Indian Head, Maryland. The DTRA wants the warhead to carry an incendiary payload for incinerating chemical or biological agents.

The design builds on the US navy's work on high-speed torpedoes (*New Scientist,* 22 July 2000, p. 26), which reduce friction around themselves by creating a gas bubble called a supercavity. The Russians have had such a torpedo since the 1990s: called Shkval, it can travel at 360 kilometres per hour.

To create a supercavity that surrounds but doesn't touch the body that created it, the object has to be travelling very fast — at least 180 kilometres an hour if it is in water. And the nose has to be flat to force fluid off the edge with such speed and at such an angle that it avoids hitting the surface of the body. But if this is to be achieved, the result is a supercavitating body with extremely low drag. Instead of being encased in water, it is simply surrounded by water vapour, which is less dense and has less resistance.

But supercavitation may not be limited to liquids. At high enough velocity a blunt-nosed body will force apart any medium it travels through, whether it be water, soil or concrete. If the cavity is large enough, the only surface in contact with the medium will be the blunt tip of the nose.

Lockheed Martin hopes the supercavitating missile will reach 10 times the depth of the current air force record holder, the huge BLU-113 bunker buster, which can break through 7 metres of concrete or 30 metres of earth.

The BLU-113 needs a thick casing to withstand the friction, but a supercavitating bomb suffers less resistance and so could have a thin casing, leaving more space for its explosive or incendiary payload.

But some experts are sceptical. 'I am still not certain that the term "cavitation" is correct for a solid,' says Nathan Okun, a naval historian. And Robert Nelson of the Union of Concerned Scientists says, 'I could believe a penetration factor increase of two but I don't believe a factor of ten.'

Michael Levi of the Brookings Institution, a Washington DC think tank, sees the need, however. 'Increased penetration is more important for chemical and biological agent destroying weapons than other bunker busters because it's essential to get detonation inside the storage area.'

Lockheed Martin will soon be putting an end to the speculation: four prototype weapons are due to be tested later this year.

Source: David Hambling, *New Scientist,* 16 July 2005, p. 30.

Clarification questions

1. What are WMDs?
2. How does the new bunker buster differ from the old design?
3. What is a 'supercavity'?
4. Why would creating a supercavity enable a bunker buster to go deeper?
5. Why are some experts sceptical about the claims for the new design?

End of chapter questions

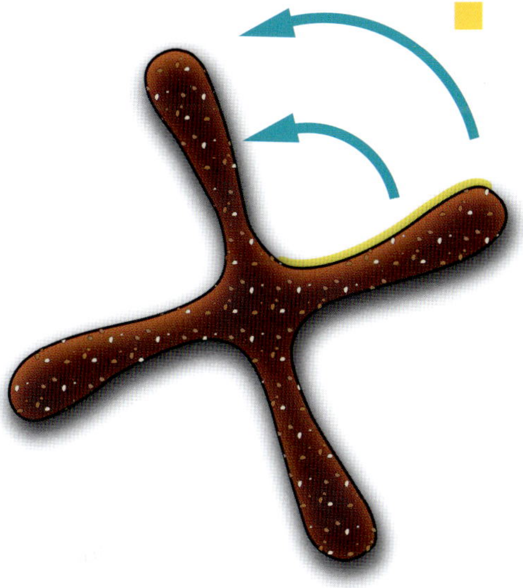

Figure 5.35:
A four-bladed boomerang

1. What is an alloy? Give an example of an alloy.
2. In what ways is a boomerang similar to an aircraft propeller?
3. Describe how a boomerang is able to fly back to the thrower.
4. Predict what would happen if you were to throw a boomerang horizontally rather than just off the vertical. Explain your reasoning.
5. A four-blade boomerang is made from plastic by a toy company. The leading edge for one of the blades is shown in figure 5.35. Copy the diagram and mark where the leading edges of the other three blades must be located in order for the boomerang to fly properly.
6. Which of the objects in figure 5.36 act as first-class levers?

Figure 5.36:
Which are first-class levers?

7. What problems do you think a trebuchet would encounter if:
   (a) the counterweight was too light?
   (b) the two arms were the same length?
   (c) the sling was placed on the shorter arm instead of the longer arm?
   (d) the pivot was stiff?
8. The range of a cannon can be altered by changing its elevation. How would a trebuchet's range be adjusted?
9. What would be the greatest range of a small cannon that fired a 20 kilogram cannonball at a speed of 120 metres per second when elevated at an angle of 30 degrees to the horizontal?
10. A fort is struck by a cannonball from a cannon that is 1 kilometre away. If the cannonball left the cannon with a velocity of 100 metres per second, at what angle is the cannon elevated?
11. You may have heard the 'fact' quoted that a battleship will move sideways due to the recoil from firing it guns. Examine figure 5.37, which shows the US *Missouri* firing nine guns at once.
    (a) How would you be able to tell from the picture if the ship was moving sideways or not?
    (b) The US *Missouri* has a mass of 58 000 tonnes and the nine guns that fire shells with an average mass of 1200 kilograms at a speed of 800 metres per second. Calculate the recoil speed that the US *Missouri* would experience when all nine guns are fired at once.
    (c) Do you think that this sideways recoil would be noticeable to observers?

Figure 5.37:
The US *Missouri* firing its guns

12. Which would best increase the range of a cannon: doubling its muzzle velocity or doubling the angle of elevation? Explain your answer.
13. Why was mustard gas more effective than earlier gas weapons such as chlorine and arsenic at subduing its targets?
14. In what ways were mustard gas and napalm similar?
15. Why was mustard gas particularly damaging to eye tissue?
16. In terms of burn severity, first-degree burns are the least critical and the shallowest, while third-degree burns are the deepest and penetrate the full depth of the skin layer. Why were second- and third-degree burns more common than first-degree burns in napalm victims?

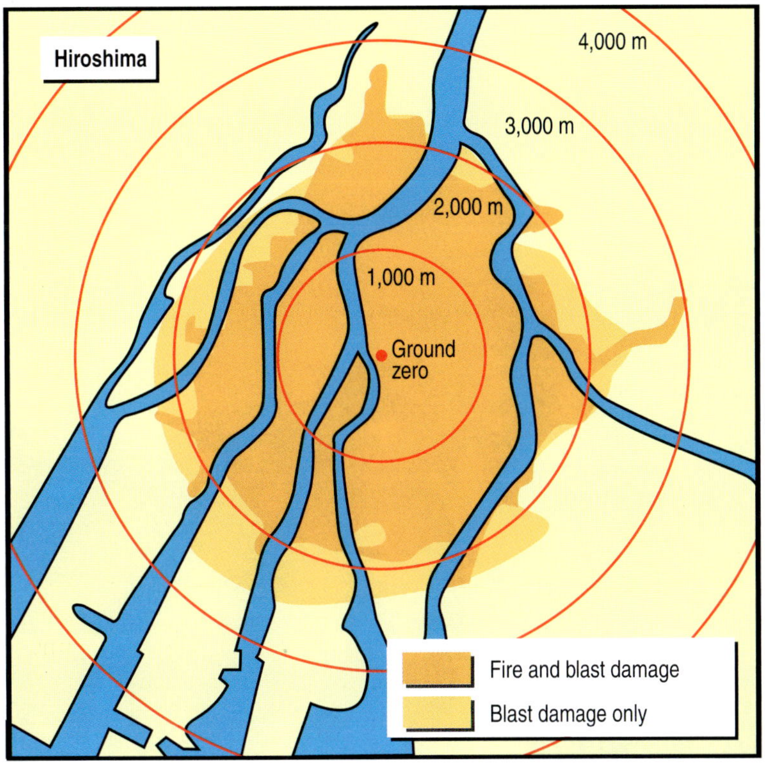

Figure 5.38:
Hiroshima damage map

17. Why was phosphorous mixed with napalm in bombs?
18. (a) Refer to the information in figure 5.38 and table 5.1.
    (i) What was the original population of Hiroshima before the bombing?
    (ii) How many of the dead were killed within one kilometre of ground zero?

    (b) As you would expect, the number of people killed gets smaller the further they are from ground zero. Contrary to expectations, though, the number of people wounded does not follow the same pattern. What explanations can you give for this apparent discrepancy?

Table 5.1: Distribution of deaths and injuries in Hiroshima

Distance from ground zero (km)	People killed	People injured	Total population
0–1.0	86%	10%	31 200
1.0–2.5	27%	37%	144 800
2.5–5.0	2%	25%	80 300
Total	27%	30%	256 300

Source: Adapted from www.atomicarchive.com/Maps/HiroshimaMap.shtml.

19. What will be the kinetic energy imparted by a bunker-buster bomb with a mass of 2000 kilograms that strikes the ground at 400 metres per second? (Kinetic energy is in joules when mass is in kilograms and speed is in metres/second.)
20. The atomic bomb that was dropped on Hiroshima was equivalent in explosive power to 15 kilotonnes of TNT (1 kilotonne of TNT — 4.184×10^{12} J). The US nuclear arsenal includes fifty B-53 nuclear warheads, each of which can produce a blast equivalent to 9 megatons of TNT. How many times more powerful is the B-53 compared to the Hiroshima bomb?

6

The brave new world of genetic engineering

In 1931, at a time when the DNA molecule that holds the genetic template of organisms had not even been isolated, Aldous Huxley wrote the novel *Brave New World*. In it, he described a future where babies were created in artificial wombs and genetically modified according to their future class and function in society. Now, nearly 80 years after Huxley's novel, the ability to manipulate the genetic make-up of living things according to our own needs and wishes is within our grasp (see table 6.1). The question is not so much *can* we do it as *should* we do it?

Table 6.1: Genetics timeline

1859	• Charles Darwin publishes *On the origin of species*, outlining the theory of natural selection
1866	• Gregor Mendel's paper on the inheritance of 'factors' in pea plants is published
1900	• The beginning of modern genetics • Mendel's results are verified by Carl Correns, Erich von Tschermak and Hugo de Vries
1905	• Nettie Stevens and Edmund Wilson describe the sex chromosomes
1910	• Thomas Hunt Morgan theorises that a particular mutation in flies is sex-linked
1944	• Oswald Avery, Colin MacLeod and Maclyn McCarty isolate DNA
1951	• Rosalind Franklin takes X-ray diffraction photos of DNA
1953	• Francis Crick and James Watson determine the structure of DNA
1961	• Nirenberg and Matthaei uncover the genetic code system in DNA
1990	• Human Genome Project started • Gene therapy first used
1996	• Dolly the sheep, the first cloned animal, was born
2000	• Fruit fly genome completed
2003	• Human Genome Project completed

Revising genetics

Before we start to consider the intricacies of genetic engineering, it might be useful to quickly revise what you have already learned about genetics and heredity in earlier science courses.

DNA

DNA (deoxyribonucleic acid) is a long molecule that is the carrier of genetic information.

Nucleotides are the subunits that make up the DNA molecule. Each nucleotide is made up of a sugar block, a phosphate block and a nucleotide base.

Just as computer code determines the way in which a computer functions, the essential blueprint for any organism — what it looks like, how it is structured, how it functions, perhaps even how it behaves — is encoded in the form of a long molecule called **DNA (deoxyribonucleic acid)**. This molecule is made up of subunits called **nucleotides**. Each nucleotide has three essential components:

- a *sugar* (or deoxyribose) block made up of oxygen, hydrogen and carbon atoms
- a *nucleotide base* made up of nitrogen, hydrogen and oxygen atoms
- a *phosphate* block made up of phosphorus and atoms of oxygen.

The nucleotide may also, depending on the type of base, contain atoms of oxygen and carbon. While the sugar and phosphate blocks in each nucleotide are the same, there are four types of nucleotide base: adenine, thymine, cytosine and guanine. These are represented by the letters A, T C and G respectively.

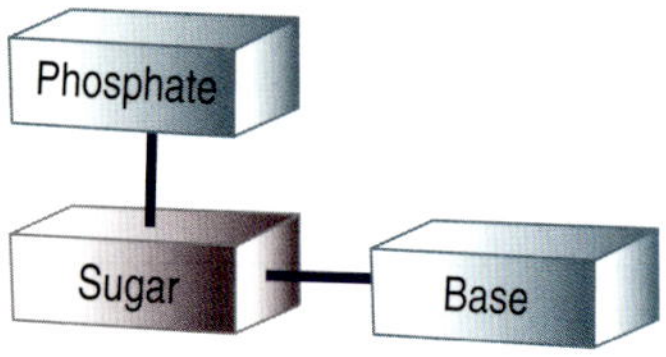

Figure 6.1:
The structure of a nucleotide

Complementary base-pairs are pairs of nucleotide bases that form the cross-links of the DNA molecule.

When the nucleotides join together to form DNA, the resulting long molecule has a characteristic 'double helix' shape, rather like a twisted ladder in which alternating chains of the sugar and phosphate blocks make up the long arms of the ladder and the pairs of bases make up the cross rungs. These base pairs will be either an A and T joined together, or a C and G. They will only pair in these combinations due to the ways in which hydrogen atoms at the ends of the bases join to form weak bonds. The base pair combinations (A — T and C — G) that form the cross-links in the DNA molecule are referred to as **complementary base-pairs**.

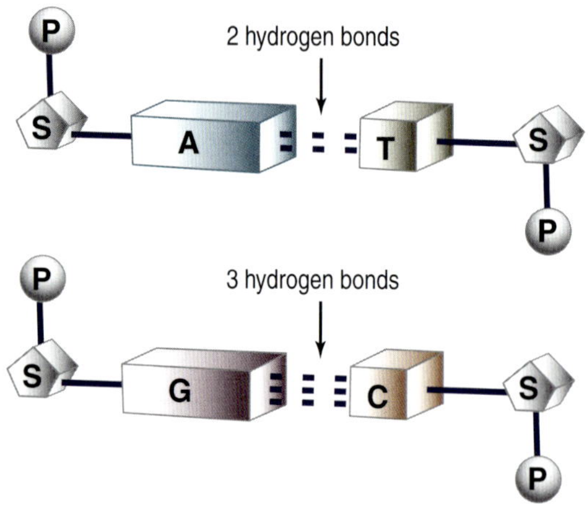

Figure 6.2:
A and T base pairs form two weak hydrogen bonds between their molecules, while C and G form three hydrogen bonds.

In groups, make a 3-dimensional model of a section of the DNA molecule from lollies or beads strung onto wire. You will need to ensure that your model has the correct shape, and that the same lollies/beads represent the same bases throughout. You should also include a key to help interpret your model.

The complete DNA molecule is very long and very thin. If you were to take the entire DNA strand from inside the nucleus of a human body cell, it would be a molecule with a length of about 1 metre but only 2 nanometres wide (0.000 002 mm), as shown in figure 6.3.

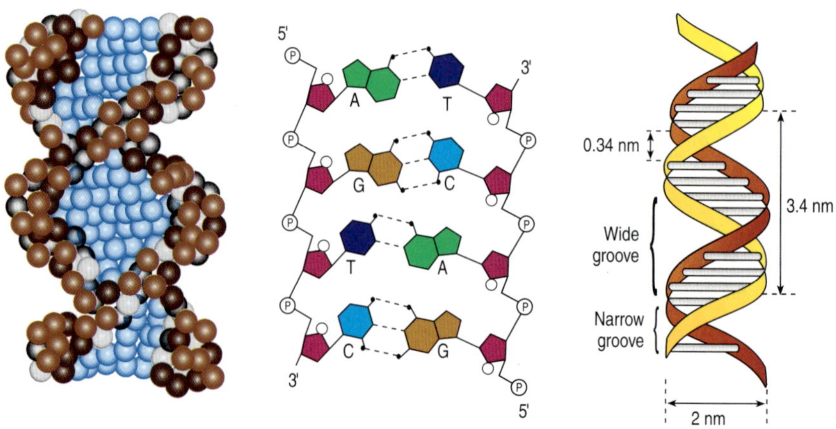

Figure 6.3:
Three representations of part of a DNA molecule

Chromosomes

Chromosomes are long thread-like structures made of DNA that are found in the nucleus of cells.

DNA is carried in the nucleus of cells as a series of tightly coiled chains making up long thread-like structures called **chromosomes**. These can generally only be seen with a microscope when the cell is about to divide, a process known as meiosis. Chromosomes were observed in dividing cells as early as 1848, but their significance was, at that time, unknown.

Different organisms have different numbers of chromosomes, but they are generally found in pairs. Humans have 23 pairs of chromosomes, making 46 in total. These pairs are numbered 1 to 22 and are known as *autosomes*. The last pair of chromosomes, pair 23, are *sex chromosomes* that determine gender. We inherit half of each pair from our parents: one copy of chromosomes 1 to 23 from each parent.

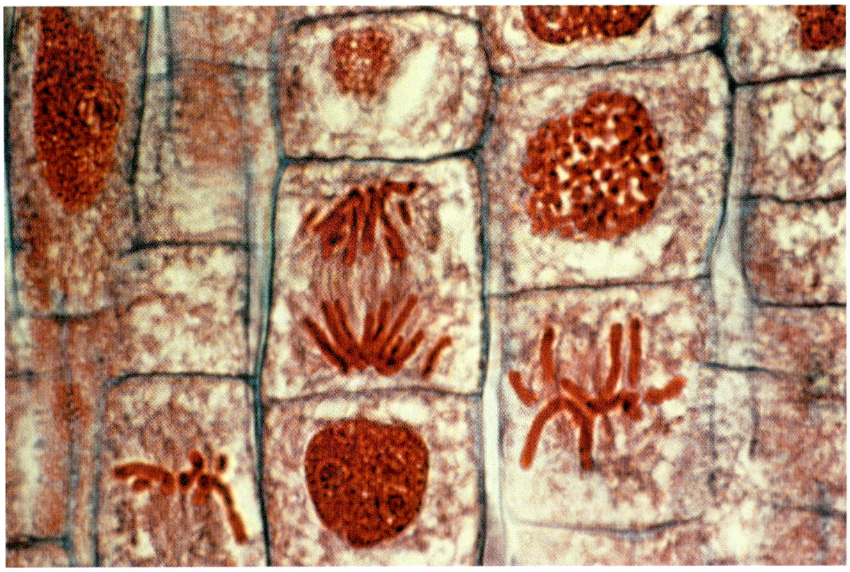

Figure 6.4:
Chromosomes become visible when meiosis is about to occur.

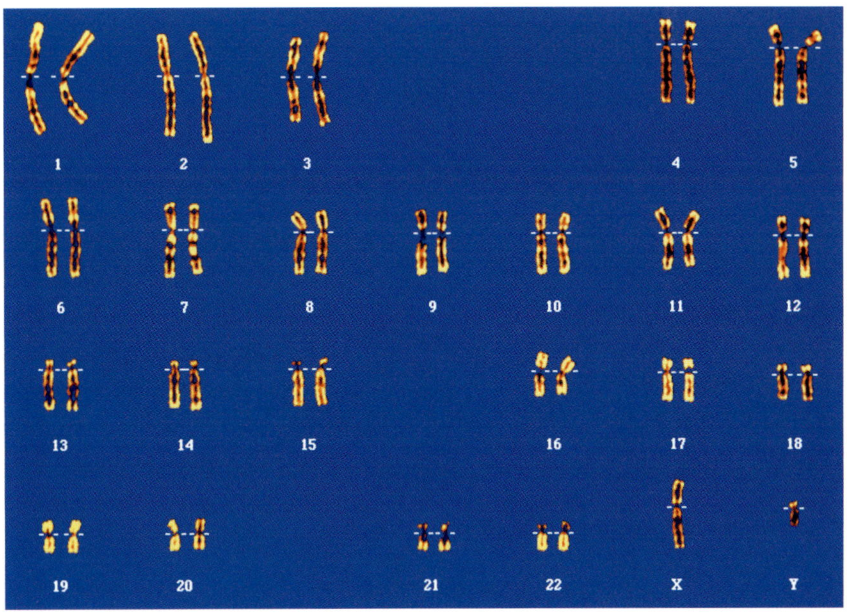

Figure 6.5:
A human's **karyotype** shows the chromosomes present in their cells. This enables extra, malformed and missing chromosomes to be identified.

A **karyotype** shows the chromosomes present in an individual's body cells.

Investigation

Extracting DNA

Aim: To extract DNA from fruit using a simplified extraction protocol

Materials: Strawberries (kiwifruit, bananas or onions may be used instead); 1 teaspoon of salt; 60 mL dishwashing detergent; chilled ethanol (isopropanol or methylated spirits may be used instead); hot water; food processor or wand mixer; knife; coffee filter paper; filter funnel; toothpick; large beaker; 500 mL conical flask; 100 mL beaker; measuring cylinder; bowl containing ice (the bowl should be large enough for the large beaker to fit in it)

Method:

1. Use the knife to finely chop the strawberries into small pieces.
2. Pour 100 mL of hot water into the large beaker. Add the salt and stir until it is dissolved in the water.
3. Add the chopped strawberries to the salt water. Use the wand mixer or food processor to blend the mixture for a minute.
4. Add the detergent, place the beaker in the bowl with ice and stir for about five minutes. The detergent helps to break down the outer membrane of both the strawberry cells and the nucleus of the cells themselves, allowing the contents of the cell (including the DNA) to enter the mixture. When this happens, enzymes that digest DNA are also released from parts of the cell. To stop the DNA being fragmented by these enzymes, the solution is iced (the enzymes are less effective at low temperatures).
5. Place the coffee filter paper in the filter funnel and put the conical flask under the funnel. Pour the strawberry mixture into the filter and allow about 50 mL of liquid to collect in the flask. Discard the filter paper and its contents.
6. Pour the filtrate into the 100 mL beaker.
7. Measure out 50 mL of chilled ethanol in a measuring cylinder. Tilt the beaker containing the filtrate and carefully pour the ethanol down the inside surface of the beaker. This will allow the ethanol to form a separate layer on the top of the filtrate liquid without mixing them together.
8. After about a minute, white precipitate will rise from the filtrate layer through to the top of the ethanol. It has a consistency similar to mucus. This is the DNA from the strawberries. (*Note:* If there is no precipitate after a minute, take a toothpick and gently stir the alcohol layer.)
9. Use a toothpick to place some of the DNA on a microscope slide and view it under a microscope. Record your observations.

What are genes?

Genes are sections of DNA carried in the chromosomes that provide instructions for the production of specific proteins.

Individual characteristics are encoded in the form of **genes** that are located on the chromosomes. This means that genes are sections of the DNA molecule. Recall from our earlier discussion that the DNA molecule is made up of two long chains of sugar and phosphate blocks with a series of bases attached to them. The bases on each chain link with the complementary bases on the other chain to form crossbars. One of these two strands is called the *template strand*, and it is the sequence of bases on this template strand that form the genes. Humans carry about 40 000 genes on the 46 chromosomes. As you can see from the karyotype in figure 6.5, chromosomes have differing lengths. The longer chromosomes are made up of longer lengths of DNA, so more genes are carried on them.

Differing combinations of these bases result in different genes being formed. On average, a gene is made up of a sequence of several thousand bases. Each sequence represented by a particular gene causes the production of a particular protein, which influences a characteristic of the organism.

The base sequences are organised as varying numbers of *base triplets* (a group of three bases). A gene could be made up of as many as 5000 base triplets. Each triplet codes for the production of a specific amino acid inside a cell. These amino acids are joined together to form the proteins just as beads are threaded together to form a necklace.

Even before the genetic code was cracked, the scientists Beadle and Tatum proposed in 1958 that each gene caused a specific protein to be produced. If the gene was changed or faulty, the protein might not be formed and this would in turn stop a particular chemical reaction occurring in the body.

There are 20 different amino acids that can be put together in combinations to form a protein. The triplet codes for these are shown in table 6.2.

Table 6.2: Triplet codes for amino acids

<table>
<tr><td>AAA
AAG</td><td>Phenylalanine</td><td rowspan="3">AGA
AGG
AGT
AGC</td><td rowspan="3">Serine</td><td>ATA
ATG</td><td>Tyrosine</td><td>ACA
ACG</td><td>Cysteine</td></tr>
<tr><td rowspan="2">AAT
AAC</td><td rowspan="2">Leucine</td><td rowspan="2">ATT
ATC</td><td rowspan="2">STOP</td><td>ACT</td><td>STOP</td></tr>
<tr><td>ACC</td><td>Tryptophan</td></tr>
<tr><td rowspan="2">GAA
GAG
GAT
GAC</td><td rowspan="2">Leucine</td><td rowspan="2">GGA
GGG
GGT
GGC</td><td rowspan="2">Proline</td><td>GTA
GTG</td><td>Histidine</td><td rowspan="2">GCA
GCG
GCT
GCC</td><td rowspan="2">Arginine</td></tr>
<tr><td>GTT
GTC</td><td>Glutamine</td></tr>
<tr><td>TAA
TAG
TAT</td><td>Isoleucine</td><td rowspan="2">TGA
TGG
TGT
TGC</td><td rowspan="2">Threonine</td><td>TTA
TTG</td><td>Asparagine</td><td>TCA
TCG</td><td>Serine</td></tr>
<tr><td>TAC</td><td>START
Methionine</td><td>TTT
TTC</td><td>Lysine</td><td>TCT
TCC</td><td>Arginine</td></tr>
<tr><td rowspan="2">CAA
CAG
CAT
CAC</td><td rowspan="2">Valine</td><td rowspan="2">CGA
CGG
CGT
CGC</td><td rowspan="2">Alanine</td><td>CTA
CTG</td><td>Aspartic acid</td><td rowspan="2">CCA
CCG
CCT
CCC</td><td rowspan="2">Glycine</td></tr>
<tr><td>CTT
CTC</td><td>Glutamine</td></tr>
</table>

As you can see from table 6.2, the alteration of even one base in a triplet can cause a change in the amino acid coded, so a different protein will be made than the one intended. For example, let us say that a particular gene is made up of the base sequence:

CGG TCA TTT GCT

This will cause the amino acids to be formed in a chain:

alanine — serine — lysine — arginine

However, if one of the bases is changed so that the gene sequence is now:

CCG TCA TTT GCT

we now get a different protein made from a different amino acid chain:

glycine — serine — lysine — arginine

A substitution error is a genetic error in which one of the bases is changed or miscopied, causing a different protein to be made in the cell.

which may have a different effect on the cell than the original protein. This type of genetic error is called a **substitution error**.

Sometimes a substitution error has very little effect at all. For example, if our original gene sequence was altered so that it read:

CGA TCA TTT GCT

we would still get the protein formed of alanine, serine, lysine and arginine because the triplet code CGG and CGA both code for the same amino acid — namely, alanine. Such cases — where a genetic substitution error has occurred in such a way that there is no effect on the protein formed — are called **silent errors**.

Silent errors are genetic substitution errors that occur in such a way that there is no effect on the protein formed.

Specific genes are found in fixed positions (*loci*) on chromosomes. For example, the gene responsible for normal colour vision is found on the X-chromosome. Some characteristics are the result of a single gene, such as the ability to see colour (which is carried on chromosome 23). Traits that are the result of a single gene are said to be **monogenic traits**. Eye colour, on the other hand, depends on four different genes on different chromosomes. We say that eye colour is therefore a **polygenic trait**.

Monogenic traits are traits that are the result of a single gene.

Polygenic traits depend on two or more genes.

How genes are inherited

Alleles are possible gene variations.

Most genes have variations known as **alleles**. These genes will have slightly different base sequences, which will cause the production of differing proteins in the cells. To better explain how this works, let us look at the gene for melanin production. Melanin is the pigment that gives colour to human skin, hair and eyes. This pigment is produced in the cells of these tissues by a chemical reaction that needs the protein tyrosinase to be present. The gene that codes for this protein is found on chromosome 11.

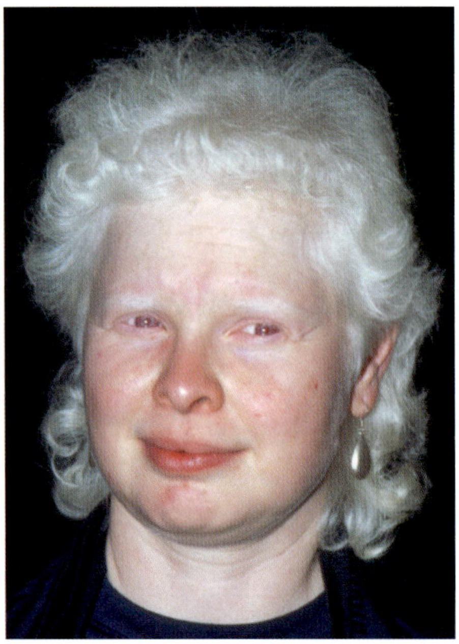

Figure 6.6: An albino has no gene for melanin production and so has no colouration in the hair, skin or eyes.

In most people, this gene is normal. However, there is a faulty version of this gene which does *not* allow the tyrosinase to be produced, so no melanin is made in the target cells. These two forms (alleles) of the gene for melanin production will produce two different proteins: *P*(tyrosinase) which causes the cells in the hair, skin and eyes to produce melanin (giving them a normal colouration), and *P*(no tyrosinase) which does not allow melanin production and causes the hair, skin and eyes to lack pigment (this is called albinism; see figure 6.6.). You will inherit one allele from your mother and one from your father on each of the chromosome pairs that contain the gene for melanin production. If the alleles from both parents for that gene are the same, you will inherit that version of the gene. That is, if both parents have the gene that causes the melanin to be produced, you will have it too (you will all have normal colouration).

A dominant allele is the characteristic that will be expressed whenever it appears in a genotype.

A recessive allele is a characteristic that is expressed only when there is no dominant allele present in the genotype.

What happens if you get the allele for melanin on one chromosome and the allele for albinism on the other chromosome of the pair? Does that mean that you will have parts of your body with no colouration? In this case, the answer is no. The cells of the hair, eyes and skin *need* only one copy of the melanin gene to start producing brown pigment, so you will have normal colouration. We say that the allele for melanin production is **dominant** and that the allele for albinism is **recessive**. A recessive gene will only become apparent when there is no dominant gene present.

An organism's **genotype** refers to the gene combination that it will be carrying for a specific characteristic.

An organism's **phenotype** refers to the characteristic that the organism expresses as a result of its genotype.

A **carrier** is an organism that carries the gene for a specific trait in its genotype but does not show that trait in its phenotype.

A **punnet square diagram** is a 3 × 3 table that allows the possible genotypes of offspring to be determined from the genotypes of the parents.

When discussing alleles, the dominant allele is usually represented by a capital letter and the recessive by the lower case of the same letter. Thus, the gene for melanin production can be represented by M and the gene for albinism is m.

The **genotype** refers to the alleles that an organism carries for a particular characteristic, while the **phenotype** refers to the characteristic that the organism actually expresses. For example, with colouration, there are three genotypes possible: MM, Mm and mm. There are only two phenotypes: normal colouration or albinism. Two normally coloured people will have the same phenotype, but may have different genotypes (one could have MM and the other Mm). A person with the genotype Mm will carry the gene for albinism (and so is called a **carrier**) and may pass it to their offspring. By using a **punnet square diagram**, it is possible to predict the chances of a carrier passing the albinism characteristic to their offspring.

If the carrier (Mm) breeds with another carrier (Mm), then there is a 25 per cent chance that their child will be am albino (see figure 6.7). If a carrier (Mm) has children with an albino (mm), then this increases to a 50 per cent chance that their child will be an albino (see figure 6.8).

×	M	m
M	MM	Mm
m	Mm	mm

Figure 6.7: Punnet square diagram showing Mm × Mm

×	M	m
m	Mm	mm
m	Mm	mm

Figure 6.8: Punnet square diagram showing Mm × mm

Identifying genes

In the past, the only tool available to geneticists was the use of pedigree charts. These were limited in the amount of information that they could provide in that they use phenotypes and logic to make a guess at a particular individual's genotype.

Genetic engineering relies on the identification of the function of specific genes carried on the chromosomes of an organism. So far, we have identified the functions of a few thousand genes and where they are situated on the chromosomes. While this might seem like a lot, it is still only a fraction of the 20 000 to 25 000 genes that are actually present.

The **Human Genome project** is a project in which several institutions around the world combined their efforts to sequence all of the human genetic code.

Gene sequencing is the process of identifying the exact order in which the bases occur on the chromosomes of an organism

In 1990, several biotechnology institutions all over the world combined their work to start decoding the genetic code of the entire human DNA molecule. This is called the **Human Genome Project** (HGP). By identifying the exact order in which the three billion or so bases occur on the chromosomes (a process known as **gene sequencing**), gene identities can be established. In 1999, the first human chromosome to be fully sequenced was chromosome 22 (which had 334 million bases making up around 550 genes), followed in May 2000 by chromosome 21(which contained only 225 genes). When the HGP was concluded in 2003, chromosomes 20, 6, 7, Y and 14 had been sequenced, and in 2004 chromosomes 5, 9, 10, 13 and 19 were completed. Once the entire human genome has been sequenced and genes identified, it should theoretically allow enormous advances in medicine, including better diagnosis and treatment of genetic disorders to occur. It has even been speculated that, by targeting specific genes that predispose a person to cancer, Alzheimer's or heart disease, human life expectancy may be greatly extended.

Technology

DNA sequencers

In the early days of genetic research, the process of determining the exact order in which the bases A, T, C and G appeared in a gene was incredibly time consuming. This procedure has become much faster with the development of instruments known as *DNA sequencers*, which use coloured fluorescent dyes to tag bases automatically. Each of the four different coloured dyes are attracted to different bases. In attaching themselves to the bases, they act as tags that are read by the sequencer. In the sequencer, the sections of the gene move past a laser that activates the dyes, causing them to emit fluorescent signals. These signals are picked up by a detector that transfers the signal to a computer which then reconstructs the base sequence. The resulting output from a DNA sequencer shows the gene as a series of coloured signal peaks with a yellow peak denoting G, a red peak for T, a green peak for A and a blue peak for C (see figure 6.9b).

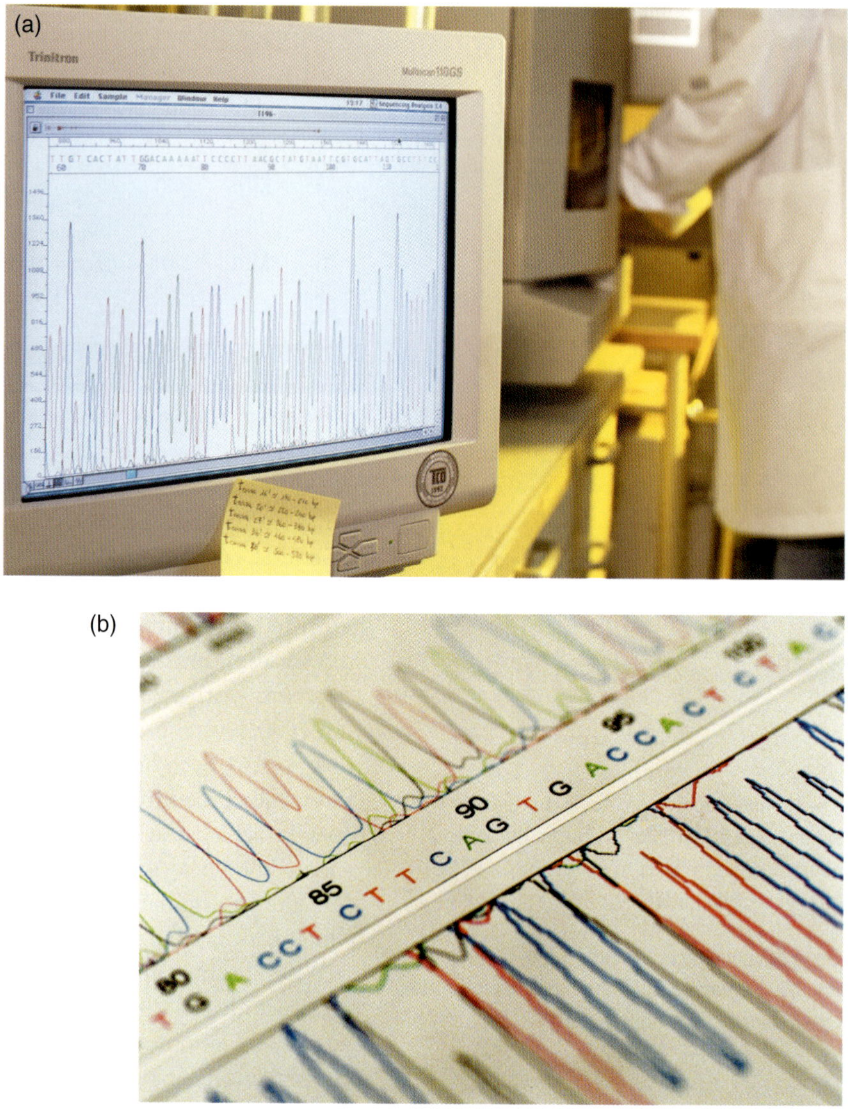

Figure 6.9:
(a) A DNA sequencer (b) output from a DNA sequencer showing the laser signals and the computer output, which identifies the base sequence

Eye on the media

A glimpse of your genetic future for $1000

The year is 2010. You walk into a lab in a shopping mall, pay $1000 and give a swab of skin cells from your cheek. A few days later you receive an email. It is a string of three billion DNA letters that hold the key to your future health.

That personal genome sequencing is within sight is remarkable. The first human genome sequence took $800 million and 11 years to complete. The 'Sanger' method used to do it has changed little since it was developed in the 1970s by Fred Sanger at the University of Cambridge. Costs have come down, but sequencing large genomes is still the preserve of large, well-funded labs.

Now two groups have developed techniques that are already 100 times faster and are set to get faster and cheaper still. The new methods do away with the need for the bacteria used in the Sanger method. Instead they use picolitre-volume beads that contain a mix of the chemicals needed to amplify DNA letters or bases. These beads are so small that millions of them can be built into a single 'chip', allowing huge amounts of DNA to be analysed at once.

Jonathan Rothberg, founder of 454 Life Sciences in Branford, Connecticut, and his team demonstrated their method by sequencing the 580 000-base genome of the bacterium *Mycoplasma genitalum* to an accuracy of 99.96 per cent in just four hours (*Nature*, DOI: 10.1038/ nature03959).

Rothberg's 70-by-75-millimetre chip can sequence over 200 000 fragments simultaneously. The 454 method uses a version of a technique called pyrosequencing. First the genome to be sequenced is cut into fragments of around 100 bases. Each bead contains one fragment which is copied many times.

Primer sequences bind to the fragments and one of the four nucleotides (A, C, T or G) is added to the mix. If the nucleotide complements the base at the first position of the unknown fragment, it releases a pyrophosphate molecule that stimulates the firefly enzyme luciferase to generate a flash of light. The nucleotides are washed away and the cycle is repeated with the next nucleotide.

'Cost will fall fast as we shrink our chips,' says Rothberg. He is predicting that miniaturisation will follow a version of 'Moore's law' — the rule of thumb that says computer chips halve in size every 18 months.

At Harvard University, Gregory Porreca and his colleagues use a similar method, but at every stage four lengths of DNA are added, each nine bases long and with one of the four nucleotides at a known position. The nucleotides are labelled with fluorescent markers of different colours. If the nucleotide binds to the fragment DNA, it gives off a flash of light.

The Harvard method is cheaper because it uses off-the shelf equipment, but is so far only able to 'resequence' known genomes. So the method can be used to spot genetic variations that predispose a patient to cancer, for example, but it cannot be used to sequence unexplored genomes (*Science*, DOI: 10.1126/science.1117389).

Cheap sequencing will give individuals and their doctors the option of consulting their DNA. 'The patient will then know what diseases he or she is genetically predisposed to and will be able to make proactive life choices to avert them,' Porreca says. 'The patient will also be able to receive treatment tailored to his or her own genetic make-up.'

Both techniques have a great deal of potential, says Daniel Turner at the Wellcome Trust Sanger Institute in Cambridge, UK, one of the centres that sequenced the human genome. 'The main limitation seems to be that the error rates of both methods are currently relatively high for individual reads.'

Source: Rowan Hooper, *New Scientist*, 20 August 2005, p. 15.

Clarification questions

1. How do the sequencing methods used by 454 Life Sciences and Harvard University primarily differ from the Sanger method?
2. What is 'Moore's law'?
3. Which of the methods used by 454 Life Sciences and Harvard University is the cheapest and why?
4. In what ways are the two methods similar?
5. What is the main social benefit of cheaper sequencing?
6. How long did Rothberg's team take to sequence a *Mycoplasma bacterium?*
7. What is the main disadvantage of the Harvard method?

Forms of genetic engineering

Genetic engineering is the technology by which the genetic make-up of an organism may be manipulated to fulfil a specific purpose.

Genetic engineering is the technology by which the genetic make-up of an organism may be manipulated to fulfil a specific purpose.

Selective breeding

Selective breeding is the process by which organisms with favourable characteristics are bred together in the hope of obtaining offspring that have those same favourable characteristics.

Selective breeding is the oldest method used by humans to manipulate the genes of organisms. As soon as humans started herds, they realised that characteristics were passed from one generation to the next, although they did not then know anything about genes!

In selective breeding, organisms with favourable characteristics are bred together in the hope of obtaining offspring that have those same favourable characteristics. So a dairy farmer may mate cows that produce high-quality milk to a bull that came from a line of good milkers in order to produce calves that would also either give good quality milk (cows) or be able to sire good milkers (bulls). A horse breeder may breed two fast horses together in order to produce a foal that is also fast.

This was all very well in theory, but in reality these breeding patterns did not always work. While modern geneticists and breeders understand that some characteristics are likely to skip generations or to be difficult to specifically breed for, many farmers in earlier centuries were more likely to blame the phases of the moon or curses when the favoured characteristics did not turn up in offspring!

Selective breeding is particularly effective when breeding for characteristics that are controlled by a single gene (monogenic traits). For example, the coarse, fast-growing coat of the Romney sheep is caused by a single dominant gene. The recessive gene gives a fine, slow-growing coat. This means that any Romney sheep that has a coarse coat has a genotype of either CC or Cc. In order to 'fix' the coarse hair in a herd (in other words, to produce a herd in which all of the sheep have the genotype CC), individual coarse-haired sheep are bred together and the coat of the offspring noted. If any pairing produced a fine-coated lamb, then the genotype of the parents would be known to be Cc and they would not be mated together again. However, as their genotype was known, they could be bred to other sheep of unknown genotype and their offspring's coats may lead the unknown genotype to be determined. By keeping careful records over a few generations, it would be possible to form a core herd of 'true-breeding' coarse-coated sheep.

Figure 6.10:
The dominant gene for coat texture in the Romney sheep is coarse.

However, the danger in breeding for specific characteristics in a herd is that, over time, all of the members of the herd would be related. The mating of related organisms is called *inbreeding*. While inbreeding is occasionally used by breeders to fix certain desirable characteristics, it may also lead to other very undesirable recessive traits being expressed. After all, any organism is the product of thousands of gene combinations — not all of which are readily apparent. In ancient Egypt, for example, where the royal family at times bred very closely (fathers marrying daughters, brothers marrying sisters and so on), there was an extremely high rate of birth defects, stillbirths and imbecility from unfavourable recessive genes that arose due to the small gene pool.

To avoid these problems, breeders will often introduce unrelated animals displaying desired traits into the herd in a process called *outbreeding*. This allows new genetic characteristics to be introduced, producing stronger offspring and less unfavourable traits.

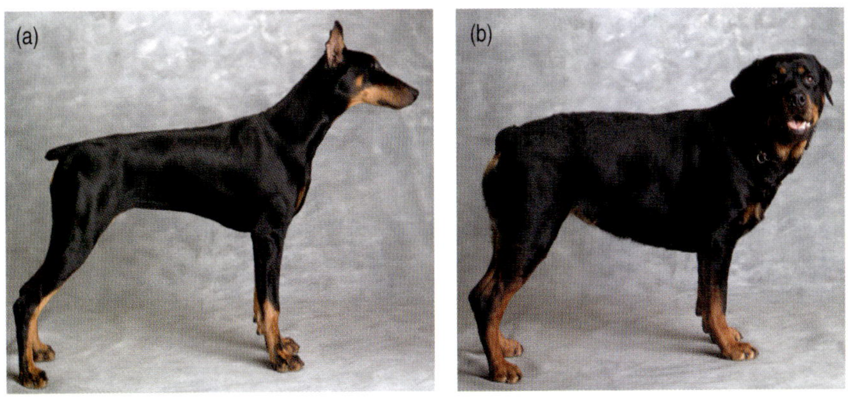

Figure 6.11:
The Doberman Pinscher and the Rottweiler originated from the same breed. The original dog was selectively bred to produce dogs with different favoured characteristics. The Doberman (a) was bred for speed while the Rottweiler (b) was bred for strength.

Cloning

Few bioengineering techniques have captured the public imagination as cloning has. However, the treatment of this technology in popular media – from Ira Levin's 1960s novel *The boys from Brazil* in which clones of Adolph Hitler are made (and inevitably turn on their maker), to films such as *The Sixth Day*, *Multiplicity*, *The Island* and *Godsend* — has generally only underlined the misconceptions that most people have about it.

Cloning is the process by which an exact genetic copy of an organism is made.

Cloning is the process by which an exact copy is made of a gene, a cell or an organism. It is a process that occurs in nature, not just in a laboratory. Identical twins have the same DNA and are, therefore, naturally occurring clones. Creatures that reproduce asexually through budding, fission and parthenogenesis are also clones.

The deliberate cloning of plants is a comparatively easy task and has been done successfully for thousands of years. This is simpler because most plants are able to reproduce naturally both sexually and asexually. By taking cuttings of an adult plant and propagating them, the cuttings will produce plants that are exact copies of the parent plant.

Figure 6.12: Dolly, the world's first animal clone, with her first lamb Bonnie, which was born from a normal mating

Cloning organisms that can only reproduce sexually under normal circumstances (such as animals and humans) has required technological development. The first complex organism to be cloned from an adult under laboratory conditions was Dolly the sheep (see figure 6.12), who was successfully created by the Roslin Institute in Scotland in 1996. Soon after, institutes in many other countries also produced successful clones. Australia's first cloned sheep (Matilda) and two cloned calves (Suzi and Mayzi) were produced in 2000.

The process of cloning is done through the following steps (see figure 6.13):

1. A somatic cell is taken from the adult organism that is to be cloned. The nucleus containing the DNA of the organism is removed from the cell. This is called the *donor nucleus*.
2. An egg cell is extracted from an organism of the same species. The nucleus of this egg cell is removed and discarded. This is now an *enucleated egg cell*.
3. The donor nucleus is transferred by injection into the enucleated egg cell.
4. Electricity is used to fuse the donor nucleus and the enucleated egg cell.
5. The new hybrid cell is cultured until it begins to divide and forms a cell cluster.
6. The cell cluster is transferred into the uterus of a surrogate organism where it implants and develops to birth.

When the newborn organism reaches adult growth, it will be an exact copy of the adult organism that provided the donor nucleus because it will have the same nuclear DNA. However, it will also carry some of the DNA of the organism that provided the egg cell. This is the *mitochondrial DNA* that was in the egg cell's cytoplasm. The mitochondrial DNA (called mtDNA) contains only incomplete genetic information and, therefore, contributes little to the phenotype of the cloned organism.

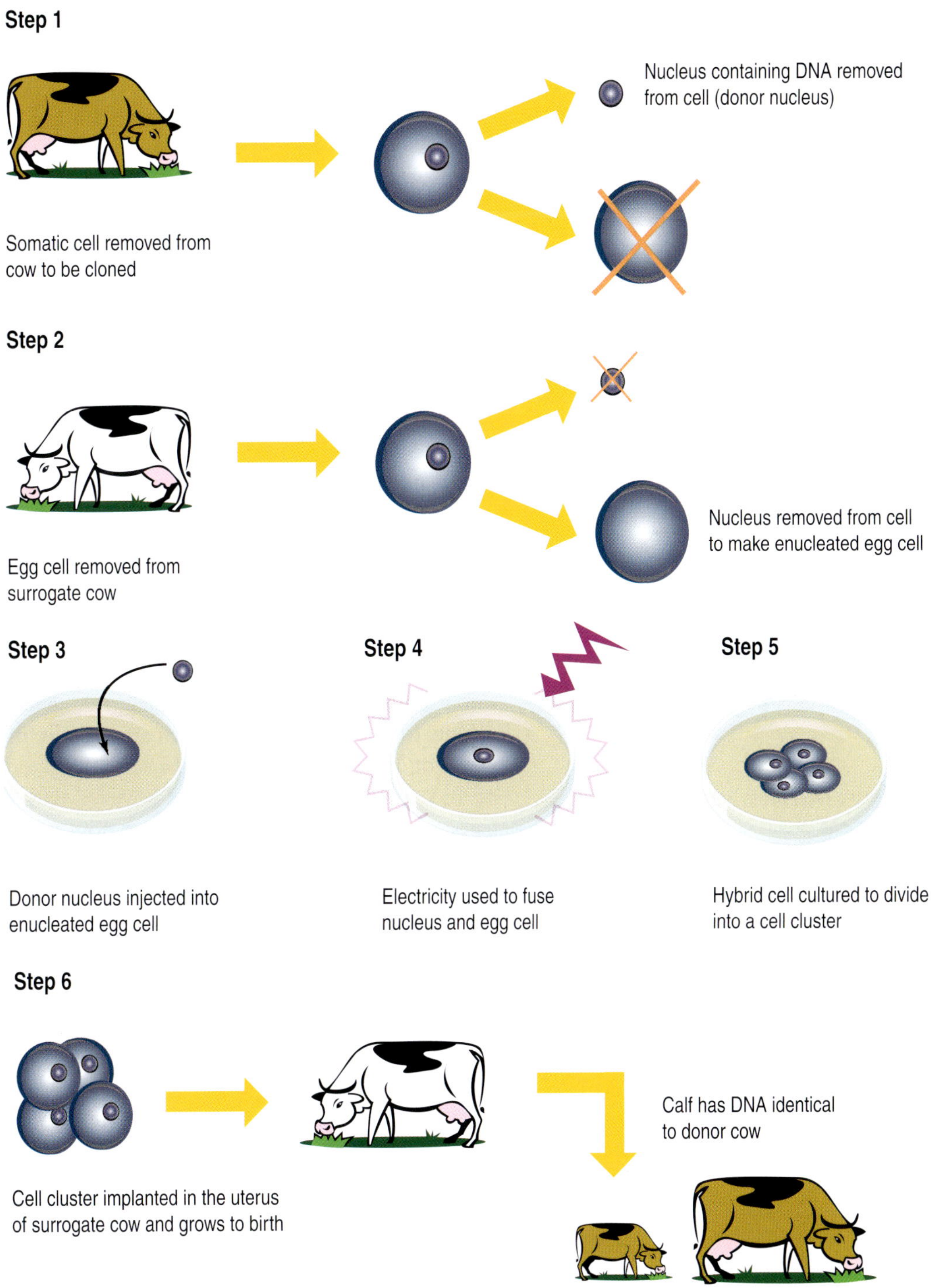

Figure 6.13:
The process of cloning an animal

One aspect of cloning animals that will require further investigation is the apparently shortened life span of the cloned animal. Dolly the sheep died in February 2003 at the age of only six, quite a young age for a sheep (which usually lives to between 10 and 16 years). This occurred only a week after the Australian clone Matilda died suddenly. It has been proposed that, while Dolly's cells were quite young, the DNA inside her cells when she was created was the same age as the donor sheep (which was six years old at the time of the experiment). The cause of these early deaths is thought to be due to a form of internal clock contained within the chromosomes themselves that controls the ageing of an organism.

Throughout an organism's life, its cells must replicate themselves to replace dead or damaged cells. Each time a cell divides, the ends of the chromosomes (called *telomeres*) become shorter. This acts as a form of clock, with each division making the telomere shorter and shorter. Eventually, cell replication is no longer possible. The more cells that fail to replicate, the faster the deterioration — in other words, advanced ageing. So the telomeres in the donor nucleus were already shortened when they were transplanted into Dolly. This meant that Dolly's cells had fewer divisions remaining than would normally be expected in a lamb, and so she reached old age faster although she matured at a normal rate.

Transgenic implantation

Under normal circumstances, genes are passed within a species from parent to offspring, and fundamental genetic differences prevent organisms of different species from combining genes by mating. In rare circumstances, matings that have produced offspring have occurred between related species. For example, horses and donkeys can produce mules (see figure 6.14), while a lion and a tiger may produce a liger. However, due to the resulting genetic and chromosomal mismatch between the parents, the offspring are sterile and therefore cannot give rise to a whole new transgenic species.

Due to the so-called species barrier, it is not possible to combine the genes of a human with a petunia or to cross a dog with a banana by sexual or asexual reproductive means. However, with the genetic technology that gained in the last three decades, it is now possible to use enzymes to 'snip' individual genes from one organism's DNA and transplant them into the DNA of an organism of a different species. Organisms that contain such transplanted genes are called **transgenic organisms** (TGOs). For example, the mice pups in figure 6.15 fluoresce green under tinted light (unlike normal mice) because their DNA contains a transplanted gene from a jellyfish that causes the fluorescence. Figure 6.16 illustrates how a transgenic mouse can be produced.

Transgenic organisms are organisms which contain gene sequences that originated in another organism.

Figure 6.14:
Sometimes related species mate to produce sterile offspring. A horse (a) and a donkey (b) may mate to produce a mule (c).

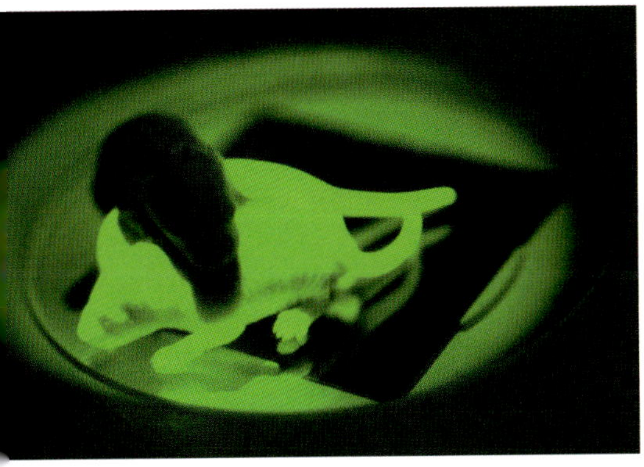

Figure 6.15:
Mice pups with the jellyfish gene showing green fluorescence when placed in particular illumination

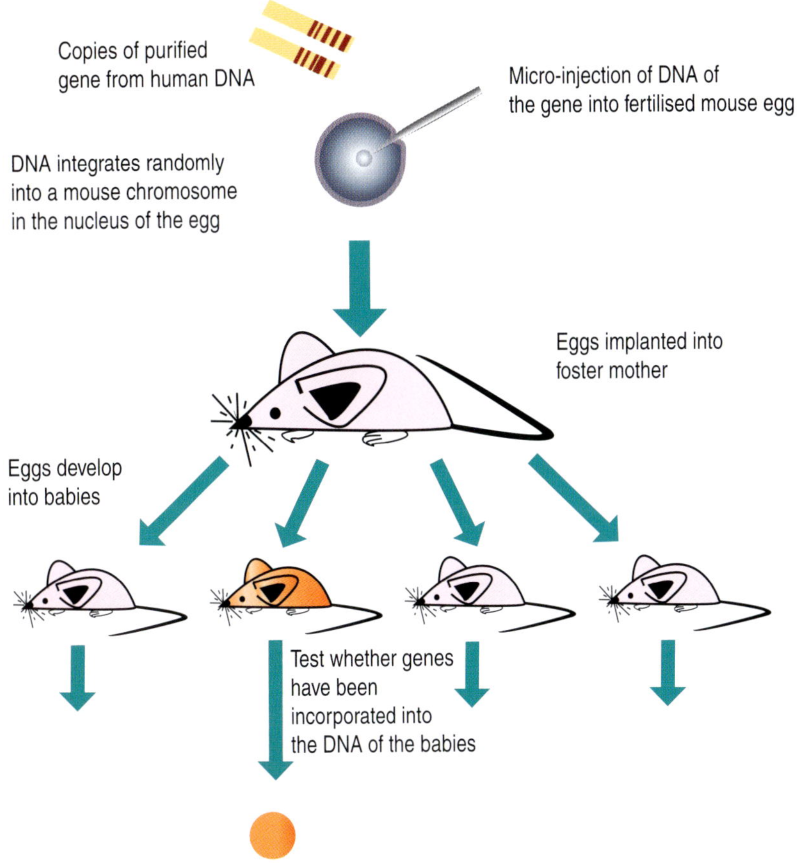

Figure 6.16:
How to make a transgenic mouse using human genes

Of course, while glowing mice might be interesting, the question you may be asking is 'What possible advantage can there be in having a glowing mouse?' Usually, these sorts of organisms are produced as a way of checking whether a particular gene-transplantation technique works or not. Genes that produce an easily visible characteristic (such as fluorescence) allow researchers to see that the transplantation method worked. The real benefits arise from the proven technique. Some of the more useful transgenic organisms include:

- bacteria that can produce human insulin, a hormone essential in the treatment of diabetes
- cows that produce milk containing human serum albumin, a vital blood factor that normally only comes from blood donations
- crops that are genetically pest-resistant without the use of chemicals or sprays.

Gene therapy

Gene therapy is a form of treatment in which specific sections of DNA are introduced into a tissue or organ in order to compensate for or replace DNA that is defective.

Gene therapy is the process by which normally coded genes are substituted into the cells of people who suffer disorders resulting from faulty or disease-producing genes. These are people who have inherited an unfavourable gene that produces proteins that do not act on the target tissues or cells in the way that the proteins produced by the correct gene would.

Cystic fibrosis (CF) is one such disorder that is currently under investigation for gene therapy. CF is an inherited disorder that occurs in 1 in 2000 births in Australia, and it is caused when a child inherits the recessive gene for it from both parents. The CF gene is found on chromosome 7. In the correct form, the gene codes for a protein that attaches to cell membranes and allows chloride ions to be transported out of the cells. The CF version of this gene, however, codes for a protein that is one amino acid short and does not allow chloride ions to be transported from the cells. This causes the airways and the pancreatic ducts to be filled with abnormally thick mucus secretions, causing lung damage and pancreatic malfunction, eventually leading to death at a young age.

At present, CF is treated by pummelling massage to help break up the mucus in the lungs and airways, and pancreatic enzymes to help with pancreas function. While this course of treatment does alleviate many of the symptoms, it does not cure it. In the case of gene therapy, normal copies of the gene are added to the target tissues, which will then enable the correct protein to be produced by the tissue cells and allow them to function normally.

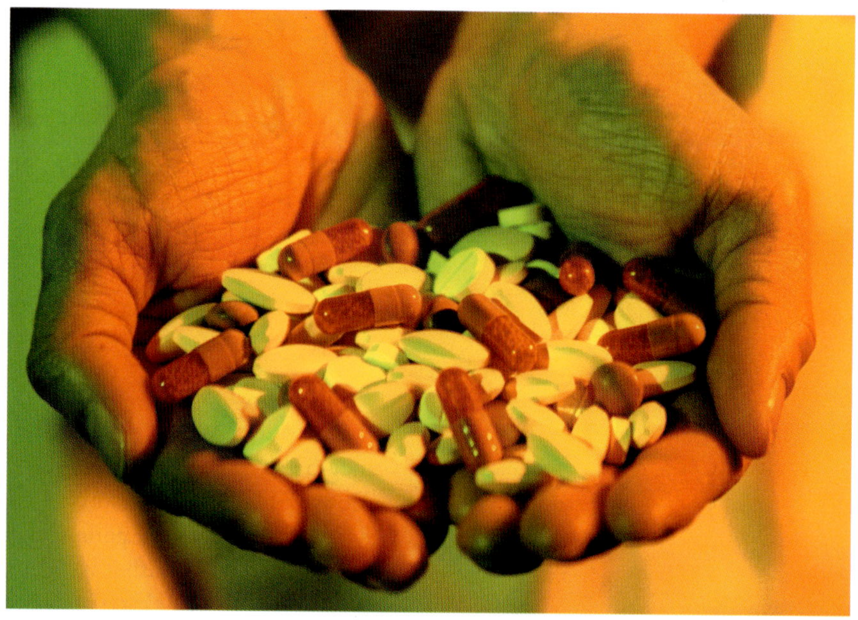

Figure 6.17:
A typical one-day supply of pancreatic enzyme tablets, as well as vitamin and antibiotic tablets, taken by a sufferer of cystic fibrosis

While gene therapy holds enormous promise, it presents numerous difficulties as well.

- *Identification of the specific gene responsible for the condition.* This is complicated by the fact that some disorders are caused by a number of genes rather than just one gene, as well as the fact that one gene may do different things in different cells, not all of which will necessarily be affected by the bad copy.
- *Identification of the target tissues.* These are the sites in the body where the effected gene is 'switched on'. For example, the gene for tongue rolling will be present in every body cell that has a nucleus, but it is only 'switched on' in the target cells — in this case, the tongue muscles. In cells where it would have no meaning (such as in your little toe), it is 'switched off'. In some genetic disorders, the identification of target tissues is quite difficult.

- *Delivering the normal copy of the affected gene to the target tissues.* This can be done in a number of ways including: injecting the DNA section into the target cells; attaching it to fatty globules called liposomes; and introducing it by injection or ingestion or, as is more usual, 'piggy backing' the gene on a modified virus that is able to enter cells easily. Some therapeutic genes have been delivered by nasal spray. The lungs and the blood have been the most successful tissues to date into which genes have been introduced.
- *'Switching on' the gene once it has been delivered to the target tissues.* While the new genes may well reach the target cells and cause the correct proteins to be produced, they are not incorporated into the chromosomes themselves. As a result, when the target cells reproduce, the faulty genes are also reproduced but the therapeutic genes are not. This has meant that while gene therapy techniques have to date led to alleviation of some genetic disorders, they have not cured them. It is hoped that, in the future, techniques will be developed which will allow faulty genes to be replaced at the chromosome level allowing for outright cures for conditions such as CF.

While gene therapy research suffered a severe setback when a teenage patient died during a gene therapy trial (apparently from the virus, not the introduced gene), some teams of researchers in France and the United States have reported small successes. In the first, the group corrected an immune deficiency in two infants. In the other, the scientists injected viruses carrying the necessary gene into three people with haemophilia. A year later, two of them were producing a blood-clotting factor that haemophiliacs lack.

Eye on the media

Can two killers make a cure?

Regulators in the US could soon be asked to approve a human trial of gene therapy for cystic fibrosis that uses a hybrid of the HIV and Ebola viruses.

Tests on monkeys show the hybrid virus delivers genes to the lungs far more efficiently than any other method developed so far. And while the idea of combining two killer viruses sounds horrifying, the risks should be low.

Despite much effort, no one has managed to develop an effective form of gene therapy to treat the lung problems of people with cystic fibrosis. The condition is caused by a mutation in the CFTR gene, and one of the problems in developing a treatment is getting a working version of CFTR into a high enough proportion of cells to restore normal lung function.

While HIV is good at adding genes to cells' genomes, it does not target lung epithelial cells. Ebola binds strongly to these cells, so Gary Kobinger at the University of Pennsylvania in Philadelphia tried adding a surface protein from Ebola to HIV. Initial tests on mice showed the hybrid virus was very efficient: the gene it carried was active in 24 per cent of airway cells after two months, a far better proportion than achieved by other delivery methods (*New Scientist,* 10 March 2001, p. 19).

But will it work in humans?

The team has now done the next best thing and used the hybrid virus to deliver a test gene, lacZ, to the lungs of monkeys. After two months the gene was active in 21 per cent of cells, Kobinger told *New Scientist*. He is preparing the results for publication.

'The gap between monkeys and humans is much smaller than from mice to monkeys,' says Ray Pickles, a gene therapy researcher at the University of North Carolina, Chapel Hill. 'This could be in human trials pretty soon.'

Kobinger now plans a more extensive monkey trial to convince the US Food and Drug Administration that human trials will be safe. The hybrid virus is created by adding HIV RNA, a stripped-down version of the Ebola surface protein RNA and the therapeutic gene to cells. The hybrid virus then self-assembles. Crucially, only the therapeutic gene is incorporated into the hybrid virus, so it cannot replicate.

All attempts in the lab to force such viruses to regain the ability to replicate have failed. Even if this did happen in the body, the resulting virus would resemble HIV, as the hybrid contains no Ebola genes — and HIV cannot infect lung cells. If it did get into the blood, where normal HIV thrives, such a virus would not replicate as efficiently as wild-type HIV. While the patient might suffer, the virus could never reach high enough levels in the body to infect others, Pickles says.

The biggest risk is that if a patient becomes infected with wild-type HIV, the wild-type virus and the hybrid could somehow merge, but such recombination has never been observed in HIV.

Despite his confidence that the hybrid will be safe, Pickles has abandoned similar research as he fears opposition from a fearful public will prevent the method ever being used. Another problem with Kobinger's approach is that patients will require repeated treatments, because the added gene ends up in surface cells that die off after a few months. Other teams are trying to find ways of targeting the progenitor cells that give rise to the surface cells.

Source: Sam Jaffe, *New Scientist*, 11 June 2005, p. 15.

Clarification questions

1. Why has a surface protein from Ebola been added to HIV to transport therapeutic genes?
2. How is the hybrid virus created?
3. What does the term 'hybrid' mean in this article?
4. What is main danger in combining these two viruses?
5. Why has researcher Ray Pickles stopped researching the use of this hybrid as a gene-delivery system?
6. How effective was the hybrid virus when used on mice?
7. What is the gene called that may cause CF in its mutated form?

Ethical arguments against genetic engineering

Ethical arguments against genetic engineering include fears about creating:

- genetically modified, or 'Frankenstein', foods
- designer genes
- clones.

GM foods or Frankenstein foods?

Genetically modified foods (GM foods) are those that contain ingredients from plants or animals that have been genetically manipulated, usually through transgenic modification. The most common sources of GM food are crops that have been modified to make them more resistant to pests, weeds and disease — factors that can decimate entire fields of plants. In Australia, Ingard cotton plants (developed by CSIRO and Monsanto) contain the Bt bacterial gene, which allows the plants to produce their own protective insecticide. This gene, taken from *Bacillus thuringiensis* bacteria, causes the cotton plants to produce a substance that is toxic to the *Helicoverpa* moths that routinely eat their way through cotton crops. Such insect-repelling crops, it is argued, protect the environment by minimising pesticide use.

Food crops such as soya beans, peas and lentils have also been subjected to transgenic experimentation. The Monsanto company has engineered a soya bean variant that is resistant to the weed-killer Roundup. These plants are able to produce an enzyme that neutralises the chemical glyphosphate, the active ingredient in Roundup. While previous methods of separating the weeds from the crops were time-consuming and expensive, farmers of these crops can now spray both the crop and the weeds simultaneously with the weedkiller without fear of killing the crop as well as the weeds.

One of the original arguments for the development of GM crops was that disease- and drought-resistant varieties could be more easily grown in the third world countries where many other non-modified crops died. It would even be possible to improve the nutritional value of these crops so that they could be enhanced with protein or vitamin A, making them an even more effective weapon in the fight against world hunger. These developments have been slow to eventuate, but other plants have been engineered for improved flavour and increased shelf life and hardiness, as well as being allergen-free. Geneticists have even created a no-tears onion and caffeine-free coffee plants.

While the vast majority of GM organisms are plants, a small number of animals have also been modified in clinical trials. Some of these include:

- cows that produce milk that contains increased amounts of casein (milk protein), which makes better-quality cheeses
- salmon that grow quickly to maturity so that they are ready to be eaten in a shorter period of time
- pigs that have been modified with genes from spinach so that their bacon is lower in fat.

However, serious concerns have been raised about the use of GM organisms in food and the long-term side effects that could arise from such 'Frankenstein foods'. As has already been shown, genes often have multiple functions or work differently in the presence of other genes. While an alien gene in a plant may allow it to ward off weedkillers or insects or disease, it is possible that it is having some other as yet unseen effect that may prove to be dangerous over time. Critics of GM foods believe that the adverse effects on consumers could include toxic or allergic reactions, antibiotic resistance and the transfer of genes to bacteria in the gut.

Other concerns arise from the creation of pesticide-resistant plants that have the potential to spread as a 'superweed'. Such plants could quickly outgrow native species and be impossible to eradicate, causing enormous ecological damage. These plants could also speed the evolution of pests until they are 'superpests' resistant to pesticide.

Plants that have been created in order to produce pharmaceutical drugs may accidentally crossbreed with food varieties with unknown consequences. This is distinctly possible in light of a 2004 report finding that theoretically unmodified varieties of major US food crops have been contaminated with transgenes. Other reports found that the pollen from GM plants could be carried tens of kilometres from their fields to fertilise non-GM plants. Given these concerns, legislation has been passed in many countries either banning the widespread use of GM products or requiring foods containing them to be labelled as such. In 1999, Australian health authorities mandated that fruit, vegetables and cereals that were genetically modified, as well as foods that could contain such GM foods, must be labelled.

Draw up a table with two columns labelled 'Advantages' and 'Disadvantages'. Under these headings, write down as many advantages and disadvantages of the use of GM products as you can think of. When you have finished, write a statement as to which side of the argument you most agree with and why.

Eye on the media

Genetic food plan axed

A multimillion-dollar project involving genetically modified food has been scrapped by the CSIRO after an altered pea produced lung inflammation in mice.

The discovery will raise new questions about the safety of genetic modification.

But supporters say the discovery of the defect has shown that safeguards work.

CSIRO has spent more than a decade and $5 million testing new techniques to make peas and beans pest- and disease-resistant by introducing alpha-amylase inhibitor protein.

As part of its risk assessment, the CSIRO asked the John Curtin School of Medical Research in Canberra to test whether the peas with the introduced protein caused an immune reaction in mice.

After two weeks, the GM peas caused an immune response and lung inflammation in mice.

The immune response increased after four weeks.

John Curtin School of Medical Research immunologist Paul Foster said the consumption of the plant by the mice had altered their basic immune response.

When the plant was inhaled it caused an inflammation of the lung, Professor Foster said. 'However; normally you would not inhale a transgenic plant,' he said.

CSIRO Plant Industry deputy chief TJ Higgins said this was only the second example of such a reaction.

'The only other example I know about was when a gene was taken from Brazil nuts and transferred into soya beans,' Dr Higgins said.

'Work was immediately stopped on those soya beans.'

Dr Higgins said the CSIRO was finalising arrangements with the Office of the Gene Technology Regulator for the disposal of the peas.

'The field sites are monitored for several years afterwards to ensure no further GM peas grow,' he said.

Dr Higgins said the same test did not have a reaction with beans, as the protein was slightly different for peas.

'This shows the regulators are doing their job,' he said.

'The regulatory system works. It ensures we have food from GM plants as safe as we get from conventional plants. Anything that has a slight risk is picked up.

'It doesn't support the view that all GM foods give rise to Frankenstein foods.'

Dr Higgins and Professor Foster said mice used in the test did not die, and their reaction to the plant had not been life-threatening.

Australian Greens senator Rachel Siewert said there were deep concerns in the community about the impacts of genetic modification.

'This one is a classic example of the problems that can eventuate when we start fiddling,' she said.

'Scientists are moving down quite dangerous lines in what they are creating, and we know it can escape into the bush as weeds and things.'

Gene Technology Regulator Sue Meek said much of the CSIRO trial had been done before the establishment of her office. However, the experiments had been conducted under the same strict guidelines.

'This allows experimental work to be conducted in the field — but tightly controlled,' Dr Meek said.

'What we have seen in this instance is how a regulatory system should work – experimental work under tightly controlled situations.'

Source: Jeff Sommerfeld, *The Courier-Mail*, 18 November 2005; p. 1.

Clarification questions

1. What is the function of the introduced gene that allows a plant to make alpha-amylase inhibitor protein?
2. What adverse effect has this modification had?
3. What other example of such a reaction is cited in the article?
4. What has been the outcome of the trials of GM beans and peas?
5. How did the immune response of the mice change with time?
6. Which of the following people mentioned in the article would you say were in favour of GM foods, which do you think were opposed and which were neutral on the subject? (Give reasons for your decisions.)
   (a) Paul Foster
   (b) TJ Higgins
   (c) Sue Meek
   (d) Jeff Sommerfeld
   (e) Rachel Siewert.

Designer genes

With rapid advances in our knowledge of the human genome and our ability to modify and change genes, it is possible that in the future we might be able to 'cure' genetic diseases in embryos by replacing faulty sections of DNA with healthy DNA. This type of manipulation is called

Germline therapy is the treatment of genetic diseases in embryos by replacing faulty sections of DNA with healthy DNA.

germline therapy and it is carried out on an egg, sperm or a fertilised embryo very early in its development. Such therapy has been successfully carried out on animal embryos under clinical conditions but, at present, it is illegal to do germline therapy on humans.

At the moment, there are only two types of advanced reproductive technologies that can legally be carried out on humans. The first involves choosing the type of sperm that will fertilise an egg: this is used to determine the sex and the genes of the baby. This technique is particularly helpful when sex-linked diseases such as haemophilia are involved. This disorder, which causes the sufferer to have uncontrolled bleeding and associated health problems, is generally found in males who have inherited the defect on their mother's X-chromosome. Female offspring would not have the disease. The second technique screens embryos for a genetic disease: only selected embryos are implanted back into the mother's womb. This is called *pre-implantation genetic diagnosis (PGD)*.

Given that we may soon be able to read the human genetic code and identify the genes that a person is carrying, a problem then arises: what do we consider to be a 'disease' that should be corrected? Often there is a tendency to define a disease as being a condition that deviates from the statistical norm, causing conditions such as obesity and high-blood pressure to be considered diseases. However, what of a person who will reach a maximum height of about 150 centimetres? This is well below average height, so does this mean that shortness is a disease?

Once it becomes possible to test quickly and reliably for thousands of genetic conditions, the biggest questions will be:

- Who decides what genes constitute diseases that should be corrected?
- Who decides whether genetic screening is done on people and embryos?
- What happens to the information that may be gathered from the genetic screening?
- Will physicians be liable for failing to test embryos or for failing to inform parents of every detail of the test results?
- Will health insurers deny policies to people with genes for diseases with high economic costs? Will life insurers? Does either have the right to?
- How can genetic profiles be kept confidential and the discriminatory use of test results be prevented?
- Since some tests will reveal information about other family members, can the privacy of these relatives be protected?

Possible outcomes of genetic screening in the future could be:

- *Genetic discrimination*. People with genetic flaws, such as a predisposition to heart disease or cancer, may be denied life insurance, health insurance and employment.
- *Differential treatment*. Employers could hire only those people whose genes indicate that they are resistant to the health hazards of the workplace, because this is a cheaper alternative to making the workplace safe for all.
- *Eugenics*. Social or political pressure may be applied to people to make child-bearing decisions on the basis of genetic information. Mating between those with valued genes may be encouraged, while mating between two people with undesirable recessive traits may be prohibited. Women carrying foetuses with genetic abnormalities may be encouraged to have abortions.
- *Genetic determinism*. Genetic determinism is the belief that behavioural and personality characteristics — such as intelligence or criminal behaviour — are mostly a function of genes. This could raise problems in legal

matters where criminals could be able to appeal on the basis that their behaviour was predetermined by their genetics and therefore beyond their control. It could also cause courses of study to be automatically excluded to those whose genetics indicate that their intelligence is insufficient to make them excel in their chosen area.

It is important to remember, however, that less than half of all diseases and disability is believed to be caused by genetic factors, and that on average each one of us carries five or so genes for disorders that can be lethal. When we can decode the genetic blueprint in individuals, we will probably find that all of us carry a large number of genes that predispose us to various conditions. This does not necessarily mean that we will get these conditions. In fact, given warning of a genetic tendency, we might be able to make lifestyle choices that allow us to avoid the diseases altogether.

Knowledge of one's genetic screen could profoundly affect the price of insurance. Insurance companies operate on a risk-assessment basis in which individuals with a greater potential of developing health complications are charged more than individuals who are deemed healthy. Actuaries look at statistics to determine how much their clients should be charged. With current advancements in genetic testing, there is the potential to discriminate against those with 'bad genes'. It is quite possible that individuals classified as being at high risk could be denied the opportunity to purchase insurance. However, insurance companies must operate as a business, looking out for themselves and their clients' best interests.

Divide into two groups:

- *Group 1.* You are representing an individual who has obtained a genetic screen and would like to purchase some insurance. Results show that the individual is at extremely high risk of developing cancer and hypertension. However, he is currently a healthy individual who does not smoke and who exercises on a regular basis. Explain why your client should be given the opportunity to purchase insurance at a reasonable price.
- *Group 2.* You are representing an insurance company. A potential client has approached your company about the possibility of purchasing insurance. After reviewing her file, including a gene scan, you realise that she falls into a high-risk category. Explain why you need to charge considerably more for providing her with insurance.

Attack of the clones

There is no doubt that cloning has its advantages. On 8 January 2001, scientists at Advanced Cell Technology announced the birth of the first clone of an endangered animal, a baby bull gaur (a large wild ox from India and South-East Asia) named Noah. Although Noah died of an infection unrelated to the procedure, the experiment demonstrated that it is possible to save endangered species through cloning. It might even be possible to bring back species that are extinct. In his novel *Jurassic Park,* author Michael Crichton wrote about bringing dinosaurs back from cell samples extracted from mosquitos in amber. In reality, it could be possible to effectively bring the thylacine (Tasmanian tiger) back from extinction by extracting cells and cloning new individuals from a preserved thylacine sample.

It has been suggested that cloned animals could serve a wide variety of needs. Entire herds made up of clones of cows that have excellent milk production or orchards of apple tree clones that produce huge numbers of apples have a lot of economic appeal. Cloned laboratory animals would allow greater control over experimental variations due to individual differences between animals. Arguments have even been made that races between cloned horses would be fairer bets, as the contest would be based upon the skill of the riders rather than the breeding of the horses!

One of the most compelling arguments for cloning arises from the possibility of cloning organs that can be created for people in need of various transplants. By extracting liver cells from a compatible donor and cloning the cells, a new liver could be made without waiting for a liver to be made available through the usual method, which relies on the death of a compatible person who is listed as a donor. Skin cells could be cloned to create sheets of skin for burns victims, a technique that is currently being investigated clinically. Using cells from the patients themselves minimises the risk of rejection, a complication that dogs many transplant procedures.

It has been theorised that couples who are infertile or who have genetic disorders could produce offspring using cloning. In this way, the offspring would be an exact genetic copy of one of the parents rather than carrying a mixture of characteristics from both parents (see figure 6.18).

Figure 6.18:
The cloned family?

It has also been proposed that, in the event of a child's accidental death, a clone of that child could be made as a form of replacement. This idea of organism replacement lies at the basis of the Missyplicity Project, a US cloning project funded by a wealthy couple who wish to have a clone made of their dog Missy, now departed.

Many see the cloning of the deceased as morally questionable and, indeed, there are as many arguments against cloning as there are to support it. One of these is that a world full of clones is a world without genetic diversity, the primary advantage of sexual reproduction in adapting to changing environments. Without genetic diversity, a population is more vulnerable to extinction.

Another objection is the large numbers of failures that are produced in order to gain one 'perfect' clone. Dolly the sheep was the only survivor out of 227 implanted clones. The rest were either miscarried or resulted in technical failures including mutations and retardation. Thus, for one Dolly, tens — if not hundreds — of lambs with abnormalities were produced that either died or were put down. When we are talking about cloning people, the implications are sinister.

While there is widespread interest in the cloning of organs and tissues for transplants, there are also fears that clones of people could be made to serve simply as donors or human 'spare-parts' whose organs could be harvested should the original person need organs (a scenario considered in the 2005 film *The Island*).

Perhaps a genetically engineered humanity of the future will arise much as Huxley envisioned all the way back in 1931: a world where castes are programmed at birth, an entire slave race of epsilon clones and a social structure whose stability is based upon the quality of your DNA.

Read the following excerpt and answer the questions after it.

> 'I shall begin at the beginning,' said the D.H.C. and the more zealous students recorded his intention in their notebooks: *Begin at the beginning*. 'These,' he waved his hand, 'are the incubators.' And opening an insulated door he showed them racks upon racks of numbered test-tubes. 'The week's supply of ova. Kept,' he explained, 'at blood heat; whereas the male gametes,' and here he opened another door, 'they have to be kept at thirty-five instead of thirty-seven. Full blood heat sterilizes.' Rams wrapped in theremogene beget no lambs.
>
> Still leaning against the incubators he gave them, while the pencils scurried illegibly across the pages, a brief description of the modern fertilizing process; spoke first, of course, of its surgical introduction–'the operation undergone voluntarily for the good of Society, not to mention the fact that it carries a bonus amounting to six months' salary'; continued with some account of the technique for preserving the excised ovary alive and actively developing; passed on to a consideration of optimum temperature, salinity, viscosity; referred to the liquor in which the detached and ripened eggs were kept; and, leading his charges to the work tables, actually showed them how this liquor was drawn off from the test-tubes; how it was let out drop by drop onto the specially warmed slides of the microscopes; how the eggs which it contained were inspected for abnormalities, counted and transferred to a porous receptacle; how (and he now took them to watch the operation) this receptacle was immersed in a warm bouillon containing free-swimming spermatozoa — at a minimum concentration of one hundred thousand per cubic centimetre, he insisted; and how, after ten minutes, the container was lifted out of the liquor and its contents re-examined; how, if any of the eggs remained unfertilized, it was again immersed, and, if necessary, yet again; how the fertilized ova went back to the incubators; where the Alphas and Betas remained until definitely bottled; while the Gammas, Deltas and Epsilons were brought out again, after only thirty-six hours, to undergo Bokanovsky's Process.
>
> 'Bokanovsky's Process,' repeated the Director, and the students underlined the words in their little notebooks.
>
> One egg, one embryo, one adult — normality. But a bokanovskified egg will bud, will proliferate, will divide. From eight to ninety-six buds, and every bud will grow into a perfectly formed embryo, and every embryo into a full-sized adult. Making ninety-six human beings grow where only one grew before. Progress.
>
> 'Essentially,' the D.H.C. concluded, 'bokanovskification consists of a series of arrests of development. We check the normal growth and, paradoxically enough, the egg responds by budding.'

Responds by budding. The pencils were busy.

He pointed. On a very slowly moving band a rack-full of test-tubes was entering a large metal box, another, rack-full was emerging. Machinery faintly purred. It took eight minutes for the tubes to go through, he told them. Eight minutes of hard X-rays being about as much as an egg can stand. A few died; of the rest, the least susceptible divided into two; most put out four buds; some eight; all were returned to the incubators, where the buds began to develop; then, after two days, were suddenly chilled, chilled and checked. Two, four, eight, the buds in their turn budded; and having budded were dosed almost to death with alcohol; consequently burgeoned again and having budded — bud out of bud out of bud — were thereafter — further arrest being generally fatal — left to develop in peace. By which time the original egg was in a fair way to becoming anything from eight to ninety-six embryos — a prodigious improvement, you will agree, on nature. Identical twins — but not in piddling twos and threes as in the old viviparous days, when an egg would sometimes accidentally divide; actually by dozens, by scores at a time.

'Scores,' the Director repeated and flung out his arms, as though he were distributing largesse. 'Scores.'

But one of the students was fool enough to ask where the advantage lay.

'My good boy!' The Director wheeled sharply round on him. 'Can't you see? Can't you see?' He raised a hand; his expression was solemn. 'Bokanovsky's Process is one of the major instruments of social stability!'

Major instruments of social stability.

Standard men and women; in uniform batches. The whole of a small factory staffed with the products of a single bokanovskified egg.

'Ninety-six identical twins working ninety-six identical machines!' The voice was almost tremulous with enthusiasm. 'You really know where you are. For the first time in history.' He quoted the planetary motto. 'Community, Identity, Stability.' Grand words. 'If we could bokanovskify indefinitely the whole problem would be solved.'

Solved by standard Gammas, unvarying Deltas, uniform Epsilons. Millions of identical twins. The principle of mass production at last applied to biology.

Source: Extract from *Brave New World,* Aldous Huxley, 1931, chapter 1.

Clarification questions

1. The book from which this extract is taken was written in 1931, long before the discovery of DNA and the cloning process. What parts of this extract are inconsistent with our present knowledge of genetics and genetic engineering processes?
2. Describe 'Bokanovsky's Process'. Is such a process possible? Explain your answer.

End of chapter questions

1. In dogs, black coat colour (B) is dominant to brown coat colour (b).
   (a) Write the genotype for a purebred dog that is being bred for brown coat colour.
   (b) Write the genotype and phenotype of a dog that is a hybrid of the two.
2. Which of the following statements are true?
   (a) A dominant trait will always be the most common in a population.
   (b) A person with red hair must have red-haired parents.
   (c) A hybrid has two different alleles in its genotype.
   (d) There are only two alleles for a gene.
   (e) It is possible to have more than one dominant gene.
   (f) Each chromosome contains a different number of genes.
   (g) If an organism has 10 chromosomes in its sex cells, then it must have five chromosomes in its body cells.
3. The sex chromosomes of a normal human foetus were found to be XX. What is the gender of the foetus?
4. Horses have 64 chromosomes in their somatic cells, while donkeys have 62 chromosomes. If a male donkey and a female horse mate, they produce a sterile offspring known as a mule.
   (a) How many chromosomes are in the horse's egg cells?
   (b) How many chromosomes are in the donkey's sperm cells?
   (c) How many chromosomes are in the mule's somatic cells?
   (d) Considering your answer to (c), explain why mules are sterile.
5. King Henry VIII divorced (and executed) wives on the basis that they did not give him sons to follow him on the throne. What was the flaw in Henry VIII's reasoning?
6. Consider the cartoon shown in figure 6.19. What points are being made by the cartoonist about Ding Duck's ability to fly?

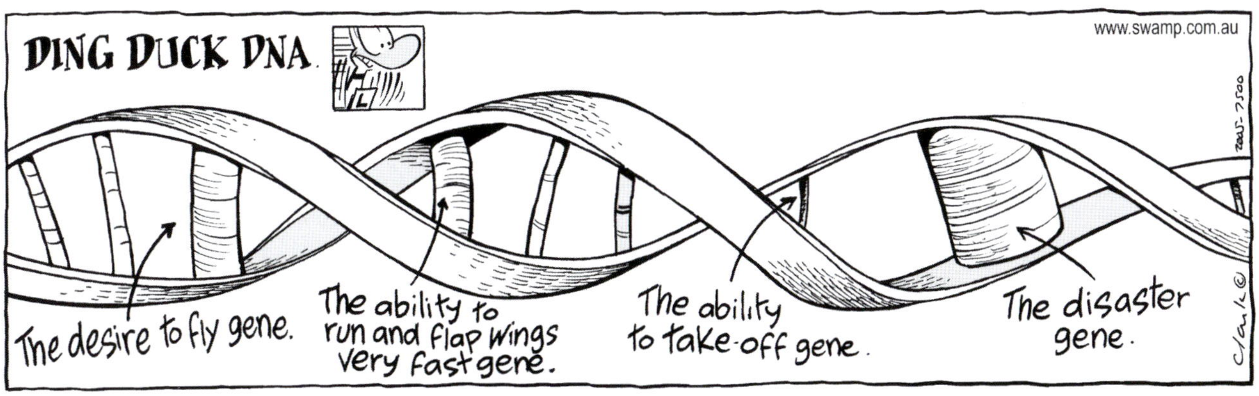

Figure 6.19:
Ding Duck DNA

7. Which of the following statements best describes the relationship between chromosomes and genes?
   (a) One gene contains many chromosomes.
   (b) One chromosome contains many genes.
   (c) One gene corresponds to one chromosome.

8. What is meant when we refer to DNA as having a 'double helix' shape?
9. What amino acid string would be produced from the following base sequence?

   AAT CGT GGG TAA CAG GAT TAC

10. Which of the following is DNA more similar to: the recipe for a cake or a list of ingredients that make the cake? Justify your answer.
11. Give four examples of asexual reproduction.
12. What is the function of telomeres believed to be?
13. What does the term 'genetic engineering' mean? Give two examples of genetic engineering.
14. How does selective breeding differ from genetic modification?
15. Explain the steps involved in the process of cloning.
16. What are the advantages and disadvantages of cloning?
17. Define the following terms:
    (a) genotype
    (b) transgenic
    (c) monogenic traits.
18. Name two ways in which genes from one organism may be transferred into the cells of another.
19. In a cloning experiment, the following steps were carried out:
    (a) The nucleus of a skin cell of a male Siamese cat was transferred into the enucleated egg of a Persian cat.
    (b) The resulting cell was then transplanted into the uterus of a Russian blue cat.
        (i) What sex will the resulting kitten be?
        (ii) Which of the three breeds will it resemble?
        (iii) Which of the three adult breeds was the surrogate?
20. Which of the following are examples of transgenic organisms or cells?
    (a) Mammals produced by cloning
    (b) Bacteria containing a plant gene
    (c) Plants produced by grafting
    (d) Wheat cells with a mouse gene
    (e) Cow cells with a gene mutation
21. Explain how gene therapy may be used to treat genetic disorders such as cystic fibrosis.
22. In gene therapy, what does the term 'target cells' mean?
23. Describe two of the benefits of the Human Genome Project.
24. How does outcrossing differ from inbreeding?
25. One sign of heat stress in cattle is an elevated rate of breathing. Consider the table below which shows the breathing rate of three different breeds of cattle when they are exposed to different environmental temperatures:

Table 6.3: Variation of breathing rate with temperature in cattle breeds

Temperature (°C)	Breathing rate (respirations/ minute)		
	Zebu	Holstein	Cross-bred
10	24	28	20
20	25	30	23
25	27	45	30
30	36	94	73
35	47	108	91

(a) Plot the data in the table on a graph with temperature in °C on the horizontal axis and breathing rate (in respirations/minute) on the vertical axis. Use three different colours of data points, one for each breed.

(b) Do any of the breeds exhibit heat stress at temperatures below 20°C?

(c) The Zebu is a tropical breed of cattle. What is the breathing rate of the Zebu cattle at about 30°C relative to the Holstein cattle?

(d) Does this graph suggest that the Holstein is better suited to tropical or temperate conditions? Explain your answer.

(e) The crossbred cattle result from crossing Zebu with Holstein cattle. Does their performance at high ambient temperatures suggest that adaptation to high temperatures is controlled in part by genes? Explain.

26. Molecular biology scientists have developed a technique that enables parents to select the sex of their future child. This technique simply separates the X-carrying sperm from the Y-carrying sperm, and then inseminating females with the preferred sex chromosomes. The procedure is banned in many countries other than for medical reasons. Potential parents with sex-linked diseases are allowed to choose to have a girl and thereby avoid the possibility of having a boy with haemophilia, for example.
    (a) Should sex selection for medical and non-medical reasons be available for parents in Australia?
    (b) What do you think are some implications of this technique?

27. You are representing an interview panel. Justify why you think the use of DNA screens is sufficient for determining whether someone is suitable and qualifies for a job.

28. Which of the following characteristics do you believe are:
    (a) completely due to genes?
    (b) partly genetic?
    (c) unrelated to genes?
        (i) Intelligence
        (ii) Height
        (iii) A tendency to shyness
        (iv) Violent behaviour
        (v) Sporting ability
        (vi) Musicianship
        (vii) Fear of heights
        (viii) Nail-biting

29. In 2000, after a sexual assault in a small Victorian town, adult male residents of the town were invited to give a cell sample voluntarily for DNA profiling. Most of the men were cooperative but some objected on the basis that the test was an invasion of their privacy. The case was solved when one of the residents, who had not yet been tested, confessed to the crime during the testing period. Discuss the issue of wide-scale DNA sampling in criminal investigations in terms of:
    (a) the rights of the individual
    (b) the community's desire to solve a crime.

7

Survive that drive!

Every day, more and more cars crowd onto our already packed highways, and with more cars comes more accidents. Car manufacturers competing for the motorist's dollar advertise increasingly grander safety systems available for their vehicles yet, despite all of the developments in car safety devices, 250 people still die on Queensland roads every year. To be sure, the issue of road safety is a complex one but these fundamental questions arise — what are we doing to ensure that every motorist survives? and is it enough?

Figure 7.1: Are Queensland's roads safe?

What price safety?

A range of safety devices can be incorporated into cars to increase the occupants' chances of being unharmed during a collision. This chapter examines the devices that you are most likely to encounter.

Seatbelts

Inertia is the tendency of an object to resist changes to its motion. Newton's First Law of Motion (the Law of Inertia) states that an object at rest will remain at rest, and an object in motion will maintain its motion, unless an outside force is exerted on the object.

The idea behind the seatbelt was to keep the wearer in contact with the seat, causing the seat and the person to move as a single unit rather than the person being thrown from the vehicle in a collision. This is particularly useful in the case of high-speed head-on collisions in which cars are very quickly decelerated to a stop. Anyone not wearing a seatbelt will be subject to **inertia**, and so will continue to travel forward in the original direction of motion, colliding with the steering column, the dashboard or the windscreen along the way. The seatbelt provides a restraining force that allows the occupant to slow down at nearly the same rate as the car.

Seatbelts were created many years before their compulsory introduction in Australia in the 1970s. In the 1940s, the two-point seatbelt — consisting of a simple belt across the lap — had been designed to keep aircraft personnel safely anchored in flight. After World War II, these seatbelts began to be used in cars, but only for people who requested to have them fitted. However, this early design had many shortfalls. If the seatbelt was

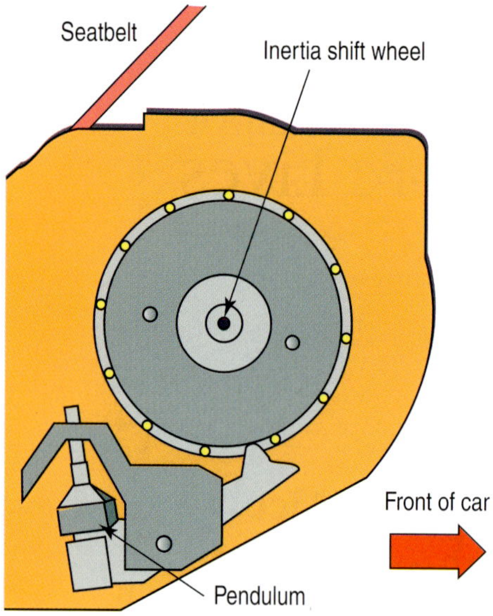

Figure 7.2: Operation of an inertia-reel seatbelt. This reel is shown in the locked position.

not secured tightly, in an accident the wearer would slide out from either under it or over it and be thrown forwards. Conversely, if the seatbelt was over-tightened, the wearer would often receive crushing internal injuries from being thrown so forcefully against the seatbelt. As a result, the seatbelt went out of favour again.

In 1959, an engineer for Volvo called Nils Bohlin came up with a new design for the seatbelt — the three-point seatbelt. This is essentially the type of seatbelt that is still used (with minor variations) in cars today. Made up of two main sections (one crossing the lap and the other crossing diagonally over the chest), the design distributed the force of impact across two of the strongest sections of the human body — the pelvic girdle and the chest. With the improved design, more people started to wear seatbelts and, by the mid 1970s, the wearing of seatbelts had been made compulsory in all Australian states. This was a major factor in the decrease of road fatalities.

One of the biggest complaints about the three-point belt, however, was that it greatly reduced free movement under normal driving conditions. The *inertia-reel seatbelt* shown in figure 7.2 reduced this complaint. While allowing the wearer some movement, it locks into place when subjected to the rapid deceleration associated with a collision. The small pendulum shown in the diagram continues to move forward under the influence of inertia and stops the reel that holds the seatbelt from turning any further, locking it in place and preventing the forward motion of the occupant.

Crumple zones

Crumple zones are sections placed at the front and rear of cars to increase the collision time and so reduce the forces acting on the passengers in a collision.

The use of **crumple zones** in vehicle construction was also pioneered by Volvo. Early vehicles were made very rigid in the belief that the more solidly they were built, the better they would withstand collision. This is indeed true. The solid vehicles were hardly dinted in collision. Alas, the same could not be said for the occupants, who were usually killed!

The problem lies in the deceleration rate experienced by both car and passengers during the collision. When the vehicle and its occupants go from travelling at around 60 kilometres per hour to a speed of 0 in a very short period of time, they experience a very large deceleration. Relative to the car, the passengers are thrown forward from rest with an acceleration roughly equal in size to the car's deceleration. On striking the dashboard, the steering column, the windscreen or even the seatbelt, they then decelerate rapidly again.

From Newton's Second Law, we know that acceleration is proportional to the force that is acting on an object, so the force exerted on the passenger by whatever they are striking is very large indeed — large enough to be fatal. In a car with crumple zones, the front and back sections of the vehicle are engineered to collapse or crumple under impact (see figure 7.3). This crumpling increases the time taken for the speed of the vehicle and occupants to be brought to rest, which means that the deceleration decreases. This in turn reduces the force acting on the passengers — once again, consistent with Newton's Second Law. Of course, the drawback is that the car resembles an accordion after the accident, but at least you have a much better chance of walking away from it afterwards to make your insurance claims!

Figure 7.3:
Crumple zones at the front and rear of cars increase the collision time and so reduce the forces acting on the passengers in a collision.

Airbags

When film stuntmen jump through the window of a skyscraper and apparently fall to the ground to make a footpath pizza, they are actually falling onto an enormous bag full of air up to a storey high. As they hit the bag, it deflates under them, and the stuntman and the bag together collapse at a reduced speed towards the ground. The airbag in a car works on much the same principle. The technology that fires the airbag, which is fitted into the steering wheel for the driver and the dashboard for the passenger, works in a fraction of a second to activate in the event of collision.

1 millisecond = 0.001 seconds

Within 20 **milliseconds** of the impact, the airbag's sensors determine the rate of deceleration of the vehicle. If that deceleration indicates a threatening situation, the controlling mechanism detonates the small explosive charge that deploys the airbag from its storage position. The airbag splits its packaging and inflates rapidly with nitrogen gas.

Figure 7.4:
Airbags increase the time interval during which the energy of a collision is dissipated.

The airbag will be fully inflated about 45 milliseconds after the initial impact. At this stage the car's occupants are only just moving forward. When their faces and upper bodies strike the airbag 15 milliseconds later, the airbag will begin to deflate and release the gas through the vent holes in its rear section. By the time 100 milliseconds have elapsed, the occupant and the airbag will be slowly sinking forward together at a controlled rate that is much slower than the rate at which this deceleration would have occurred during an unprotected collision. The increased time leads to a smaller acceleration and smaller forces acting on the occupant — Newton's Second Law again. As you can see, all this happens very quickly in only one-tenth of a second, which is less than human reaction time. In fact, many people have remarked that they were unaware of being in a collision until they found themselves looking at the bottom of an airbag!

Investigation

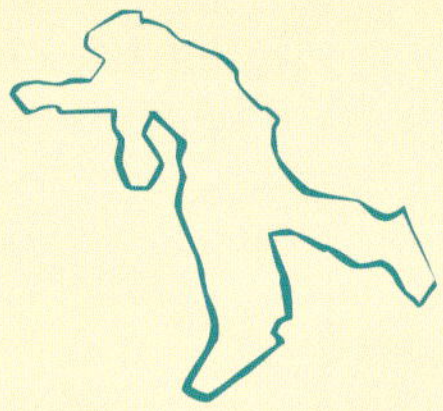

Simulating airbags

Aim: To simulate the way in which airbags lessen the damaging effects of impact on vehicle occupants

Materials: One garbage bag; two eggs with intact shells, rubber band; air pump or vacuum cleaner with an inflation setting

Method:

1. Pump air into the garbage bag until it is firm. (Take care not to over-inflate the bag and cause it to burst!) Twist the opening shut and seal it with the rubber band.
2. Drop the egg onto the inflated bag from a height of about 1.5 metres (see figure 7.5). Observe what happens to the egg.
3. Inflate the garbage bag again. Place the rubber band loosely around the opening in such a way that air escapes easily when the bag is pressed. (You may need to experiment with this method to get it working properly.)
4. Drop the second egg onto the bag from the same height as the first, and observe what happens to the egg.

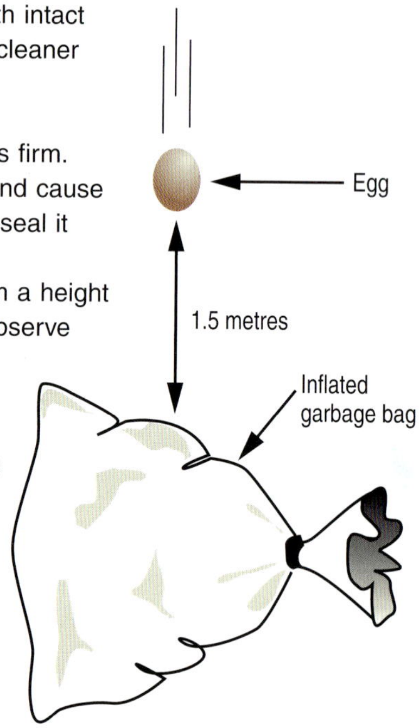

Figure 7.5: Dropping an egg onto an inflated garbage bag

Discussion question

Based on what you have observed here, why do you think it is important that the airbags in a car inflate quickly but then deflate again as they are hit by the car occupant rather than staying inflated throughout the impact?

Collapsing steering wheel

Some of the most horrific injuries used to occur to drivers when they were thrown onto the rigid steering-wheel column. This would cause crushing injuries to the chest, rupture of internal organs or even impalement. A modern steering column collapses when the degree of crumpling in the front of the car reaches a certain level.

Dropping engine block

The engine block in the front of the car was often implicated in front-seat occupant injuries in earlier years. As the front of the car crumpled, the engine block would be pushed backwards through the firewall (which prevents engine heat from being transferred into the occupant cavity) and into the legs of the people in the front seat. This would cause crushing injuries and entrap the passengers in the vehicle. In modern car designs, the crumpling of the front of the car will cause the supporting struts holding the engine to collapse, so the engine drops down towards the road.

Safety cell

An **alloy** is made of either a combination of metals or a metal mixed with non-metals. For example, brass is an alloy of copper and zinc.

The passenger safety cell is a network of strong **alloy** beams that are built into the car's structure in such a way as to surround the entire passenger cavity. Unlike the parts in the crumple zones, these sections are extremely rigid, designed to protect the cavity from outside intrusion.

While many different safety devices exist, not all of them are fitted as standard equipment (i.e. at the basic car cost) on the vehicle. How much safety do you get for your money these days?

1. Find out which safety features/devices are included as standard or available as optional extras for a variety of cars. (Possible sources of information include the internet, car manuals, car yards and car advertisements.) Where a safety feature is available as an optional extra, find out how much extra it will cost. Place your information in a table such as that shown below:

Car	Safety feature								
	Seatbelt pre-tensioners	ABS	Airbags (front)	Curtain airbags	Airbags (side)	EBD	BA	Height-adjustable seatbelts	Centre rear lap-sash seatbelts
Suzuki Marvel	S (front)	S	S	Optional ($1200)	Optional ($2000)	NA	S (auto)	S	NA

S = standard NA = not available

2. From your research, which car would you say is the safest to be in if you were involved in an accident?
3. Did the number of standard safety features vary with the cost of the car?
4. Were there any cars that could not be fitted with a particular safety device? If so, why was this?

Eye on the media

Respect safety

Car buyers should pressure importers and manufacturers to change their outlook on new vehicle safety levels, according to the RACQ.

The Club wants Australian drivers to have access to a full range of lifesaving active and passive safety aids, such as dual front airbags, anti-lock (ABS) brakes, **traction** control, electronic stability control and side or head airbags.

The tyre's ability to grip a surface and so push the car forward is referred to as **traction**.

Car makers point to cost, lack of customer demand and complexity in supply and distribution, tactics which undermine a buyer's ability to buy a better blend of safety. Current practices include removing safety equipment from models being imported, restricting safety equipment to higher specification models and bundling optional safety equipment together, or with non-safety related items, making the options more costly.

RACQ Technical Services executive manager Steve Spalding said the 'practice of 'de-speccing' vehicles for the Australian market was undesirable.

'De-speccing is when imported models arrive in Australia with less safety equipment than is found in the same car sold in overseas markets,' Mr Spalding said. 'The local car company specifically asks the factory to omit the equipment.

'RACQ has identified numerous examples of de-speccing on Australian cars, compared with the same cars sold in Europe and New Zealand.

'We want importers and manufacturers to fit safety equipment as standard or, as a compromise, to make such equipment available optionally, unbundled and across the full model range at affordable prices.'

He said customers should not face lengthy delays or unnecessary costs when ordering safer vehicles.

'Importantly, if customers want ABS brakes, they should be able to tick that box without having to pay for another safety feature or luxury item as well,' Mr Spalding said.

'The small car segment, for example, would greatly benefit from additional airbags and ABS brakes, while four-wheel-drives, which are more prone to roll-overs than sedans, could be made safer with side and curtain airbags.'

Many of the best sellers in the Australian market lack key safety features.

Front side airbags on the base Commodore Executive sedan are only available with air conditioning, a scenario repeated for a passenger airbag in the Holden ute. Traction control is only available from Acclaim sedan models upwards.

The base XT Falcon offers side airbags and traction control as stand-alone options and the Falcon ute can be specified with a passenger airbag or ABS. However, Ford's manual Fiesta LX cannot be bought with ABS at all.

Toyota, 2004 market leader, provides a mixed performance. Its Corolla Ascent bundles safety features together; Camry Ateva's front side airbags are only available in conjunction with a sunroof; the CV Rav4 does not have ABS at all; while the Cruiser offers an impressive but costly safety pack. The GX Prado 4.0 offers ABS, but the GX 2.7 does not.

The Mazda3 Neo bundles ABS, electronic brake distribution, brake assist, front side airbags and dual row curtain airbags together as an option, while other 3-models have those features as standard.

RACQ also has concerns about the lack of basic safety features, such as airbags and ABS, available to fleet and commercial operators buying 'tool of trade' vehicles. Mr Spalding said the Club encouraged all motorists to make safety a key consideration when selecting their next vehicle.

Source: Chris Bishop, *The Road Ahead,* June/July 2005, p. 3.

Clarification questions

1. What point is the author trying to make in this article?
2. Explain in your own words what 'de-speccing' means.
3. The article refers to the practice of 'bundling optional safety equipment together'. Why does the author consider this to be a bad thing?
4. What is included in the bundle for the Camry Ateva?
5. What reasons do car makers give for not making the full range of safety aids standard in all cars?

Why accidents happen

An accident occurs when the driver of a car is unable to avoid a hazard. It may be that the driver saw the hazard but did not have sufficient time to manoeuvre, or the vehicle for one reason or another was not adequately able to respond to driver control. As you can see from table 7.1, a variety of factors can lead to a crash. The main causes of crashes (apart from a decision on the part of the driver to disregard road rules) tend to be speeding, driver impairment, and inattention or inexperience.

Table 7.1: Assessed contributing factors to crashes*

	Fatal crashes		All reported crashes	
	Number	Proportion of fatal crashes	Number	Proportion of all reported crashes
Alcohol/drugs	107	38%	2503	11%
Disobeyed traffic rules†	83	29%	8945	41%
Inattention	74	26%	6397	29%
Inexperience	48	17%	4452	20%
Speed	45	16%	1108	5%
Fatigue§	37	13%	1148	5%
Age	31	11%	1163	5%
Other	31	11%	2996	14%
Rain/wet road	16	6%	1916	9%
Road conditions	12	4%	1077	5%
Negligence	12	4%	444	2%
Other driver conditions‡	10	4%	1426	6%
Vehicle defects	4	1%	632	3%
No street lighting	4	1%	77	0%
Total crashes	284	100%	22 083	100%

* More than one contributing factor could be attributed to a crash and therefore this table may not reflect crash totals.
† Disobeyed traffic rules does not include alcohol/drugs, inexperience, speed and inattention.
‡ Driver conditions do not include inattention, negligence, inexperience, fatigue or age.
§ Includes fatigue by definition crashes

Source: Queensland Department of Transport, *Road traffic crashes in Queensland: 2003*, p. 38.

Stopping distance is the distance that a car travels between the driver seeing a hazard and the car coming to a halt.

Reaction distance is the distance a car travels between the driver seeing the hazard and actually starting to brake.

Braking distance is the distance required for the brakes to bring the car to a halt.

The distance a car travels between the driver seeing a hazard and the car coming to a halt is called the **stopping distance**. The stopping distance itself is equal to the sum of the **reaction distance** (the distance that the car travels between the driver seeing the hazard and actually starting to brake) and the **braking distance** (the distance that is required for the brakes to bring the car to a halt). Each of the accident factors mentioned above influences the stopping distance because they affect the reaction distance, the braking distance or both.

Factors affecting reaction distance

Speed is the rate at which an object travels a distance.

We use the term **speed** to describe the rate at which a vehicle travels. If we say that a car has a speed of 60 kilometres per hour, we mean that, in a one-hour period of time, that car would travel a distance of 60 kilometres. In a two-hour period of time, the same car would travel 120 kilometres. We can say in general that the speed is equal to the distance the vehicle travels divided by the time of travel, and we can represent this with a formula:

$$\text{speed} = \frac{d}{t} \text{ (where } d \text{ is distance and } t \text{ is time).}$$

Converting metres per second into kilometres per hour

Speedometers are usually marked to measure speed in kilometres per hour, but the equations that we use to describe the motion of a vehicle require its speed to be described in metres per second. We will therefore need to be able to convert from one set of units to another.

The easiest way to convert kilometres per hour into metres per second is to *divide by 3.6.* To convert metres per second back to kilometres per hour, *multiply by 3.6.* For example, a speed of 72 kilometres per is equal to 20 metres per second (i.e. 72 ÷ 3.6), while 10 metres per second is equivalent to 36 kilometres per (i.e. 10 × 3.6).

So why is 3.6 the magic number? Think of it this way. Each kilometre has 1000 metres, each hour has 60 minutes and each minute has 60 seconds, so each hour has 3600 seconds. Therefore:

$$\begin{aligned}\text{1 kilometre per hour} &= \text{1000 metres per 3600 seconds}\\ &= (1 \div 3.6) \text{ metres per second}\end{aligned}$$

In other, words, we divide by 3.6. We can similarly show why we multiply by 3.6 to convert from metres per second to kilometres per hour.

Calculating reaction distance

The reaction distance ($d_{reaction}$) is equal to the speed of the vehicle when the hazard was seen multiplied by the reaction time of the driver:

$$d_{reaction} = \text{speed} \times t_{reaction}$$

The reaction time ($t_{reaction}$) of the driver is equal to the period of time that elapses between the driver first seeing the hazard (or pedestrian) and then starting to brake. Many drivers believe their reaction times to be much faster than they really are. The reaction time of a typical driver surprised by a hazard varies widely between about 0.8 and 3 seconds, averaging out at about 1.5 seconds. However, the reaction time of a driver who is fatigued, under the influence of drugs or alcohol or having a lapse in concentration will be much larger than this, so their vehicle will travel much further between them seeing a hazard and putting their foot on the brake.

For the average driver who reacts in 1.5 seconds while their car is travelling at 60 kilometres per hour (i.e. 16.7 metres per second), the reaction distance can be calculated as:

$$d_{reaction} = \text{speed} \times t_{reaction} = 16.7 \times 1.5 = 25 \text{ metres}$$

A tired driver travelling at the same speed who has a reaction time of 3 seconds will travel twice as far before starting to brake:

$$d_{reaction} = \text{speed} \times t_{reaction} = 16.7 \times 3 = 50 \text{ metres}$$

Obviously, increasing the speed of a vehicle will also increase the reaction distance.

Eye on the media

Drinking within limits

Staying under 0.05 blood alcohol content is different for men and women.

As a general rule, to stay under the legal driving limit for the concentration of alcohol in your blood and breath, men can have two standard drinks in the first hour and one every hour after that, while women can have one standard drink in the first hour and one every hour after that.

However, a Queensland transport spokesperson said this was only a guide as the number of drinks a person could have before going over the limit depended on a number of factors such as body weight, height, proportion of fat, metabolism, previous meals and medications.

The standard drink guide for women is different from men because women are usually smaller and therefore reach the legal limit more quickly.

Some different beers and wine can vary in alcohol content and in glass size so always read the label on the bottle or ask the bartender to find out the exact number of standard drinks.

'No matter how much or how little you have had to drink, alcohol will slow your reaction time, reduce concentration, increase confidence, affect your vision and hearing, and relax you, making you more likely to fall asleep at the wheel,' the Queensland Transport spokeswoman said.

She said drivers should be aware that at a level of 0.05 per cent concentration of alcohol in your blood and breath, your risk of being involved in a crash was twice what it normally is if you had not been drinking.

'At 0.15 per cent concentration of alcohol in your blood and breath, your chances of a crash are 25 times more than if you had not been drinking,' she said.

'The concentration of alcohol in your blood and breath will start to rise as soon as you start to drink and can take from 30–60 minutes after you have stopped drinking until it reaches its highest concentration.

'The only way to reduce the level after drinking is to wait. You could still be over the legal driving limit the next morning if you had a large number of drinks the night before.

'Play it safe by catching a cab or public transport or arrange a lift with a friend who hasn't been drinking. That way, you are not risking your licence or your life by drinking and driving.'

Source: *The Courier-Mail,* 29 June 2005, p. 43.

Clarification questions

1. What factors influence the rate at which a person metabolises alcohol?
2. Why are women on average unable to metabolise alcohol as quickly as men?
3. What is the effect of alcohol on a driver?
4. How much more likely is a driver with a blood alcohol level of 0.15 to have a crash than one with a blood alcohol level of 0.05?
5. The average human has five litres of blood in their body. How much alcohol in litres is in the bloodstream of a person who has a blood alcohol level of 0.05?
6. What advice does the article give on how to avoid a drink-driving charge?
7. How can you lower your blood alcohol level after drinking?

Eye on the media

Road accidents linked to exposure to petrol fumes

Petrol station workers are more than twice as likely to have an accident while driving home as on the way to work. This is the first confirmation of a link between low-level exposure to petrol fumes and road accidents.

Even motorists inhaling petrol fumes at a self-service station may be more likely to have a crash, according to Jung-Der Wang at the National Taiwan University, Taipei, who led the study.

His team looked at the number of road accidents experienced by a group of 20 000 workers at the same company. Nearly half worked on petrol station forecourts and had therefore been exposed to petrol fumes, the rest had office jobs or similar and were the control group. There was no demographic difference between the two groups.

The employees had 626 injury-causing road accidents between 1991 and 2000, with the forecourt workers having 61 per cent more accidents than the office workers. But even more telling was the timing of the extra accidents. There was no statistical difference between their accident rates on the drive to work. But on the way home, the accident rate of the forecourt workers was 2.4 times that of the office workers. The study will appear in a future issue of *Accident Analysis and Prevention.*

Though the study did not monitor the extent of exposure to petrol fumes, the link was clearly demonstrated by a change in accident rates after 1997. That was when Taiwan's Environmental Protection Administration made it a legal requirement for petrol stations to install vapour recovery systems. These suck petrol vapour back into the storage tank during refuelling, cutting vapour release by 90 per cent. After these devices were installed, the accident rate of forecourt workers dropped to the same level as in the general population.

Vapour recovery devices are mandatory in areas of the US with a local pollution problem. And though they are not yet common in Europe, a European Union directive states that all petrol stations should install them as part of its drive to cut emissions of volatile organic compounds. The UK is not planning to implement the directive until at least 2010.

Sean Semple of the University of Aberdeen, UK, has studied the neurotoxic effects of solvents on painters at a naval dockyard in Scotland. 'There were a number of stories of how after a 12-hour shift they were involved in accidents,' he says. 'And there were a number of occasions when they were taken in by the police, because they thought [the painters] were drunk!'

But Semple doubts that ordinary motorists are at risk. 'I would think short-term exposure to solvents would have to be quite high to have any effect.'

Source: Mick Hamer, *New Scientist,* 16 July 2005, p. 16.

Clarification question

Compare this article with the previous one 'Drinking within limits' and then discuss the dangers of petrol fumes for the average driver compared with the dangers of alcohol consumption.

Acceleration is the rate at which an object changes its velocity.

Factors affecting braking distance

The rate at which an object changes its velocity is referred to as its **acceleration** (a). If the velocity of an object increases, the object is said to have *positive acceleration*. It is said to have *negative acceleration* (also called *deceleration*) when the velocity decreases.

The average acceleration of an object can be determined by using the equation:

$$a = \frac{(v - u)}{t}$$

where u is the object's initial velocity at the beginning of a time interval, v is the object's final velocity at the end of the time interval and t is the length of the time interval.

Once the driver has started braking, other factors come into play. The braking distance of the car (d_{braking}) is related to the speed of the vehicle and the rate at which the car is able to bring the car to a stop:

$$d_{\text{braking}} = -\frac{\text{speed}^2}{2a}$$

Although the acceleration in this case is a negative number (being a deceleration), we will end up with an overall positive value for the braking distance.

Once again, the speed of the vehicle is implicated. In fact, it has an even bigger effect on the braking distance than it did on the reaction distance. If you double your speed, you will have four times the braking distance!

The deceleration rate depends upon a large range of things including the mass of the car, the condition of the brakes, the type of brakes, the road surface, the slope of the road and the condition of the tyres. (We will look at some of these in more detail later.)

As a result, the braking rate varies enormously between situations and cars. Given good road conditions, a Toyota Corolla can brake at –6 metres per second per second, a Mazda Protege at –7 metres per second per second and a Ferrari Maranello at –9 metres per second per second.

Driving too fast

No doubt you have at one time or another seen an advertisement aimed at drivers telling them to keep to the speed limit and not to drive too fast. But exactly how fast is *too* fast?

Consider this example: the driver of a car travelling through a suburban street sees a child chase a ball into the street a distance of 25 metres in front of the car. If the driver's reaction time is 0.9 seconds and the car is capable of decelerating at 8 metres per second per second, will the child be run over if the car's speed is (a) 50 kilometres per hour; or (b) 65 kilometres per hour?

(a) Given that speed = 50 kilometres per hour; t = 0.9 seconds; and a = –8 metres per second per second, to find d_{stopping}, the speed must be converted into metres per second:

50 kilometres per hour = 13.9 metres per second

Then the values for reaction distance and braking distance must be found:

$$\begin{aligned} d_{\text{reaction}} &= \text{speed} \times t_{\text{reaction}} \\ &= 13.9 \times 0.9 \\ &= 12.5 \text{ metres} \end{aligned}$$

$$d_{braking} = -\frac{speed^2}{2a}$$

$$= -\frac{13.9^2}{2 \times (-8)}$$

$$= 12.1 \text{ metres}$$

Reaction distance and braking distance are now added to find the total stopping distance:

$$d_{stopping} = d_{reaction} + d_{braking}$$
$$= 12.5 + 12.1$$
$$= 24.6 \text{ metres}$$

Therefore, the car travelling at 50 kilometres per hour stops a bare 0.4 metres from the child. That was close!

(b) We will now do the same thing with an initial speed of 65 kilometres per hour:

65 kilometres per hour = 18 metres per second

$$d_{reaction} = speed \times t_{reaction}$$
$$= 18 \times 0.9$$
$$= 16.2 \text{ metres}$$

$$d_{braking} = -\frac{speed^2}{2 \times a}$$

$$= -\frac{18^2}{2 \times (-8)}$$

$$= 20.3 \text{ metres}$$

$$d_{stopping} = 16.2 + 20.3$$
$$= 36.5 \text{ metres}$$

This means that not only does the car hit the child, it does not come to a stop for another 11.5 metres afterwards!

The problem is that 65 kilometres per hour does not seem very fast, and many people routinely drive at 60 kilometres per hour or greater through suburban streets with a legal limit of 50 kilometres per hour. Of course, the physics of the situation is often not obvious to people who are a little late for work or who are trying to get to the train station in time. So, the simplest answer to our question is: any speed over the posted limit is too fast!

Investigation

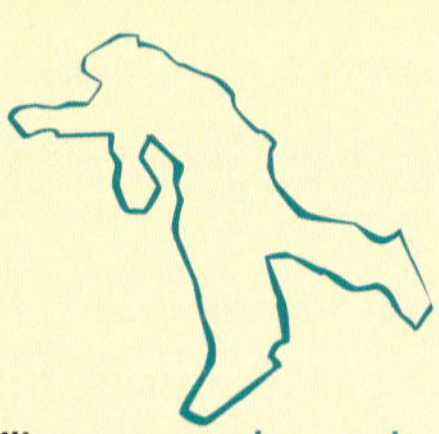

A **dilemma zone** is a region before a set of traffic lights where a driver's course of action (to either accelerate or brake) on seeing a yellow light is not clear.

The dilemma zone

When a traffic light turns yellow, the drivers in the cars approaching the intersection must make a decision — whether to continue travelling through the intersection or to start braking. If the drivers continue, they run the chance of breaking the law by running a red light. On the other hand, they may not be able to brake hard enough to bring the car to a halt before entering the intersection. The distance before the traffic lights in which either course is possible is called a **dilemma zone**.

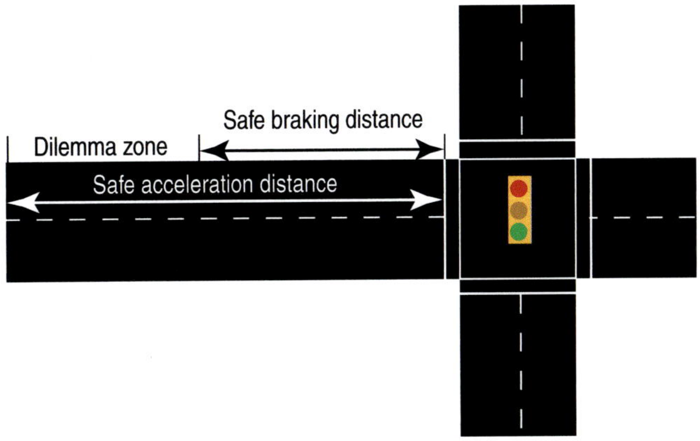

Figure 7.6:
The dilemma zone

Aim: to identify whether there is a dilemma zone at a set of traffic lights
Materials: Pedometer, stopwatch
Method:
Part A (on the street)

1. Find an intersection that has a traffic light and a safe crossing.
2. When the traffic is stopped at the lights, use the pedometer to measure the width of the intersection in metres.
3. Record the speed limit for the street (you will need to convert this from kilometres per hour into metres per second).
4. Measure the duration of a yellow light at the intersection. You should do this for at least five cycles, and find the average length of a yellow light in seconds.

Part B (at school)

1. Using the intersection width and the speed limit, calculate the time required for a car to travel through the intersection.
2. Use this time and the duration of a yellow light to determine the car's largest possible distance in front of the traffic light if the car is to clear the intersection before the light turns red (the safe acceleration distance shown in figure 7.6). Assume that the car will be travelling at the speed limit.
3. Use the deceleration for an average car (approximately –7 metres per second per second) to determine what distance the car requires to come to a stop. Once again, assume that the car is travelling at the speed limit. You will need the equation:

$$d_{\text{braking}} = -\frac{\text{speed}^2}{2 \times \text{deceleration}}$$

4. If the distance calculated in step 2 is greater than the distance calculated in step 3, there is a dilemma zone for this intersection. This zone will lie between the two distances.

Discussion questions

1. Does your chosen intersection have a dilemma zone?
2. Would you expect most intersections to have a dilemma zone?
3. How would decreasing the duration of the yellow light affect the length of the dilemma zone?

Road conditions

Friction is the force that acts between two contacting surfaces resisting their motion past each other; caused by interaction of surface irregularities and molecular bonds.

In order to understand how the condition of a road surface affects the deceleration rate of a car, we will need to look at the concept of friction. In general, **friction** is a force that arises when objects or surfaces are moving past each other while in contact. Although the processes that create friction are very complicated, friction is believed to be due mainly to two effects: irregularities in the objects/surfaces and the bonds between the particles in the objects/surfaces.

The first cause of friction arises from the fact that all surfaces have irregularities. On some things such as bricks, sandpaper or Monte Carlo biscuits, the surface irregularities are easy to spot. However, the surfaces of some materials (such as highly polished glass or metal) seem perfectly smooth to the touch. It is only when these apparently smooth materials are examined under a microscope do the minute bumps and ripples that pepper their surfaces become visible (see figure 7.7).

Figure 7.7:
When viewed under a microscope, this polished metal surface appears rough when magnified 300 times.

As two surfaces come into contact, their lumps and bumps also make contact and, when one surface moves across the other, the little bumps catch on each other. This makes it harder to move the surfaces relative to each other, and it also causes the surfaces to heat up. You can easily experience this phenomenon by rubbing your hands together. The little friction ridges (also called fingerprints) that cover your hands catch on each other, and you soon notice the heating effect.

The second cause of friction operates on an even smaller (molecular) scale than the interaction between the bumps. As the bumps and ridges of the two surfaces push against each other, bonds form between the particles of the two materials. These bonds are quite strong and break only when a large enough force is applied to pull them apart. As surfaces move across each other, bonds are made and broken between the particles, causing a resistance to the motion (friction).

The three types of friction we are concerned with are static friction, sliding friction, and rolling friction. The differences between the types of friction lie in the way the bumps on the surface interact.

Static friction must be overcome as you first start to push one surface across another. When two surfaces are placed into contact, a fair amount of interlocking occurs between the peaks of the two surfaces, and bonds form between the particles of the materials. Remember, these bonds are very strong. As you start to push one surface across the other by applying a force, the combination of particle bonds and peak locking acts to exert a resisting force in the opposite direction, so the surfaces do not move. As you exert more force, the peaks start to fracture and the bonds begin to break. Eventually, if you apply sufficient force, enough of the bonds and peaks will be damaged for the two surfaces to start moving past each other.

Sliding friction takes effect once the static friction has been overcome by the applied force and the materials are actually moving past each other. Sliding friction is smaller than static friction in magnitude. Sliding friction acts in such a way as to resist the motion of the surfaces. So an object sliding down a hill has a friction force acting upon it that is directed in the opposite direction (that is, up the hill).

Rolling friction is actually a combination of static friction and sliding friction, and it allows easy movement of an object across rolling surfaces — a handy fact that many ancient civilisations made use of. It is believed that the great stone blocks used to build the pyramids in Egypt and the ziggurats in South America were at times transported by being pulled over a series of rollers made from logs. This would have been a faster and more efficient method than dragging the blocks over the ground (and safer for the workers too).

The **coefficient of friction** (μ) is a measure of the friction between two surfaces in contact with each other.

Each of these types of friction relies upon the nature of the surfaces that come into contact, which is usually an indication of how smooth or rough they are. In each case, this is put in terms of a value called the **coefficient of friction (μ)**. Note that the friction coefficient depends upon which pair of surfaces are in contact. The coefficient of friction for aluminium sliding across wood will differ from that for aluminium sliding across ice. The coefficient of friction also differs for the same pair of surfaces according to whether the friction experienced is static or sliding (see table 7.2). Usually, we denote the coefficients of static, sliding and rolling friction as μ_s, μ_k and μ_r respectively. The coefficient of friction has no units, and you will usually find that the friction coefficient has a value less than 1. A coefficient of friction that is equal to 0 indicates that there is no friction at all acting between the surfaces.

Table 7.2: Values of static and sliding friction for surface pairs

Surface pair	μ_s (Static friction)	μ_k (Sliding friction)
Rubber on dry concrete	0.9	0.7
Rubber on wet concrete	0.6	0.4
Rubber on ice (wet)	0.14	0.1

Notice what happens to the value for static friction in table 7.2 as the surface changes. Remember that the way a wheel grips a road for traction and control depends upon the static friction acting between the rubber wheel and the road surface. The higher the value of static friction, the better grip that the rubber tyres have on the road. A dry road is optimal for safe travelling. When the road gets wet, the static friction value drops and the tyres have less grip on the road. If the road is covered in wet ice (as can happen in many parts of Australia such as the Snowy Mountains), the tyres have very little grip on the road surface. As a result, the car is very sensitive to steering and it is easy to lose control of the vehicle. In snowy conditions, you will usually see drivers attaching chains to their tyres. These allow the tyres to gain traction on the icy surfaces much as spiked shoes allow arctic explorers to walk on frozen surfaces.

Investigation

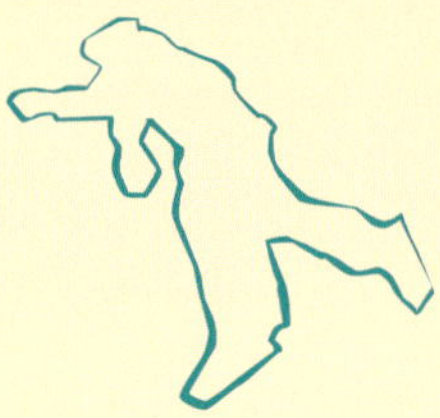

Looking at friction

Aim: To investigate the relative size of the friction coefficients for a variety of surface pairs

Materials: A 500-gram wooden block; two pairs of axles; four carpenter's staples; four toy car wheels with rubber tyres (from a hobby shop); Newton scale; Plasticine; flat concrete surface; water; olive oil

Method:

1. Mount the wheels on the axles and attach the axles to the wooden block as shown in figure 7.8. The wheels should be able to move freely.
2. Place the wheeled block on the concrete surface. Attach the Newton scale to the block and gently pull the scale until the block *just starts* to move forward. Note the scale reading — this is the value of static friction between the tyres and the dry concrete.

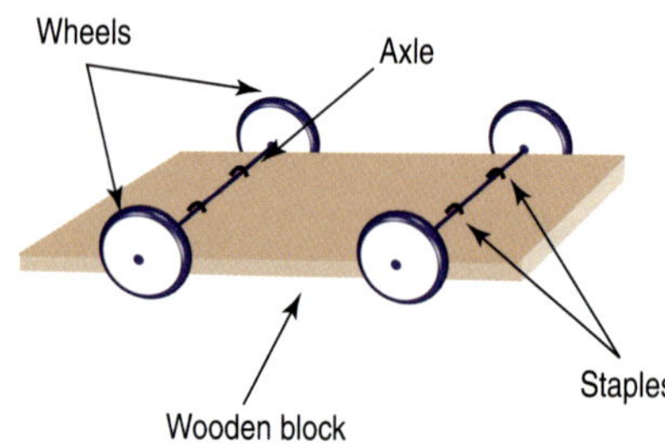

Figure 7.8: Fitting the wheels to the block

3. Pull the wooden block with the Newton scale at a steady speed across the concrete. Make sure that the wheels are rolling freely during this stage. Note the scale reading — this is the value of the rolling friction between the tyres and the concrete.
4. Use the Plasticine to jam the wheels of the block. Drag the block on its immobilised wheels with the Newton scale and note the reading — this is the sliding friction value.
5. Pour water onto the concrete and repeat steps 2 to 4.
6. Pour oil onto the concrete and repeat steps 2 to 4.
7. Enter your results in a copy of the results table.

Results table:

Surface	Static friction (N)	Rolling friction (N)	Sliding friction (N)
Dry concrete			
Dry concrete			
Oily concrete			

Discussion questions

1. Which combination generates the greatest friction value?
2. Which generates the least friction?
3. What does this imply abut the surface conditions for safe driving?

Eye on the media

Roads to ruin

It's time again to condemn those poorly designed or neglected roads that make motoring a misery. RACQ wants the name, location and details of Queensland's shabbiest roads, plus suggestions on how to fix them, to help the Club lobby governments for a fix.

The Club's previous Unroadworthy Roads Survey was conducted four years ago and received 1400 responses, which identified 500 unroadworthy roads, most of them in regional or country areas.

Worst offender was the Barkly Highway, between Mount Isa and the Northern Territory border, which was characterised as rough, narrow, badly aligned, flood prone and having poor shoulders.

RACQ traffic and safety executive manager John Wikman said in the case of the Barkly Highway, several new bridge crossings and the elimination of some single-lane sections had improved the flow of freight, tourists and industry along the stretch.

We are aware of several other projects scheduled or underway on at least six of the top 20 list of State or Federal government controlled roads identified in the 2001 survey,' Mr. Wikman said.

'We'd like to see more money tipped into roads which rank as the worst offenders, and that includes those from 2001 which are still considered substandard or poorly maintained.

'Better roads yield substantial benefits in terms of safety and economic efficiency.'

'The Federal Government's National Road Safety Strategy estimates that half of the targeted reduction in fatality rate per 100 000 population by 2010 will come from improvements to roads.'

He said an example of 'unroadworthiness' would be narrow, undulating sealed roads with broken edges or steep drop-offs. There might also be no line markings or so many potholes that your vehicle would be constantly jarred.

Queensland roads have consistently ranked lower in independent surveys than their interstate counterparts.

Source: Chris Bishop, *The Road Ahead*, June/July 2005, p. 27.

Clarification questions

1. Why was the Barkly Highway considered unroadworthy?
2. What are some of the characteristics of 'unroadworthy' roads as mentioned in this article?
3. What has been done to improve the Barkly Highway?

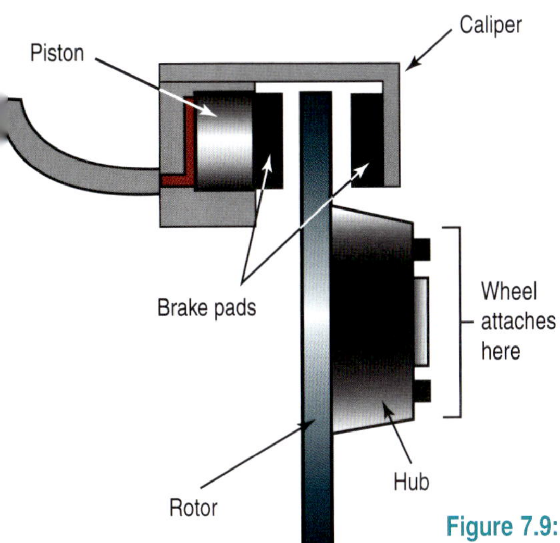

Figure 7.9: A disc brake

Braking

Usually, a driver's first reaction to seeing a hazard is to start braking. The most common type of brake fitted in cars is the disc brake (see figure 7.9). The wheels of the car are mounted on the hub, which is attached to a steel rotor. On either side of the rotor is a brake pad, which is a thick pad of high-friction material. When the brake pedal is depressed, the piston causes these two pads to be pressed against the rotor. This clamping effect slows the rotation of the rotor and thereby slows the wheel.

When a car is brought to a stop by braking, there are two different types of friction acting: the sliding friction acting between the steel rotor and the brake pads ($\mu_k \approx 0.6$) and the rolling friction between the tyres and the road surface ($\mu_r \approx 0.01$). Of course, a car can be slowed simply by taking your foot off the accelerator and allowing the rolling friction to bring the car to a halt but, as you can see from the relative sizes of the friction coefficients, it would take a much greater distance to stop this way than by using the brakes.

Unfortunately, when faced with the need for a sudden stop, many drivers instinctively stamp on the brake as hard as they can, inadvertently causing the wheels to 'lock'. This happens because the heavy pressure exerted on the brake pedal is transmitted to the piston controlling the force with which the callipers press the pads against the rotor. The pads are pressed so tightly to the rotor that the rotor stops turning altogether. As a result, the wheels stop rotating and the car travels on tyres that are locked in one position. Locked wheels have no traction, and the car slides. At this point, the deceleration of the vehicle depends solely upon the sliding friction between the tyres and the road surface. This deceleration is a lot less than that provided by properly engaged brakes and so the car travels a lot further by the time it comes to rest. In wet conditions, the value of sliding friction is reduced even further. This is especially dangerous if you have been tailgating the car in front of you and the car stops suddenly!

When the brakes and wheels have locked, the best solution is to pump the brakes. You do this by taking your foot off the brake pedal, allowing the wheels to start turning again and then reapplying the brakes. If the wheels lock again, you repeat the process.

While this procedure does work, it usually does not occur to the driver of a car to do it when all they can see is the rear end of the car in front of them getting very close, very quickly.

Anti-lock braking systems (ABS) will pump the brakes as an automatic process, much more quickly and efficiently than a panicking person would. The heart of an ABS is a small computer that monitors the rate at which the wheels of a car are turning. When it senses that the wheels have stopped rolling but the car is still moving forward, it releases the pressure on the brake pads and allows the wheels to turn before it reapplies the pressure. An ABS may repeat this release-and-reapply process up to 15 times in a second, allowing the car to come to a controlled stop rather than an uncontrolled slide.

Hydroplaning occurs when water is not carried from underneath the tyres quickly enough by the wheel tread, causing the car to ride on a thin cushion of water rather than making solid contact with the road.

An ABS is particularly useful in wet conditions, where large amounts of water on the road surface may cause the car to **hydroplane**. Hydroplaning occurs when water is not carried from underneath the tyres quickly enough by the wheel tread, causing the car to ride on a thin cushion of water rather than making solid contact with the road. When drivers feel their cars starting to slip or become very unresponsive, their instinctive reaction is to brake, and this can all too easily turn into a slide due to locked wheels.

Cracking down on speed

There are a number of different ways in which the speed of a car can be determined. Some of them use a calculation of the car's average speed while others allow the instantaneous speed of a vehicle to be measured.

So far this chapter has concentrated on *average speed*, which is calculated from the total distance covered over a specific period of time. This average value does not necessarily reflect the speed of a car at a specific moment in time. For example, if you drove the 985 kilometres between Brisbane and Sydney in 14 hours, your average speed would be 70.4 kilometres per hour. Obviously, it is unrealistic to believe that the car travelled at exactly 70.4 kilometres per hour for the entire journey, particularly one this long! Typically, there would have been frequent stops for the bathroom, fuel, chips and coffee. There would have been sections of highway where you travelled at 110 kilometres per hour, small towns where you reduced your speed to 50 kilometres per hour, and inevitable delays where you were caught behind slow cars pulling caravans up a mountain range on a single-lane road. The journey would have been one long sequence of starting, stopping, accelerating, cruising and braking.

Instantaneous speed is the speed of travel at a specific moment in time.

When considering if you are going to be booked for speeding or if you are travelling too fast for safety, it is usually the **instantaneous speed** that you are thinking about. When you are driving down the highway and your speedometer indicates that you are travelling at 92 kilometres per hour, it is indicating your instantaneous speed. As you brake, your instantaneous speed changes. When you stop for lunch, your instantaneous speed is zero. So while your *average* speed for a particular journey may have been 52 kilometres per hour as you travelled to school, your *instantaneous* speed at any particular moment in time might have been 0, 40 or even 110 kilometres per hour!

Figure 7.10: The speedometer indicates the vehicle's instantaneous speed.

Speedometer

The speedometer is located in the driver's console of a vehicle and is either connected by a cable to the gearbox or electronically connected to the transmission. The speedometer indicates the instantaneous speed of the vehicle and is calibrated specifically for the car it is in, based upon the recommended tyre size. When drivers fit non-standard tyres to their vehicles, many do not realise that this will have an effect on the speedometer's reading. As a result, they may believe that they are travelling at exactly the speed limit when in fact they could be travelling faster or slower than this.

VASCAR

VASCAR stands for Visual Average Speed Computer And Recorder, and it is a device used by police and road authorities to determine a vehicle's average speed. It works by establishing the time a vehicle takes to travel a set distance. Two landmarks on a set stretch of road are selected as reference points and the distance between them entered into the VASCAR. As a car passes the first of the landmarks, the VASCAR is activated by an observer, who then gives a further signal when the car has passed the second landmark. The device then performs a fairly easy calculation of speed (distance divided by the time interval) and then produces a reading of the car's average speed. When used by the police, the number plate of the offending vehicle is radioed further down the road to other officers, who will direct the driver to pull over and then issue them with a speeding ticket. While the system works well in theory, it is notoriously unreliable in reality as error is possible due to three main factors:

1. Reliance on the reaction time of the officer in depressing the buttons. This becomes a real problem over very short distances between reference points.
2. Parallax caused by the fact that an officer can sit in only one location, either across from the first landmark or across from the second. As a result, the view to the other end is skewed, making the exact moment of the vehicle's passing difficult to gauge. This is called **parallax error**.
3. The difficulty faced by officers who need to remain hidden for the entire time while retaining a clear view of the road. If a driver entering the test zone sees the controlling officer early enough, the driver can brake and slow the vehicle enough to extend the transit time, which in turn lowers the value of the average speed calculated.

Parallax error is the misjudgement of distance that occurs as a result of a shift in the position of the observer.

Amphometer

An amphometer (or Digitector) is a device that measures the time elapsed between the activation of two sensor cables.

The human error inherent in the VASCAR system led to the development of the **amphometer** (or **Digitector** as it is also known). This device functions on essentially the same principle as the VASCAR, except that two cables are placed across the road at a measured distance and they are activated by a vehicle rolling over them. These cables between them send a reading of time elapsed for the vehicle's transit to a central processor, which then calculates average speed much as the VASCAR does. The amphometer's main advantage is that because human reaction time is no longer a factor, the distances between the cables can be quite small, allowing the driver very little time for evasion.

Speed guns

Speed guns are handheld devices that use the reflection of soundwaves or light waves to determine the speed of an object at which the guns are directly pointed.

These days, you are far more likely to encounter **speed guns** than at any time in the past. One type of speed gun, the *radar detector*, relies on the emission of radio waves that are then reflected back to a detector from a vehicle passing anywhere up to 500 metres away. When the radio waves are reflected from a moving vehicle, they experience a change in frequency, a phenomenon known as Doppler Shift (this is discussed in chapter 3). The faster the vehicle is travelling, the bigger the change in frequency seen by the detector. The device is calibrated so that a particular size of frequency shift correlates to the speed of the vehicle, which is then displayed. Many of these devices are connected to hidden cameras so that when a speeding vehicle is detected, a picture is taken of the vehicle displaying its number plate. This allows a ticket to be issued automatically by post to the registered owner.

A later version of the radar gun is the *laser radar gun* (see figure 7.11), which detects speed by measuring the frequency shift of a fine laser beam rather than radio waves. This is usually a handheld device that is aimed at vehicles (hence the term 'gun'), and it is able to obtain accurate readings at distances of up to a kilometre away.

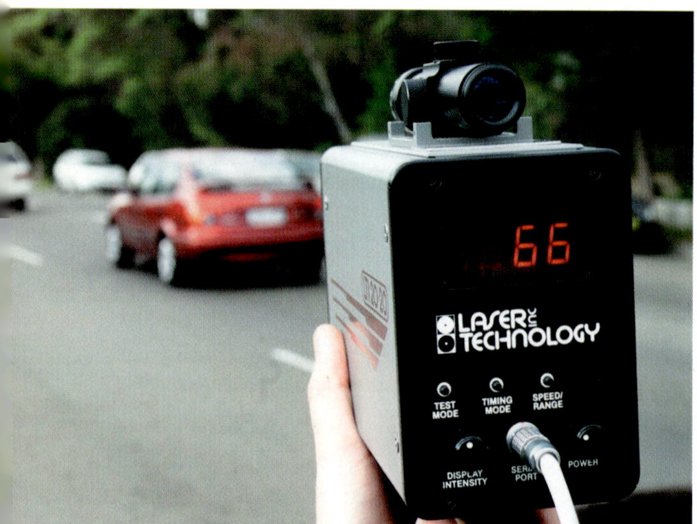

Figure 7.11
Police use laser guns to measure the instantaneous speed of vehicles.

Ironically, it is the high precision of these devices that has caused complaints against them. There was a court case many years ago in which a lawyer, defending his client against a speeding fine, claimed that the radar gun used to charge his client was defective. To prove it, he took the device outside and pointed it at a tree, which the gun registered as travelling at 65 kilometres per hour! While this story is often trotted out to show how radar-gun readings are inaccurate, those who have since repeated the demonstration quickly realised that the gun was measuring the speed of the *leaves* of the tree, which were being whipped towards the gun by the wind. So the error lay not in the radar gun but in the human hand that held it — something all too common in science!

Investigation

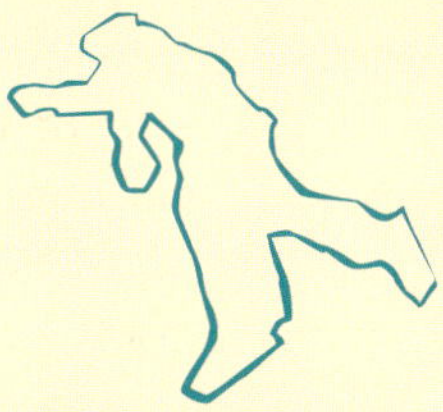

Measuring average speed

Aim: To simulate the way in which a VASCAR determines a vehicle's average speed

Materials: Pedometer, stopwatch

Method:

Part A (on the street)

1. Find a straight section of road where there is a clear view for at least 30 metres.
2. Select a landmark near the road that is highly visible from a distance. From this landmark, use the pedometer to measure a distance of 30 metres. Do not walk on the road to do this — keep to the footpath!
3. Standing at the 30-metre mark, look towards the landmark you have chosen. When a car passes the landmark, start the timer. Stop the timer when the car passes you. *Note:* To get a more accurate measurement for the time, you could have more than one person timing the same car. If you do this, note each timer's reading and find the average time for each car.
4. Note the time taken for the car to travel the 30-metre distance.
5. Reset the timer and repeat the timing process for at least 20 cars.

Part B (at school)

Calculate the average speed for each of the vehicles that were timed. This will be equal to the distance marked out divided by the time the cars took to travel it.

Discussion questions

1. What proportion of the cars timed were travelling over the speed limit for this section of road?
2. What difficulties did you experience while doing this experiment?

Safety campaigns

In an effort to make people more aware of the dangers involved in road use, a number of advertising campaigns have been devised by government and council transport departments. These vary enormously in approach.

Road safety campaigns are common in Queensland.

Figure 7.12:
Car safety advertisements

Look at the advertisements in figure 7.12 and answer the following for each one:

1. What safety issue is the advertisement addressing?
2. What consequence does the advertisement imply will happen?
3. How is the advertisement designed to be effective? Does it use: humour? language devices? threats?
4. Is the advertisement to be used on a billboard or in a newspaper? How can you tell?
5. How does the layout including the use of pictures catch your attention?

Eye on the media

Speed ads on the mark

Now, consider the following letter that was written to the editor of a Queensland newspaper:

The article that you published in your paper last week included interviews with a large slice of the community who were appalled by the Queensland Transport *Speed Kills* ads on the TV. Many of them felt that the realism of scenes, where young people lie dead on the street after an accident or where a young mother and her baby are struck and killed by a speeding car, was 'upsetting' and 'sickening'. Well, let me tell you that, as a paramedic with the fire department who has to arrive week after week to real-life scenes of the carnage that speeding drivers cause, these ads are nowhere near as upsetting or sickening as the real thing! If people find that they are having such a visceral reaction to these ads, all I can say is — good! If this is what it takes to make people think twice about pressing that accelerator pedal down, I say congratulations to Queensland Transport for finally having the guts to produce such a shocking hard-line campaign!

K. D., Ashgrove

Comprehension questions

(a) Can you think of any advertisements that might fit the description of those mentioned in this letter?

(b) Is the author in favour of or against these advertisements?

(c) Discuss road safety advertisements you have seen that you have found confronting or disturbing. What aspects of these advertisements presented you with the biggest problems? How effective do you think these graphic and emotional advertisements are?

Independent inquiry: The ultimate safety vehicle?

Using the information in this chapter and other sources such as journals, magazines and the internet, investigate the possible ways in which cars can keep their occupants safe regardless of the wishes of the occupants themselves. You may consider these ideas for a start:

- automatic/robot cars that drive themselves
- accident-avoidance devices
- cars that immobilise themselves unless certain conditions are met (for example, unless seatbelts are done up or an in-built breathalyser detects an acceptable blood alcohol level).

When conducting your inquiry, consider the advantages and disadvantages of each possibility as well as exploring what social and financial repercussions may be involved.

End of chapter questions

1. Why is an amphometer more likely to be used than a VASCAR?
2. What factors affect the rate at which a car is able to brake?
3. A holidaying physics teacher drives her old Volkswagen from Brisbane to Cairns, a distance of about 1100 kilometres. Her average speed was 80 kilometres per hour. She trades in the Volkswagen and purchases a brand new Toyota Corolla. She drives back to Brisbane at an average speed of 100 kilometres per hour. Calculate her average speed over the journey.
4. What is the average speed of a vehicle (in kilometres per hour) that travels:
   (a) 400 kilometres in five hours?
   (b) 100 metres in 4 seconds?
   (c) 200 kilometres in three hours and then takes two hours to cover the last 60 kilometres?
5. Calculate the acceleration of a bus that takes 5 seconds to reach a speed of 12 metres per second from a stationary start.
6. You are standing in the bus travelling to school when the bus suddenly brakes. In what direction will you be thrown: towards the front of the bus or towards the back of the bus? Explain your answer.
7. Which of the following factors could contribute to the value of:
   (a) reaction distance of a car?
   (b) braking distance of a car?
      (i) Driver tiredness
      (ii) Car size
      (iii) Driving under the influence of alcohol
      (iv) Rain
      (v) Bald tyres
      (vi) Talking on a mobile telephone while driving
8. How do seatbelts protect their wearers in the case of a collision?
9. If you were to compare the performance of a car on Earth with that of an identical car on the moon, which would you expect to have the greatest acceleration? Explain your answer.
10. Consider the photos of the Formula One car and the drag racer included here.
    (a) Describe the ways in which they differ.
    (b) List the individual design features of each car and explain what function or purpose each feature serves.

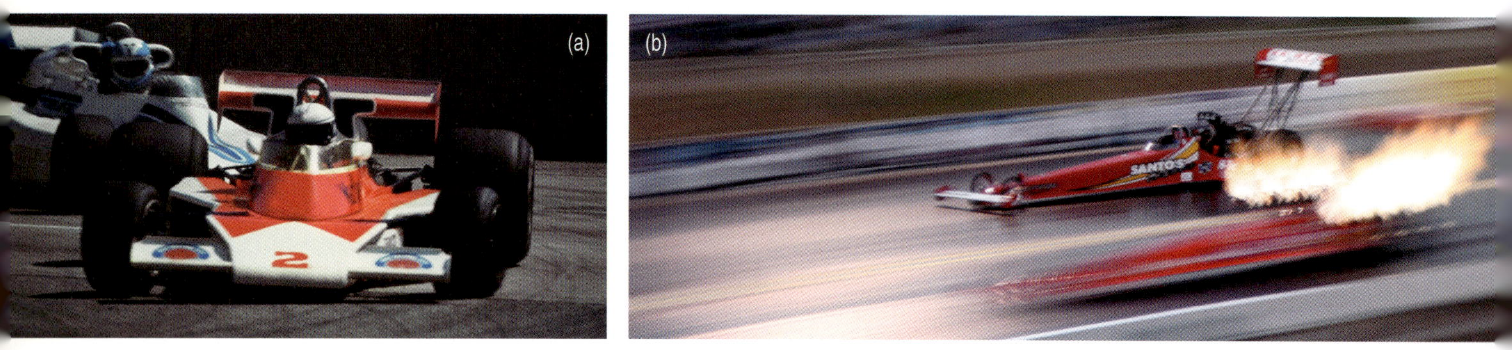

Figure 7.13: A Formula One car (a) and a drag racer (b).

11. It has been found that shorter people (under 165 centimetres in height) often suffer more collateral damage from the activation of an airbag in a collision than taller people do. How could you account for this?
12. What is whiplash? What design feature in a car minimises the effects of whiplash?
13. Drivers are generally advised to travel with a '2-second following distance' between their car and the one ahead of them.
    (a) What does this mean?
    (b) According to this guideline, what distance should separate two vehicles travelling at 60 kilometres per hour?
    (c) Why does this rule of thumb hold regardless of the speed of the vehicles?
14. The driver of a car travelling at a steady speed of 20 metres per second sees a pedestrian step onto the crossing in front of him. The driver takes 0.6 seconds to hit the brakes, which then accelerate the car at a uniform rate of –8.2 metres per second per second. Find the car's:
    (a) reaction distance
    (b) braking distance
    (c) stopping distance.
15. On 6 May 1954, Roger Bannister became the first man to run one mile in under four minutes, recording a time of three minutes and 59.4 seconds. If one mile is equivalent to 1.609 kilometres, what was Bannister's average speed in metres per second?
16. What speed will a cyclist be travelling at if he accelerates uniformly from rest at a rate of 2.2 metres per second per second over an 8-second period?
17. Convert the following:
    (a) 20 kilometres per hour into metres per second
    (b) 100 kilometres per hour into metres per second
    (c) 15 metres per second into kilometres per hour
    (d) 3 kilometres per second into kilometres per hour.
18. A car is cruising past a house at a speed of 15 metres per second when a dog sitting outside decides to chase it. If the dog is able to maintain an acceleration of 1.7 metres per second per second, how long will the dog take to catch the car? *Note:* a useful equation is: $d = ut + \frac{1}{2}at^2$.
19. A truck takes 0.5 seconds between activating one cable of an amphometer and activating the second cable placed 10 metres away. If the speed limit in the area is 50 kilometres per hour, will the driver be recorded as speeding?
20. Which of the following objects would you expect to have the smallest amount of inertia and which would have the greatest?
    (a) A cricket ball after being struck by a bat
    (b) A ping-pong ball hit by a paddle
    (c) An 18-wheel truck travelling at 80 kilometres per hour
    (d) A bullet fired from a gun
21. In a head-on collision between two cars, the drivers of both vehicles are thrown forwards and then backwards. Explain why this happens.

8

What's your poison?

Consider the substances in figure 8.1. Which of them would you consider to be a drug? Considering that the World Health Organization defines a drug to be 'any chemical substance that affects the central nervous system, causing either physiological or psychological changes', then *all* of the substances in figure 8.1 would have to be considered to be drugs of one kind or another. So why is it that some drugs are made illegal while others are not?

Drugs and the human brain

Figure 8.1:
Can you spot the drug?

While the general term 'drugs' tends to make us think immediately of cocaine, marijuana or ecstasy, or even of medicinal chemicals such as painkillers and antibiotics, few of us realise that the chemicals inside beer, cigarettes, coffee and even chocolate also qualify as drugs because they have an effect on our mood, thinking patterns, behaviour and physiology. Before we start looking at what these chemicals are and what they do, it is necessary to look at their prime target: the brain and the nervous system.

Brain chemistry

Neurons are nerve cells.

A **soma** is the cell body that contains the nucleus.

Dendrites are spiky extensions from the soma that receive impulses from other neurons.

An **axon** is a long extension from the soma that transmits electrochemical impulses to other neurons.

The brain is the control centre of the body. In an average adult, it weighs between 1.3 and 1.4 kilograms, making up around 2 per cent of a human's body mass. It is 17 centimetres long, 14 centimetres wide and 9 centimetres high, with a spongy texture. It contains 100 billion **neurons**, each of which is connected to as many as 10 000 other brain neurons. Each neuron has three main sections:

- **soma** — the cell body that contains the nucleus
- **dendrites** — spiky extensions from the soma that receive impulses from other neurons
- **axon** — a long extension from the soma that transmits electrochemical impulses to other neurons. It is surrounded by a myelin sheath that insulates it, much as plastic insulates the outside of an electrical wire. At the end of the axon is the axon foot.

The structure of the neuron is shown in figure 8.2.

A **synapse** is the small gap between the dendrite of one nerve cell and the corresponding axon of the next nerve cell.

Neurotransmitters (or **neuromediators** or **transmitter substances**) are chemicals that travel across the synapse between the axon of one neuron and the dendrite of the next

The impulses that travel from one neuron to another do so by passing from the axon foot of one neuron to the dendrite of another neuron. However, the dendrites and the axons are not actually touching. They are separated by a small gap called a **synapse** over which the electrochemical impulses cannot jump. Instead, as the impulse enters the axon foot, it causes the production of specialised chemicals called **neurotransmitters** (also called **neuromediators** or **transmitter substances**). The neurotransmitters travel across the synapse to the dendrite of the next neuron where they cause an electrochemical impulse that then travels down *that* neuron and so on (see figure 8.3). Most of the effects of psychotropic drugs (drugs that affect the mind, mood or other mental processes) are caused by their influence on the production, reception or detection of these neurotransmitters.

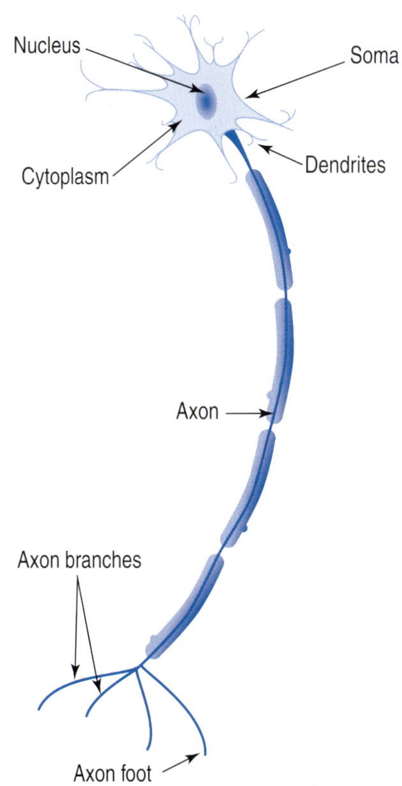

Figure 8.2:
Structure of the neuron

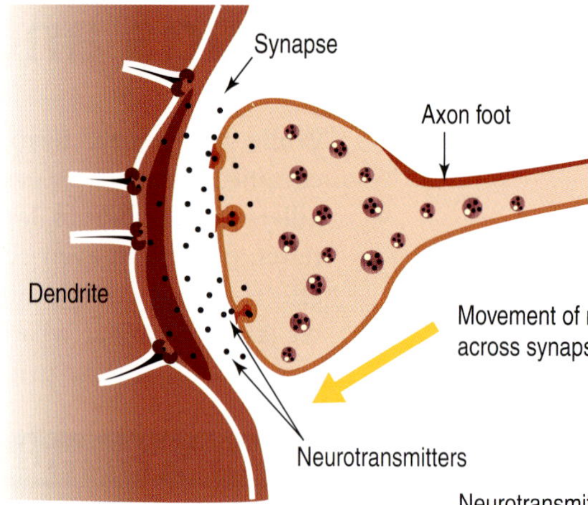

Figure 8.3:
Neurotransmitters travel across the synapse between the axon of one neuron and the dendrite of the next.

Amine neurotransmitters are a family of neurotransmitters made up of compounds that have nitrogen atoms attached to them. This group includes the neurotransmitters acetylcholine, dopamine, serotonin and norepinephrine.

Acetylcholine

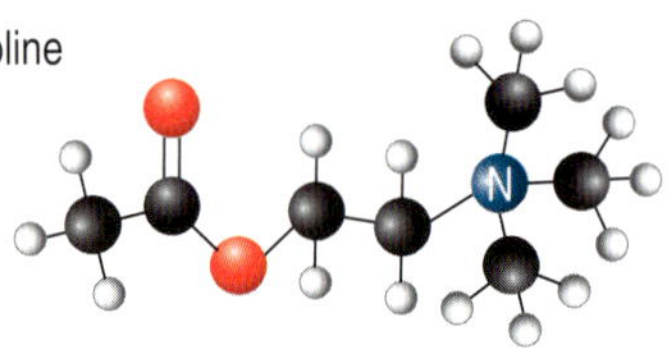

Dopamine

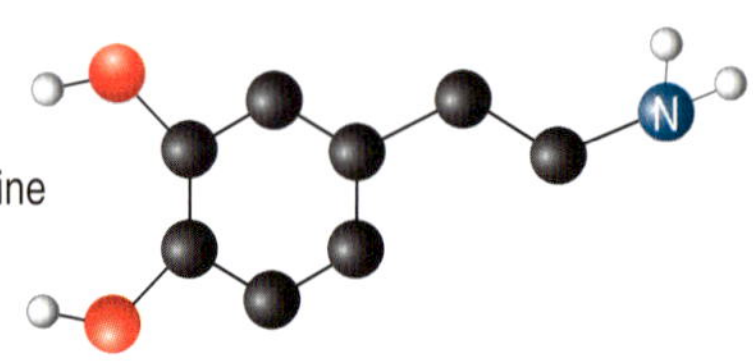

Serotonin

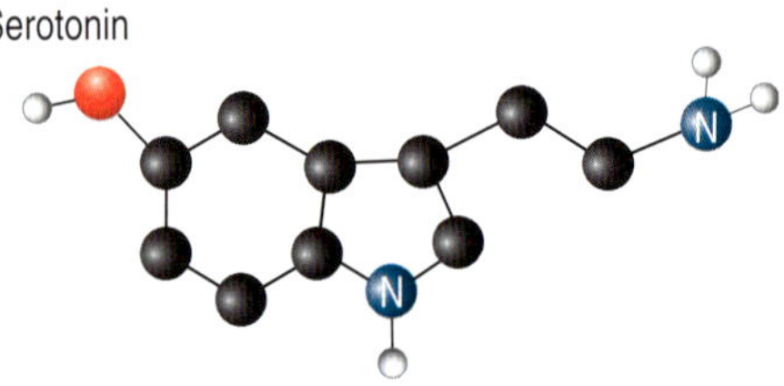

Norepinephrine

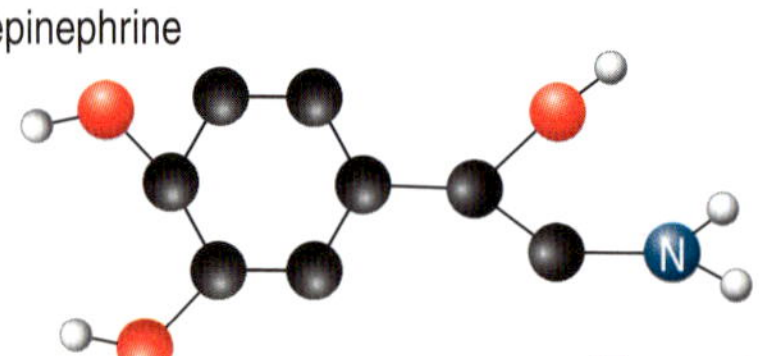

Figure 8.4:
Members of the amine neurotransmitter family have a nitrogen atom attached to them.

Neurotransmitters

Most of the 35 or more neurotransmitters are made from amino acids derived from dietary protein. In many cases, vitamins and minerals such as vitamin B_6, folic acid, niacin, iron and vitamin C are required for their production. The most important neurotransmitters are a group known as the **amine neurotransmitters** (see figure 8.4). All the neurotransmitters in this family (serotonin, norepinephrine, dopamine, and acetylcholine) are compounds that have nitrogen atoms attached to them.

- *Acetylcholine* is produced from the vitamin choline (found in meat, eggs and nuts) and is the most widely used neurotransmitter in the body. It is responsible for the stimulation of muscles and helps to schedule sleep, as well as being needed for the formation of memory and learning. Acetylcholine is used more when the body is under greater stress. Botox, which is used to prevent facial wrinkles, acts by blocking acetylcholine in the facial muscles, causing them to be paralysed. It is believed that there is a link between acetylcholine and Alzheimer's disease based on the fact that people suffering from this disease have up to 90 per cent less acetylcholine in their brains than non-sufferers.
- *Norepinephrine* (also known as *noradrenaline*) is produced from the conversion of the dietary amino acid *tyrosine* (found in dairy products, soy products and bananas). This neurotransmitter has a stimulating effect on the body and the brain much the same as that of adrenalin, causing us to come into a state of high alert when surprised or endangered, increasing our heart rate and our blood pressure and focusing our attention. Like acetylcholine, norepinephrine is also important for forming memories.
- *Serotonin* is produced with the help of the amino acid tryptophan. This neurotransmitter affects emotion and mood. While normal levels of serotonin have a calming effect in the brain, decreased levels of serotonin are associated with depression, aggression, obsessive-compulsive disorder and sleep disorders.

- Like serotonin, *dopamine* plays a role in the regulation of mood and is found only in the central nervous system. Excessive amounts of dopamine are found in the frontal lobes of schizophrenics while inadequate amounts of dopamine are found in sufferers of Parkinson's disease. Dopamine's main function appears to lie within the brain's association of certain behaviours or substances with reward mechanisms and motivation. It is generally believed that all substances that trigger dependencies in human beings increase the release of dopamine in a specific area of the brain, the *nucleus accumbens,* which is the brain's 'pleasure centre' (see figure 8.5).

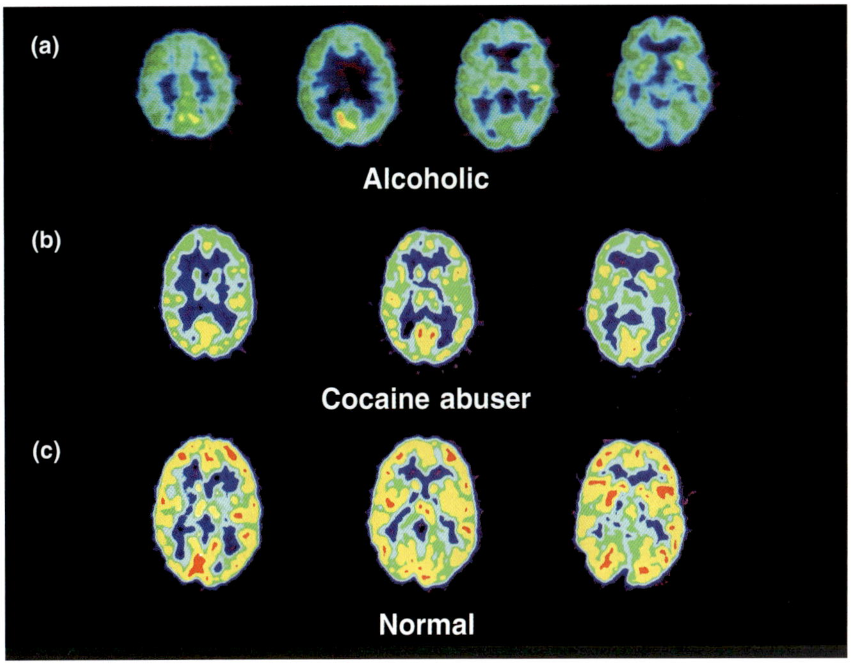

Figure 8.5:
Positron emission tomography (PET) scans showing the brain's response to an avalanche of dopamine in the *nucleus accumbens* of people with dependencies. The red areas indicate large numbers of dopamine receptors. These images show that people who are alcoholic (a) or cocaine-dependent (b) have fewer dopamine receptors than do people without such addictions (c).

While not a member of the amine neurotransmitter family, *endorphin* (short for *endogenous morphine*) is an important neurotransmitter that is involved in pain reduction and is associated with the brain's pleasure centres. Strangely enough, it is also the chemical that allows bears and other animals to hibernate during the winter, slowing heart rate, respiration and metabolism.

Drug classifications

Different drugs affect the brain in the different ways. Some substances imitate natural neurotransmitters and take their place on their receptors. Other substances increase the production or secretion of neurotransmitters, while still other substances block a neurotransmitter, preventing it from making contact with its receptor. The drug-classification system divides frequently abused drugs into categories according to how they imitate or interfere with the chemicals responsible for transmitting messages within your brain and between your brain and body.

Depressants

As the name implies, depressants are drugs that slow down the central nervous system. This results in a much slower rate of transmission of impulses from neuron to neuron. While this may give a feeling of relaxation and a release from inhibitions, the decreased speed of impulses between the neurons in the brain causes a slower reaction time and impaired coordination. This in turn leads to a slurring of speech, poor balance and stumbling or weaving when you walk.

Alcohol

Alcohol is a chemical term that describes a group of organic chemicals. The one that is found in alcoholic drinks is *ethanol* (chemical formula: C_2H_5OH), a clear and flammable liquid that is soluble in water. Alcohol's effect on the brain as a depressant is the result of several different processes. First, alcohol binds to the receptors for a number of different neurotransmitters including serotonin, glutamate and acetylcholine. This causes a sedative effect. Second, alcohol increases the amount of dopamine in the brain by interfering with the processes involved that 'mop up' excess dopamine in the synapses. Third, alcohol has a dehydrating effect on the tissues of the brain. We will examine the effects of prolonged alcohol use in more detail later.

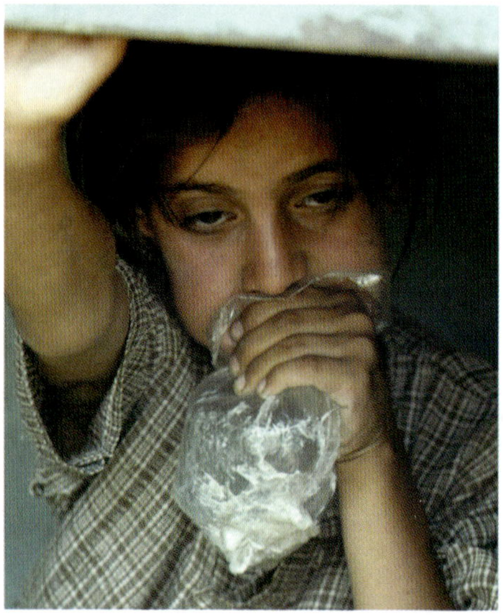

Figure 8.6: An Iraqi orphan inhales glue from a plastic bag. Typically, sustained use of inhalants irreparably damages the cognitive centres of the brain.

Inhalants

Inhalants are volatile substances that are breathed in (or 'huffed') to achieve an intoxicating effect similar to that of alcohol. They can be solvents (paint thinners, petrol, glue, nail polish remover, liquid paper), gases (aerosol propellants, nitrous oxide, propane) or nitrates (amyl nitrate). The most heavily implicated component of these substances is *toluene* (chemical formula: C_7H_8). Toluene, like alcohol, binds to the acetylcholine receptors (causing sluggishness and sedation; see figure 8.6) and raises the levels of dopamine (which give a perception of pleasure and reinforce addiction).

Narcotics

Narcotics are a form of depressant in that they slow down the transmission of impulses in the central nervous system but have the added effect of providing pain relief.

Opiates (heroin, morphine, opium, codeine)

Opiates are a family of drugs that are derived from the opium poppy and used medicinally as painkillers and sedatives. They can produce drowsiness, nausea and constipation; depending on the amount of drug taken, they can also depress respiration. Opiates are able to work because they closely resemble endorphins, the neurotransmitters that the brain manufactures to bind to the *opioid receptors* in the nervous system and block the perception of pain. Opiates, pretending to be endorphins, can attach themselves to the opioid receptors and produce the same effect.

Opiates can also cause a feeling of euphoria (a 'high'). This feeling is produced when the opiates stimulate the production of dopamine and suppress the action of the enzymes that would normally break it down again once it has reached a receptor. Because this happens in the nucleus accumbens, it stimulates a feeling of pleasure as well as a desire to repeat the experience (the beginning of addiction).

Stimulants

In contrast to depressants and narcotics that slow the transmission of impulses through the central nervous system, stimulants are drugs that speed up impulse transmission. The effects of stimulants include euphoria, a sense of wellbeing and the perception of enhanced clarity and speed of thought. However, they can also cause anxiety, paranoia, increased heart rate and blood pressure, reduced appetite, restlessness, insomnia and a feeling of being 'shaky.'

Cocaine

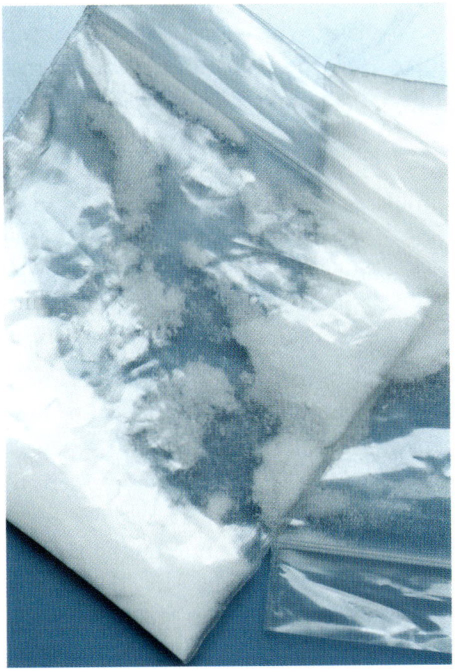

Figure 8.7: Cocaine powder

Cocaine (chemical formula: $C_{17}H_{21}NO_4$) is extracted from the leaves of the coca plant, which is found naturally in the Andes Mountains of South America. Most of the world's cocaine comes from Colombia, Peru and Bolivia. Originally it was used in medicine for its ability to numb nerves before dental work, but it has found great popularity as a recreational drug. Cocaine can be inhaled nasally (snorted), smoked (when in the form of *crack cocaine*), or injected (see figure 8.7). All these methods allow the drug to enter the bloodstream quickly.

Cocaine acts by blocking the reuptake ('mopping up') of certain neurotransmitters such as dopamine, norepinephrine and serotonin. By binding to the substances that normally remove the excess of these neurotransmitters from the synapses, cocaine prevents them from being reabsorbed by the neurons that released them and allows them to increase in concentration in the synapses. As a result of the increased amounts of the neurotransmitters, the user of cocaine experiences euphoria (from dopamine), feelings of confidence (from serotonin) and increased energy (from norepinephrine).

If cocaine is used often enough, the neurons will adapt to the increased levels of dopamine by manufacturing more dopamine receptors. This makes the receptors more sensitive to decreased (or rather normal, non-drug induced) levels of dopamine, giving rise to feelings of depression, cravings and listlessness when cocaine is not present. Cocaine's effect is also quite short, prompting the user to repeatedly administer cocaine to re-experience its effects.

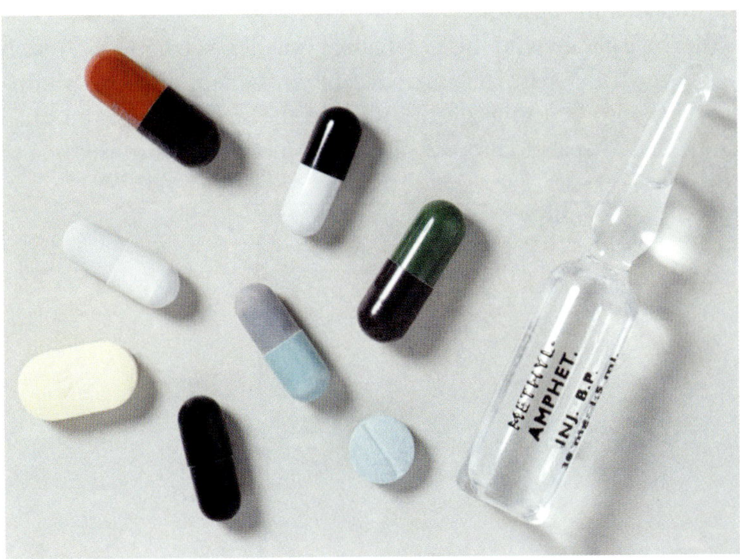

Figure 8.8: Amphetamines in a variety of forms

Amphetamines

Amphetamines, (also known as *speed*), are synthetic stimulants. They are usually found in powder form and used as capsules, sniffed through a narrow tube, prepared for injection, swallowed in a drink or smoked with tobacco (see figure 8.8). Amphetamines were issued to combat troops in World War II to keep them fighting in the absence of food. Later, other amphetamines were used as weight-control drugs due to their ability to suppress the appetite, control weight, and increase the metabolism. Because they increase levels of the neurotransmitter norepinephrine, resulting in increased energy levels and alertness, amphetamines have also been used to treat disorders including narcolepsy and attention-deficit hyperactivity disorder (ADHD) and to combat fatigue.

Like cocaine, amphetamines increase the concentration of dopamine in the synaptic gap but do it in a different way. Amphetamines are similar in structure to dopamine. They enter the axon feet of neurons, move into the spaces where dopamine is stored and force the dopamine out of the neuron and into the synapse. As with cocaine, repeated use of amphetamines causes an increased tolerance so that users need larger quantities of the drug in order to get the same effects.

Nicotine

Nicotine is recognised as one of the most frequently used addictive drugs. It is found in the leaves of the tobacco plant, which originated in South America. When extracted in a pure form, nicotine is a colourless liquid that turns brown when burned. When tobacco is smoked or chewed, nicotine (as well as a variety of other compounds) is ingested. When smoked, nicotine is delivered rapidly to the brain — within 10 seconds of inhaling.

On entering the body, nicotine stimulates the adrenal glands and causes epinephrine to enter the bloodstream. When nicotine enters the central nervous system, it binds to the acetylcholine receptors. The combined effect is an increase in blood pressure, breathing rate and pulse as well as a sudden release of glucose, putting the body in an alert state. This may explain why many smokers find that they think more clearly when they have a cigarette. Strangely, nicotine can also have a sedative effect on users if they are feeling nervous or threatened.

Nicotine also causes dopamine to be released in the pleasure/reward centres with the same effect that is observed in users of cocaine — a feeling of pleasure or euphoria and a motivation to continue the behaviour. Because the immediate effects of nicotine wear off after a few minutes, the desire a smoker feels for nicotine returns more quickly than does the desire for other drugs felt by most drug users.

Caffeine

Caffeine is the common name for the chemical *trimethylxanthine* (chemical formula: $C_8H_{10}N_4O_2$). It is found in coffee, tea, cola drinks, energy drinks, stay-awake tablets, diet pills and even chocolate. It is one of the most widely consumed stimulants in the world.

In its pure form, caffeine is a white crystalline powder (see figure 8.9) with a very bitter taste that is particularly noticeable in coffee. When it is consumed, caffeine is absorbed through the walls of the stomach and the small intestine. From there it passes into the bloodstream and travels to the brain.

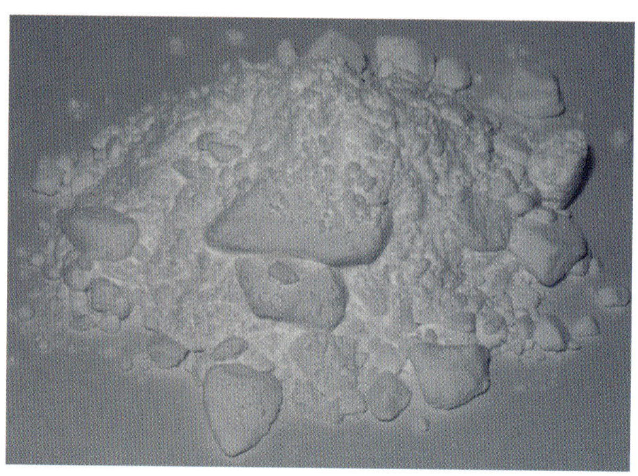

Figure 8.9: Caffeine in its pure form

Investigation

Extracting caffeine from tea

Aim: To extract caffeine from tea

Materials: Tea leaves; distilled water; sodium carbonate; dichloromethane; anhydrous sodium sulfate; stirring rod; 2 x 250 mL beakers; separating funnel; filter funnel; hotplate; piece of cheesecloth or muslin; filter paper; evaporating dish; Bunsen burner; tripod; gauze mat; matches

Method:

1. Place 30 grams of the tea leaves in a 250-millilitre beaker.
2. Add 100 millilitres of distilled water and 5 grams of sodium carbonate to the beaker and stir the contents with a glass rod.
3. Boil the contents of the beaker on a hot plate for 10 minutes.
4. Allow the tea solution to cool to room temperature and then filter the cooled solution through muslin or cheesecloth into a clean beaker. Pour the filtrate in the beaker into a separating funnel.
5. Pour 10 millilitres of dichloromethane into the separating funnel with the tea solution. Two layers will form. Gently swirl the two layers for about 5 minutes — *do not shake the layers together.*
6. Pour the lower organic layer into a 250-millilitre beaker, leaving any emulsion layer (the thicker section between the two layers).
7. Repeat steps 5 and 6 three more times, decanting the lower layer into the same 250-millilitre beaker each time.
8. Add anhydrous sodium sulfate to the beaker. (The sodium sulfate will remove any water and water-soluble salts that are retained in the dichloromethane or accidentally transferred during decantation.)
9. Using a funnel and filter paper, filter the contents of the beaker to remove the solid sodium sulfate
10. Transfer the solution to an evaporation dish and place the evaporation dish over a Bunsen burner with a tripod and gauze. Heat the dish until a dry powder forms. This will be crude caffeine (usually slightly green in colour).

Discussion questions

1. Pure caffeine powder is white and crystalline in appearance (see figure 8.9). Your powder may be slightly green. How can you explain the difference in their appearance?
2. Why was it necessary to do the separation step four times instead of just once?
3. Looking at the method you have used, in which liquid is caffeine more soluble: water or dichloromethane? Explain your reasoning.
4. What does the word 'anhydrous' mean? How has the appearance of the sodium sulfate changed between being placed in the solution and being filtered out again? What do you think has happened here?

Caffeine operates using the same mechanisms that amphetamines, cocaine, and heroin use to stimulate the brain although to a lesser effect. One of the chief brain chemicals that caffeine affects is *adenosine,* which influences sleep patterns. When adenosine is created in the brain under normal circumstances, it attaches to adenosine receptors and causes drowsiness by slowing the rate of impulse transmission and dilating (widening) the blood vessels. This allows more oxygen to be supplied as sleep causes the breathing rate to drop. Caffeine mimics adenosine and binds to the adenosine receptors, so the adenosine cannot reach the receptors. As the receptors do not detect adenosine, the rate of impulse transmission between neurons is not slowed down. In fact, it speeds up a little.

This increased rate of impulse transmission triggers the adrenal glands to produce epinephrine (adrenaline), which causes increased heart rate, higher blood pressure, greater blood flow to the muscles and increased energy levels. By blocking adenosine reception in the brain, caffeine makes you feel more alert.

Caffeine, like cocaine and heroin, also manipulates dopamine production in the pleasure centres of the brain. This leaves the consumer feeling good and more likely to later desire repeating the pleasurable effect.

The strength of the effect of caffeine on the brain depends on the amount that is consumed. Table 8.1 indicates the amount of caffeine in common substances.

Table 8.1: Amount of caffeine in common substances

Substance	Serving size	Average caffeine content (mg)
Instant coffee	1 cup	65
Drip-filtered coffee	1 cup	110
Tea	1 cup	30
Cola drink	1 can	40
Hot chocolate	1 cup	4
Chocolate bar (milk chocolate)	50 g	5
Chocolate bar (dark chocolate 85% cocoa)	50 g	60
Headache tablets	1 tablet	30–65
Stay-awake pills	1 tablet	100–200

If you consume between 75 and 150 milligrams of caffeine a day (the average person's rate of consumption), then the effects are fairly mild: a slight increase in metabolism, higher breathing and heart rate, increased energy levels and increased concentration. Larger doses of between 250 and 750 milligrams of caffeine daily, particularly for people who are usually non-users, can have more detrimental effects: headache, restlessness, nausea, muscle tension, rapid heart rate, disturbed sleep, nervousness and unsteady hands. Like all other drugs, it is possible to overdose on caffeine — a condition called *caffeinism*. The symptoms of caffeinism vary according to individual tolerances but they generally occur with consumption of over 750 milligrams of caffeine. Over-consumers can experience anxiety attacks, hear ringing in their ears and see flashes of light in a similar way to migraine sufferers. They also have elevated levels of blood sugar similar to diabetics. Caffeinism can even kill, with the lowest dose known to kill an adult being 3200 milligrams.

Like other drugs that manipulate the dopamine concentrations in the nucleus accumbens, caffeine can be addictive; regular use of more than 350 milligrams of caffeine a day causes physical dependence on it. When you stop consuming caffeine, the characteristic symptoms of withdrawal are produced — severe headaches and feelings of irritability and fatigue. Many users resume caffeine consumption simply to avoid the withdrawal symptoms!

Eye on the media

Good news for fans of the black stuff

Drinking coffee might have least one health benefit: reducing your risk of liver cancer.

A team led by Manami Inoue of the National Cancer Centre in Tokyo analysed a 10-year study of 90 000 people in Japan. They found that middle-aged and elderly people who drank coffee daily had half the rate of a common liver cancer compared with less frequent consumers (*Journal of the National Cancer Institute,* vol. 97, p. 293). And the more coffee people drank, the lower the risk. It is not clear if decaffeinated coffee has the same effect, as few people in the study drank it. 'There is caffeine and antioxidants in coffee. Maybe both or either are protective,' says Inoue.

Meanwhile, a large US study found no association between drinking caffeinated tea and coffee and the incidence of colon or rectal cancer (*Journal of the National Cancer Institute,* vol. 97, p. 282). But people who drank decaf had less rectal cancer — perhaps because they lead healthier lives, suggest the researchers.

Despite their ubiquity, the long-term effects of coffee, tea and caffeine remain uncertain (*New Scientist,* 29 January 2000, p. 30). Inoue says people should not change their drinking habits on the basis of her study alone. When it comes to liver cancer, it's more important to be screened for hepatitis, she says. 'That is the biggest risk factor.'

Source: *New Scientist;* 26 February 2005; p. 21.

Clarification questions

1. What does the word 'ubiquity' (in the last paragraph) mean here?
2. What is the biggest cause of liver cancer?
3. Does decaffeinated coffee have the same effect as normal coffee?
4. Who were the group tested during this experiment?

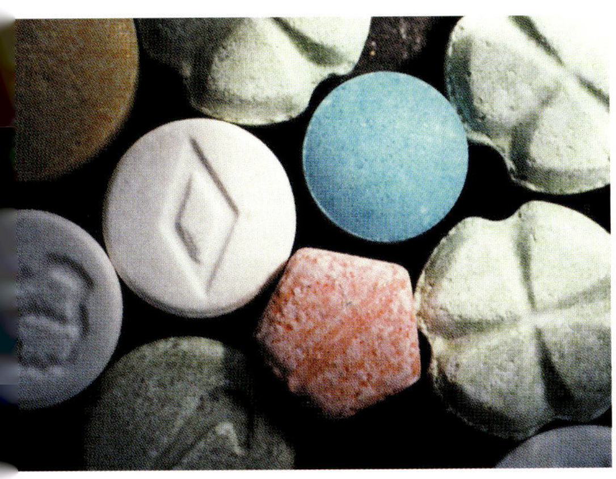

Figure 8.10: A variety of ecstasy tablets

Ecstasy

Ecstasy (see figure 8.10) is another name for methylenedioxymethamphetamine, or MDMA for short (chemical formula: $C_{11}H_{15}NO_2$). It is a synthetic drug that acts as a stimulant and a hallucinogen. Like amphetamines and cocaine, ecstasy prevents the absorption of excess amounts of neurotransmitters, thus increasing their levels in the synaptic gap and their effect on the neurons on the receiving end of the gap.

The initial effect of ecstasy is an increased release of serotonin that causes the user to experience increased energy, euphoria and the suppression of certain inhibitions in relating to other people. Later, the increased serotonin levels trigger the release of substances that deactivate the excess serotonin. It has been found that these deactivating substances may actually reduce serotonin levels to *below* normal levels, causing an ecstasy user to experience depression and lethargy for periods up to several weeks after the drug use.

Like all psychoactive drugs that produce a sensation of pleasure, ecstasy increases the release of dopamine into the pleasure centres. Some of this excess dopamine is indirectly created in response to increased levels of serotonin.

Eye on the media

Ecstasy downer

If you have a family history of depression, you should probably think twice about taking ecstasy. The drug appears to trigger depression in people with a genetic susceptibility to the condition.

Ecstasy tricks neurons into dumping large quantities of the mood-regulating neurotransmitter serotonin into the brain. It is thought to work by causing serotonin transporters to reverse direction, pumping serotonin out of neurons instead of sucking it back in.

There are two versions of the serotonin transporter gene, long and short. The short version has already been associated with depression brought on by stressful events. David Rubinsztein and colleagues at the University of Cambridge wanted to know whether there might be a similar link with ecstasy use.

They studied 66 chronic ecstasy users, 30 cannabis smokers and 28 people who had never used illicit drugs. This revealed that of the people with two copies of the short gene, ecstasy users were much more likely than cannabis smokers and abstainers to suffer from clinical depression (*American Journal of Psychiatry,* vol. 162, p. 609). This raises the possibility that studies of the risks from ecstasy use underestimate the danger for some individuals, Rubinsztein says.

Source: *New Scientist,* 12 March 2005, p. 5.

Clarification questions

1. How many people took part in the study?
2. How do the long and short versions of the serotonin transporter gene differ in the way they function?
3. Which genetic characteristic does this article indicate causes a greater likelihood of clinical depression resulting from ecstasy use?
4. How does ecstasy work?

Hallucinogens

Hallucinogens are drugs that are characterised by the random speeding up and slowing down of impulses in the central nervous system. This causes a distortion of the signals between neurons that is then interpreted as distortion in the perception of reality, causing hallucinations. Mild hallucinations tend to be experienced as sensual enhancement: touch seems more sensitive, pain is magnified, hearing is altered, vision can be enhanced or blurred. Thinking ability may be affected, with short-term memory problems, a false sense of time, an inability to focus, a reduced ability to learn and a tendency to be easily distracted. Some of the more serious effects are paranoia, the triggering of schizophrenia, psychotic breaks and uneven heart rate leading to heart attack.

Figure 8.11: LSD blotters

LSD

Lysergic acid diethylamine (chemical formula: $C_{20}H_{25}N_3O$), generally known as LSD or acid, is a very strong hallucinogen. In its pure form, it is colourless, odourless and tasteless. A typical dose of LSD is only one-tenth of a milligram — about the size of a grain of sand. LSD is usually soaked into an absorbent material such as blotting paper (see figure 8.11) or a sugar, and then swallowed or allowed to dissolve in the mouth.

The main effect of LSD is to intensify senses and feelings. It produces visual and hearing distortions and causes the surroundings to be brightly coloured. It can also produce synaesthesia, a condition where the senses are merged — for example, the perception that colours have taste or sounds have texture.

In the brain, LSD affects a number of receptors. Like most of the drugs already discussed, it increases dopamine levels in the nucleus accumbens and thereby causes euphoria. It also stimulates the body into producing epinephrine (adrenalin) and puts the body into a heightened alert state, increasing heart rate, blood pressure and energy levels. Additionally, LSD binds to most of the serotonin receptors much as ecstasy does, causing relaxation of inhibitions and heightening pleasure. LSD's hallucinogenic properties are believed to be caused by its effect on receptors in the brain known as 5HT2A receptors, allowing sections of the brain that would normally be dormant during wakefulness to be stimulated, creating a dream-like state.

LSD's effects normally last from six to 12 hours. It is typical for LSD users to encounter problems getting to sleep after a 'trip' and to have feelings of depression and listlessness, with many reporting that they do not feel completely normal until they have had one or two full nights of restful sleep.

Marijuana (cannabis)

Marijuana is composed of the leaves, stems, flowers and seeds of the hemp plant, *Cannabis sativa*. The active chemical in marijuana is tetrahydrocannabinol (chemical formula: $C_{21}H_{30}O_2$), usually abbreviated to THC. When smoked, the dried marijuana buds and leaves are rolled into a cigarette called a *joint* or a cigar called a *blunt*. It can be smoked in a pipe or in a water pipe called a *bong*. Marijuana can also be stewed to form a tea and mixed into foods such as biscuits and cakes. THC can be consumed in the form of *hashish* or *hash*. Hash is made by extracting the resin from the leaves and flowers of the marijuana plant and then pressing it into cakes (see figure 8.12). It is stronger than marijuana, and may contain five to 10 times as much THC.

Figure 8.12: (a) Marijuana (b) Hashish

In the brain, THC connects to the *cannabinoid receptors* that are found in clusters throughout the brain and are particularly dense in the parts of the brain that influence pleasure, memory, thought, concentration and sensory and time perception. As a result, heavy THC consumption impairs a user's ability to form memories, recall events and shift attention from one thing to another. THC also disrupts coordination and balance by binding to receptors in the cerebellum and basal ganglia, the parts of the brain that regulate balance, posture, coordination of movement and reaction time.

Heavy marijuana use can lead to a condition known as *Amotivational Syndrome* (or 'burnout'). Sufferers of this condition experience apathy, decreased libido, shortened attention span, poor memory and judgement, impaired ability to concentrate, decreased communication skills and an inability to retain interpersonal relationships.

Effects on health

When questioning why some drugs are legal while others are not, perhaps it should be considered whether their legality is in some way related to their potential for harm. Surely, the illegal drugs are much more harmful than the legal ones? There is no doubt that the effects of drugs such as heroin on the human body can be devastating. Figure 8.13 shows a series of photographs of a heroin addict taken over a nine-year period. The first photo was taken when the subject was 29 years old, the last when she was 38.

Figure 8.13:
Photos showing the deterioration of a heroin addict over nine years

On the other hand, the long-term effects of legal drugs such as alcohol and nicotine can also be harmful. A survey conducted by the National Drug Strategy found that two-thirds of those surveyed thought that 'drug problems' were associated with (in decreasing order of magnitude) cannabis, heroin, cocaine, alcohol, amphetamines and tobacco. In fact, alcohol and smoking are the top two causes of drug-related deaths in Australia — far in excess of any of the illegal drugs — while practices such as chroming (which was not even mentioned by the survey respondents) are claiming an increasing number of young victims. How is this the case when, so far as effects on the brain are concerned, alcohol and nicotine appear to be no worse than any other drug?

Smoking hazards

At present, the nicotine content of the average cigarette cannot exceed 1.5 milligrams. The main problem with smoking lies not so much with the nicotine content but rather with the effects of the other substances that are found in the cigarettes that deliver the nicotine into the body. When the cigarette burns and the smoker inhales the smoke, the smoker is also inhaling over 4000 other substances including:

- *Tar.* This is a thick, sticky yellow-brown substance that stains the smoker's fingers, teeth and lungs. It is made up of particles consisting mostly of nitrogen, oxygen, hydrogen and carbon dioxide. A number of carcinogens (cancer-causing substances) are present in tar.
- *Carbon monoxide.* This is a poisonous gas that is also found in car exhaust fumes. When carbon monoxide enters the blood, it takes the place of oxygen, cutting down oxygen supply to the body. When combined with nicotine, it is believed to cause heart disease and damage blood vessels in the eyes.
- *Hydrogen cyanide.* This chemical directly damages the cilia, the tiny hairs that line the breathing tract and protect it from bacteria and damaging particles. The damage leads to an accumulation of toxic agents in the lungs, increasing the likelihood of developing lung disease.
- *Ammonia.* A strong chemical sometimes found in household cleaners, ammonia damages the cilia.
- *Formaldehyde.* Commonly used for preserving dead people in morgues and lab specimens, this chemical also damages the cilia.
- *Agricultural chemicals and additives.* A wide range of carcinogenic chemicals – including herbicides, fungicides and insecticides — is routinely used in growing tobacco.
- *Radioactive compounds.* The radioactive compounds found in highest concentration in cigarette smoke are polonium-210 and potassium-40. Both are carcinogens

Every year, around 19 000 Australians die from diseases caused by smoking: cancer, lung disease/emphysema, heart disease and stroke.

Cancer

While cancer has long been connected with smoking, it was only recently that science uncovered the exact mechanism by which this process occurred. Some of the components of cigarette smoke, most notably the carcinogen benzpyrene, damage a critical gene in the DNA of cells — the *p53 gene*. As discussed in chapter 6, when cells divide to reproduce themselves, the DNA sequence inside those cells is also copied into the new cells. During this process, it is common for small errors to arise just as spelling mistakes occur when you are copying something from the board in class into your notebook. If left uncorrected, further reproductions will lead to more errors until the damaged cells divide uncontrollably, forming cancerous tumours. One of the key jobs of the p53 gene is to correct these errors within the cell at an early stage. If the p53 gene is damaged, then the DNA errors accumulate and increase the risk of cancer developing.

Lung disease / emphysema

Lung disease occurs when the lungs have been damaged (see figure 8.14). Cigarette smoke leads to lung disease in a number of different ways:

- Chemicals in the smoke can paralyse and even destroy the cilia that line the upper airways. These small hairs normally have a waving motion that causes incoming particles and bacteria to be passed upwards again, preventing them from moving down into the sensitive areas of the lungs and causing infection. When the cilia are unable to function, the damaging particles and bacteria can continue unchecked into the lungs.

- Once in the lungs, the smoke particles are an irritant. The lungs react to this irritant by increasing the amount of mucus lining their inside surface and by narrowing the air passages. Unfortunately, the thickening of this mucus layer makes it much harder for oxygen molecules to reach the air sacs where it is transferred into the bloodstream.
- Tar in the cigarette smoke forms a coating over the inside surfaces of the lungs. Initially, this only interferes with oxygen exchange. Over time, however, this layer thickens and hardens, causing the lungs to become less elastic and unable to inflate properly.
- The carbon monoxide in cigarette smoke bonds with the haemoglobin in the blood cells that would normally allow oxygen to be carried. While the carbon monoxide is attached, blood cells cannot capture oxygen.
- Chemicals in cigarette smoke actively destroy the air sacs that absorb oxygen and get rid of carbon dioxide.

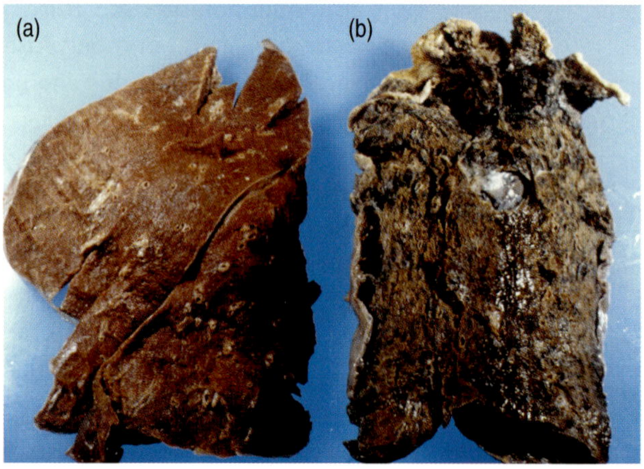

Figure 8.14:
(a) Healthy lungs
(b) Lungs of a smoker

Heart disease and stroke

Cigarette smoking is a risk factor for *arteriosclerosis*, which is the hardening (*sclerosis*) of the arteries. One of the most common forms of arteriosclerosis is **atherosclerosis** (see figure 8.15). This is caused by a build-up of plaque (calcium and blood products) in the inner lining of an artery. As these plaque deposits continue to form, the artery wall begins to harden (see figure 8.16).

Atherosclerosis is the narrowing and hardening of arteries.

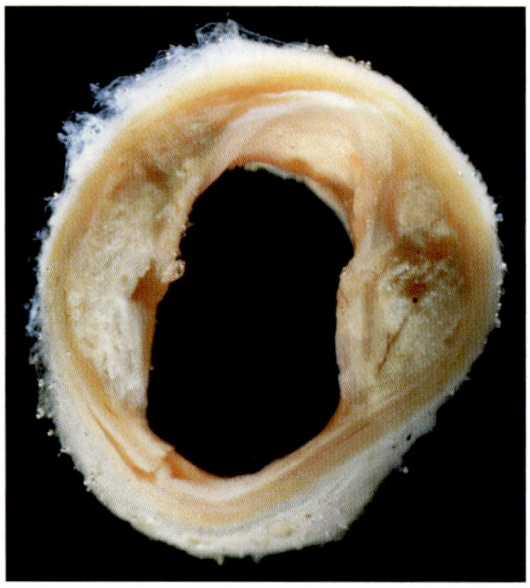

Figure 8.15:
Atherosclerosis affects the coronary arteries, which supply the heart. In addition to physically blocking the artery with fatty plaques (visible at left and right), atherosclerosis may lead to internal bleeding and clot formation. Cigarette smoking is a factor contributing to atherosclerosis.

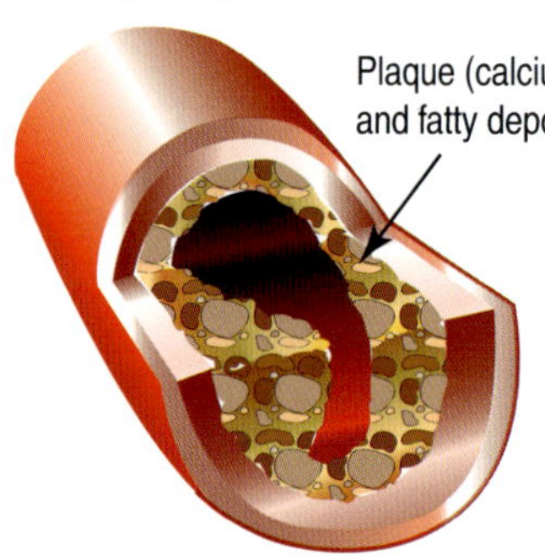

Figure 8.16:
Hardening of the arteries is caused by the build-up of plaque.

A **heart attack** is a biological event that occurs when the narrowing or blockage of the blood vessels that supply the heart interrupts the supply of oxygenated blood to the heart muscle.

Fibrillation is when a heart beats rapidly but ineffectively, creating an inadequate supply of oxygen to the body.

The blood vessels may even become completely blocked, with disastrous effects. A **heart attack** occurs when the narrowing or blockage of the blood vessels that supply the heart interrupts the supply of oxygenated blood to the heart muscle. Starved of oxygen, some of this muscle dies. This may cause the heart to go into **fibrillation** so it beats rapidly but ineffectively, creating an inadequate supply of oxygen to the body. If the brain is starved of oxygen for long enough, death follows. Most heart attacks occur when the plaque lining a cardiac blood vessel cracks open. This causes a blood clot to form and stick to the plaque, resulting in a bulge that blocks the artery. The sticky blood cells of a smoker will tend to clump together, forming clots that can also block the blood vessels even if plaque layers are still quite narrow.

As well as the possibility of damage to the heart by blockages, the heart muscles of a smoker must work much harder because the blood must be pushed through much narrower pathways. This causes the heart itself to become enlarged and suffer greater stress, resulting in increased blood pressure with all its attendant health risks.

Eye on the media

Smoking's ugly face

Young women are being urged to face the facts — smoking makes them ugly.

In a bid to stop young women smoking, the Queensland Cancer Fund will launch a shock campaign focusing on the negative effects smoking has on their looks.

They say most young women are unconcerned about the long-term health risks associated with smoking, so the campaign will focus on immediate effects such as wrinkles, bad breath and yellow teeth.

'It's probably better to focus on the things that have an immediate impact on younger people because they live for the now, not for the future,' the cancer fund's Susan Greenbank said.

'Part of it is that we think we're pretty invincible. We keep thinking that we'll quit and we won't be an older smoker so we won't face the health problems associated with smoking.'

The campaign is targeted at women aged 18 to 24 and will be released in the next six to 12 months.

In 2004, 13.3 per cent of 14- to 19-year-old females and 27.9 per cent of 20- to 29-year-old women smoked daily.

Using television, radio and billboard advertisements, the $600 000 Queensland Health-funded program will follow a similar campaign recently launched in the United Kingdom.

The British version includes ads featuring women with damaged and yellow teeth and raw gums, and a poster of a young woman with a 'cat's bum mouth'. Other negatives include less elastic skin, degraded collagen levels, a sallow complexion, and saggier and aged-looking skin.

The campaign may also help counteract the effect smoking in movies has on the young. In a survey of 1500 people aged under 25, the Cancer Council of

New South Wales found almost half (49 per cent) of secondary school students believed smoking by celebrities encouraged young people to start smoking.

Source: Anooska Tucker-Evans, *The Sunday Mail*, 10 July 2005, p. 33.

Clarification questions

1. What effects does the article suggest that smoking has on physical appearance?
2. Why is it felt that this campaign will be more successful than those that emphasise the health risks of smoking?
3. Why do you think that the percentage of female smokers increases from the 14–19 year age bracket to the 20–29 year age bracket?
4. According to the article, how are young people affected by watching actors smoking in movies?

Alcohol

The blood alcohol concentration (BAC) is a measure of the amount of alcohol in grams that is present in a 100 mL sample of blood.

Alcohol abuse is an increasing problem in our society. It is estimated that dealing with the effects of this abuse costs approximately $7.5 billion a year. While many tend to think of alcoholics as old men, the biggest consumers of alcohol are people under 25 years of age who binge drink. Alcohol affects people differently according to their rate of consumption, gender, body weight, medication, tiredness, metabolism and state of health. While medical authorities recommend that, for health, men should consume no more than 210 grams of alcohol in a week and women no more than 140 grams, there is *no* level of alcohol that is safe for every person at every time. Table 8.2 shows the usual effects on the human body according to the amount of alcohol (in grams) present in each 100 millilitres of blood. This is referred to as the **blood alcohol concentration (BAC)**.

Table 8.2: Stages of alcohol intoxication

Blood alcohol concentration (BAC) (g/100 mL)	Symptoms
0.01–0.05	• Behaviour may be normal according to tolerance • Legally able to drive if on an open license
0.03–0.12	• Increased sociability and self-confidence • Reduced inhibitions and concentration • Some motor-control impairment
0.09–0.25	• Impaired memory and perception • Comprehension problems • Increased reaction time • Impaired balance and coordination • Blurred vision • Drowsiness
0.18–0.30	• Confusion • Disorientation • Dizziness • Staggering • Slurred speech • Increased pain threshold • Apathy • Lethargy

0.25–0.40	• Stupor; inability to stand or walk; vomiting; incontinence
0.35–0.50	• Unconsciousness • Depressed body temperature and heart rate • Incontinence • Possible death
0.45+	• Death

A standard drink contains 10 g of pure alcohol in the form of ethanol.

Alcohol consumption is measured in terms of the **standard drink**, which contains 10 grams of pure alcohol in the form of ethanol. Different types of drinks contain varying amounts of alcohol and so are rated in terms of their equivalence to a standard drink. Figure 8.17 shows the number of standard drinks in a number of beverages.

Figure 8.17: How many standard drinks?

Table 8.2 lists the dangers of consuming too much alcohol in a single sitting, the worst being death. However, prolonged alcohol abuse can also lead to serious disorders such as liver disease and diabetes.

Investigation

How alcoholic is this?

Aim: To distil the alcohol from an alcoholic beverage and determine its alcohol density

Materials: Distillation apparatus; 100 mL of wine; 150 mL beaker (to catch the distillate); 20 mL measuring cylinder; Bunsen burner; boiling chips

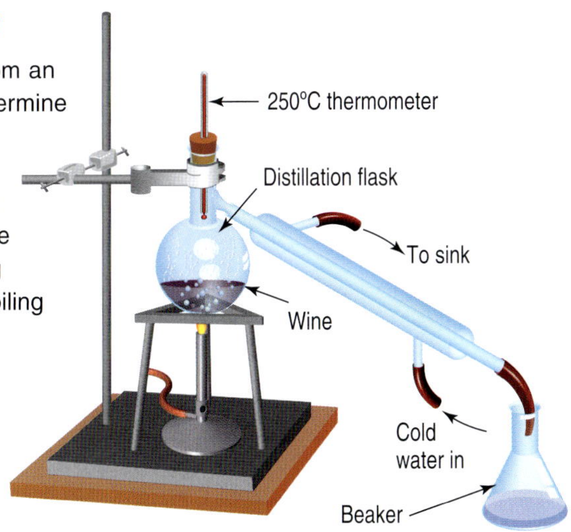

Figure 8.18: Set-up of distillation apparatus

Method:

1. Assemble the distilling apparatus as shown in figure 8.18.
2. Remove the thermometer from the distillation flask and use the funnel to pour 100 mL of wine into it. Add a boiling chip and reseal the flask with the thermometer in place.
3. Turn on the water so that it flows through the cooling jacket of the condenser tube. Have your teacher check your set-up.
4. Light the Bunsen burner under the distillation flask and adjust the burner until it has a blue flame. Make sure that the flame does not form a hot spot at the bottom of the flask.
5. After boiling begins, reduce the heat so the wine boils at a moderate rate.
6. As the temperature of the wine approaches 80°C, watch the condenser closely. At a temperature of between 80°C and 85°C, a clear liquid will start to come out of the condenser and enter the catchment beaker. The temperature should remain fairly constant as the distillate comes off. This distillate is the alcohol from the wine.
7. When no more liquid comes into the catchment beaker and the temperature begins to rise again, remove the beaker and turn the Bunsen burner off. (Leave the distillation apparatus in place until it has cooled before disassembling.)
8. Measure the volume of alcohol obtained by distillation. This will be equal to the alcohol content of the wine as a percentage (as we used 100 mL of wine).

Discussion questions

1. Check the label of the bottle from which your wine sample came. What is its alcohol content? How does your experimental value for the alcohol percentage compare?
2. Give reasons why they may differ.
3. How did the distilling apparatus separate the alcohol out?
4. Try the experiment again on a different type of alcoholic beverage and compare your results with those you obtained for the wine.

Hangover/dehydration

While some people rarely have hangovers, most people who have drunk excessive amounts of alcohol will have a hangover the next day. The symptoms of a hangover include headache, nausea, diarrhoea and depressed mood. The headache is due to the dehydration of the body and the brain membranes. This occurs because the liver, which breaks down alcohol, needs water to flush away the toxins. Because alcohol is a diuretic, it causes increased urination that drains the body of fluid. To get the water that it needs to do its job, the liver diverts fluid from the other organs including the brain.

Alcohol's diuretic effect causes important salts to be flushed away as well. Due to the disturbed balance of water and salts in the system, diarrhoea may result. Much of the nausea arises from the toxic effects of the additives (*congeners*) that give a drink its taste, colour and smell. The only cure for a hangover is time — even better, not drinking too much in the first place!

Liver disease

When the body metabolises alcohol, 90 per cent of the alcohol is broken down by the liver. During this process, the liver produces large amounts of toxins including acetaldehyde. Under normal circumstances, the liver is able to flush these toxins away. However, after prolonged use of alcohol, the liver falls behind and starts to accumulate toxins, becoming damaged as a result. The liver is responsible for removing many other toxins from blood apart from alcohol, it makes critical chemicals and stores important substances such as glycogen. Therefore impaired liver function causes the deterioration of the body until eventually the liver fails and death follows.

Diabetes

When alcohol enters the body, it is converted into glucose that can be used by the body for energy. The body responds to glucose by releasing insulin from the pancreas. Insulin breaks down excess glucose in the bloodstream to form glycogen, which is stored in the liver. As alcohol intake increases, more insulin is produced. Over many years of alcohol intake, the pancreas loses its ability to produce enough insulin to break down the resulting glucose, causing increased levels of glucose in the blood. This is diabetes, which can lead to blindness, organ failure, loss of extremities, coma and even death if not managed properly.

Cocaine

One of the chief risks in using cocaine is that, regardless of what form the cocaine takes or how frequently it is used, it can cause heart attacks or strokes. Such 'sudden deaths' have occurred to both first-time users and regular users. Other health problems include heart arrhythmia, seizures and respiratory failure. Cocaine interferes with the appetite, so many chronic users can become malnourished.

There are other problems that can arise depending on the method by which the cocaine is used. Snorting cocaine on a regular basis quickly leads to nosebleeds, loss of sense of smell and a continuously runny nose. Eventually, the structures in the nose are eaten away and the nose collapses. Swallowing cocaine reduces the blood flow to the stomach and bowel. This causes cramps and nausea and, in the long term, may eventuate in ulcers or gangrene of the bowel (in which the bowel is eaten away). Injection of cocaine can cause a dangerous allergic reaction and, as with any injected drug, there is the risk of contracting HIV or hepatitis.

A particularly dangerous combination is cocaine taken with alcohol. The liver combines these two substances to form a compound called *cocaethylene*. This compound heightens the euphoric effect of the cocaine but dramatically increases the risk of sudden death.

Heroin

As heroin is generally injected into the bloodstream, early damage appears in the circulatory system in the form of collapsed veins, abscesses and the destruction of the lining of the heart and the heart valves. As the circulatory system works in combination with the respiratory system, there are often lung problems leading to pneumonia, bronchitis or respiratory failure.

If heroin is used during pregnancy, there is usually an effect on the foetus. Recorded effects have so far included spontaneous miscarriage, low birth weight, problems with physical or mental development, and the birth of babies with a heroin addiction.

One of the biggest dangers with heroin is that street heroin varies in purity and in the combination of additives that may be mixed with it to pad it out for increased dealer profit. Injection of extremely pure heroin can lead to accidental overdose. Many of the additives do not dissolve properly in the bloodstream and clog the blood vessels that supply vital organs, leading to infection or even the death of these organs.

Ecstasy (MDMA)

MDMA is notorious for interfering with the body's ability to regulate its temperature. As the body temperature increases rapidly, the heart, liver or kidneys can shut down and the muscles start to break down. Some medical practitioners have compared the condition of the muscles after a high dose of MDMA to melting.

MDMA can interfere with its own metabolism. As a result, some users have taken more of it in the belief that the dose that they had taken up to then 'wasn't working'. This can cause the body systems to rapidly shut down as a result of overdose. Like the other members of the stimulant family, MDMA can cause heart problems and seizures.

Due to variations in the way they are made, ecstasy tablets may contain other additives such as ephedrine, ketamine, caffeine, cocaine or methamphetamine, all of which can cause a heart attack when combined with MDMA. As ecstasy is seen as a 'party drug', it is often consumed with alcohol, creating a potentially lethal combination.

LSD

As LSD is not addictive as such, it tends not to be used habitually and many users stop its use voluntarily. While it does not damage the health of the body in the same way as the long-term use of addictive drugs, its effects on the psychological state of the user are potentially devastating. Many users experience flashbacks to earlier trips that can occur up to two years after the last use. These occur suddenly and can debilitate the person while they occur — especially if they are driving a motor vehicle at the time. Due to LSD's effect on the brain chemistry, it may cause long-term psychoses in some users and trigger an existing potential for schizophrenia or depression in vulnerable people. Because of its hallucinogenic nature, accidental deaths occur when users believe that they can fly or walk through solid objects, or that they are invulnerable.

Cannabis

Because smoking is the most common method of cannabis use, many of the risks associated with using it are similar to the risks associated with smoking tobacco. These risks include heart disease, lung damage, cancer and stroke. A recent US study found that marijuana smokers have more health problems than cigarette smokers. Perhaps this is so because marijuana users usually inhale more deeply and hold their breath longer than tobacco smokers do, thereby increasing the lungs' exposure to carcinogenic smoke.

It has also been theorised that marijuana smokers experience more health problems because the active chemical in it (THC) suppresses the immune system, allowing the body to be more easily invaded by bacteria and viruses and less able to protect itself from cancer. There has been some evidence that marijuana use interferes with male fertility, causing malformation of sperm.

Legal issues

Consider the following hypothetical situation: Someone you know well has late-stage cancer that gives them a great deal of pain. The painkillers prescribed for them make them sleep all the time, and this is not how they want to spend their last weeks. You have read on the internet that marijuana is excellent for controlling pain in cancer sufferers without the heavy sedation effects of normal medications. While marijuana is an illegal substance, you know someone who can get some for you so you can give it to the patient.

In groups, discuss what you would do if you were in this position, keeping all legal and moral dimensions in mind. Compare your decisions with another group and discuss reasons for differences in opinion.

The legality of a particular substance's use and the penalties involved for its possession depend on the substance's legal classification rather than its medical classification. After all, banning all hallucinogens would also necessitate the banning of things such as chocolate, which contains anandamide (a chemical similar in effect to THC).

The idea of legality varies from country to country. Medications containing pseudoephedrine or codeine are freely available in Australia but are controlled substances in some other countries. In Amsterdam, hash cookies can be bought in cafes!

Having seen that legal drugs can have effects and dangers that may not be markedly different from those of illegal drugs, we must now consider the historical reasons governing the legality and illegality of drugs in this country. In 19th century Australia, there were no illegal substances as such. Many therapeutic substances contained opiates and could readily be bought from pharmacists, doctors and grocers. The original Bonnington's Irish Moss, used in the treatment of throat infections and cold symptoms, contained opium and morphine. Laudanum, a derivative of opium, was often used by women who suffered from depression. Coca-cola, when it was first released as a tonic, contained small amounts of cocaine that gave the drinker a boost. (The boost was later provided by caffeine instead of cocaine.) The users of these products were generally middle-class, middle-aged women or health professionals. What changed?

Early regulation of drugs occurred by way of poisons laws that imposed requirements on the sale and labelling of some drugs. These laws were a response to the misuse of drugs such as arsenic in poisonings, suicides and homicides, and they did not affect drug availability or prevent their legal sale. Over the last 150 years or so, some drugs have gone from being 'available poisons' to substances whose use or unauthorised sale have become matters for criminal prosecution. A number of different historical factors have been suggested to account for the making of some substances illegal.

Anti-Chinese prejudice

During the gold-rush years, some 50 000 Chinese immigrants came to Australia. They were fleeing the environmental, economic and political difficulties in China and were attracted by the lure of gold. Opium dens started to arise on the goldfields, places habituated by both Chinese and European clients. From the goldfields, opium dens found their way into the cities, as did many of the Chinese immigrants. The Chinese were generally peaceful and industrious but hostility was felt against their communities, particularly because of their different customs and traditions. Anti-Chinese leagues were established and public disapproval of these immigrants led to the cry to ban opium dens.

In 1895, South Australia (followed quickly by New South Wales, Queensland and Victoria) passed laws that banned opium smoking and the sale, trafficking, manufacture and possession of opium suitable for smoking. Although Caucasian Australians were substantial consumers of cocaine, opiates and other drugs, these laws specifically targeted smokeable opium — a substance used primarily in the Chinese community. In 1905, the newly formed Commonwealth government banned the importation of opium suitable for smoking. By 1910, it had been made an offence to be in possession of a prohibited import such as opium without reasonable excuse (such as medicinal purposes).

Increasing domination of the health services by the medical profession

As the social status of doctors rose and the profession started to consolidate as a group in the early part of the 20th century, they gained more power. Certain drugs with undeniable medicinal benefits (such as cocaine and heroin) were targeted as being misused by the general public.

In order to prevent the misuse but still allow use of the drug when needed in medicine, it was legislated that such drugs could be provided only on the approval of qualified doctors. This approval took the form of a prescription that was filled by the doctor himself or by approved pharmacists, a similar method to what we have today. All active ingredients such as morphine and cocaine were removed from commonly available products. It has also been suggested that this measure was implemented in order to make the provision of medicines the sole domain of the medical profession and, therefore, a monopoly.

International influences on Australian legislation

The Hague Convention for the Suppression of Opium and Other Drugs (1912) was an agreement between 42 countries to restrict to medical and legitimate authorities the manufacture, sale and use of opium, heroin, morphine and cocaine. This convention led Australia in 1914 to restrict the importation of opium, morphine, heroin and cocaine to medicinal purposes, and requiring all importers to obtain a licence.

The 1925 Geneva Convention on Opium and Other Drugs restricted for medical and scientific purposes the manufacture, importation, sale, distribution, exportation and use of cannabis, as well as medicinal opium, cocaine, morphine and heroin. This convention influenced the Australian government to prohibit the importation of cannabis in 1926, with the first Queensland penal controls on unauthorised cannabis use introduced in 1937.

Growth of recreational drug use in the 1960s and 1970s

Until the 1960s, drug dependence was not viewed as a major social problem in Australia. However, during the 1960s and 1970s, the 'Flower Power' generation was encouraged to 'tune in, turn on and drop out'. This philosophy encouraged the use of hallucinogens such as marijuana and LSD as a way of altering consciousness, in the belief that this would allow people to 'connect with each other' on a peaceful level. This rebellion against their society — which the young generation saw as stale, war-torn and anti-individualist — led to increased pressure being placed on governments by the more conservative sections of society to control what they saw as an anarchic movement. It was believed that the control of hallucinogens would help in this, so they were made illegal.

The economic benefits received by the government from the sales of tobacco and alcohol

The true extent of tobacco's health dangers was only understood well after its use was firmly entrenched in the community. The widespread use of cigarettes, and the fact that they are made by large and powerful companies that have gone to great lengths to protect their business interests, have made it very difficult for health measures to be introduced to improve community health.

Below is a list of street names for a variety of drugs. Using the internet, determine the more common name for the drug and its classification. Enter your findings in a table such as the one below:

Street name	Common name	Classification
Bush	Cannabis	Hallucinogen

Street names:

- Acid
- Blow
- Charlie
- Coffin nail
- Dexies
- Ganj
- Gary
- Golden brown
- Hooch
- Horse
- Trips
- Tweed
- Whippets
- Whiz

End of chapter questions

1. Describe the effect of marijuana on the brain.
2. What is the function of the nucleus accumbens?
3. Name four neurotransmitters and describe their functions in the brain.
4. In preparing for an all-night study session, Foolish Fred took two stay-awake tablets and drank 10 cups of percolated coffee. Estimate the amount of caffeine that he has consumed and describe the effects they will have on him.
5. How do stimulants differ from narcotics?
6. What effect does adrenalin (epinephrine) have on the body? Explain in terms of this how sudden shocks may lead to heart attack in some people.
7. How does crack cocaine differ from normal cocaine?
8. Spiders were fed a number of different substances: benzedrine (an amphetamine), caffeine, chloral hydrate (a sedative) and marijuana. The webs produced by the spiders under the influence of these substances are shown in figure 8.18. Looking at these webs, do you think that these substances have similar effects on spiders as they do on humans? Explain your answers.

Figure 8.19: Web (a) was produced by the control spider, which was not given any substance. The other webs were produced by spiders under the influence of (b) benzedrine, (c) caffeine, (d) chloral hydrate and (e) marijuana.

9. Consider the following facts:
   - After World War II, half of Australia's population were smokers.
   - In 2001, one in four males and one in five females aged 14 years and over were regular smokers.
   - Smoking-related diseases such as lung, mouth and throat cancer, heart disease and lung disease are the greatest killers of people over 50.

   What reasons can you think of that would explain the decreasing smoking rate but increasing death rate from smoking-related illnesses?
10. Explain why the average woman needs only two standard drinks to be at the legal driving BAC of 0.05 while the average man needs three standard drinks.
11. Consider the following facts:
   - The legal blood alcohol limit for drivers in Australia is 0.05 per cent.
   - The liver can metabolise (break down) alcohol in such a way that it lowers the blood alcohol percentage by 0.015 per cent every hour.
   - The consumption of one standard drink raises the blood alcohol level by 0.016 per cent.

   (a) Write an equation for calculating the approximate blood alcohol percentage (BAP) of a person based upon the number of standard drinks consumed (N) and the number of hours (h).
   (b) Bill has consumed eight standard drinks in two hours at the family barbecue. How many hours must he then go without further alcohol consumption before he is legally able to drive home?
12. Look at the poster in figure 8.20, produced as part of a drug-awareness campaign.
   (a) Why is ice being packed around the patient? What aspect of ecstasy use is being treated here?
   (b) Are all of the claims about the effects of taking ecstasy true? Have any of the effects been exaggerated?
   (c) What is the purpose of the light that is being shone in the patient's eye? What information does it give to the medic?
   (d) Who is the poster's intended audience? Explain what aspects of the poster tell you this.
   (e) What places would you put this poster to ensure that it reached its target audience? Justify your reasoning.

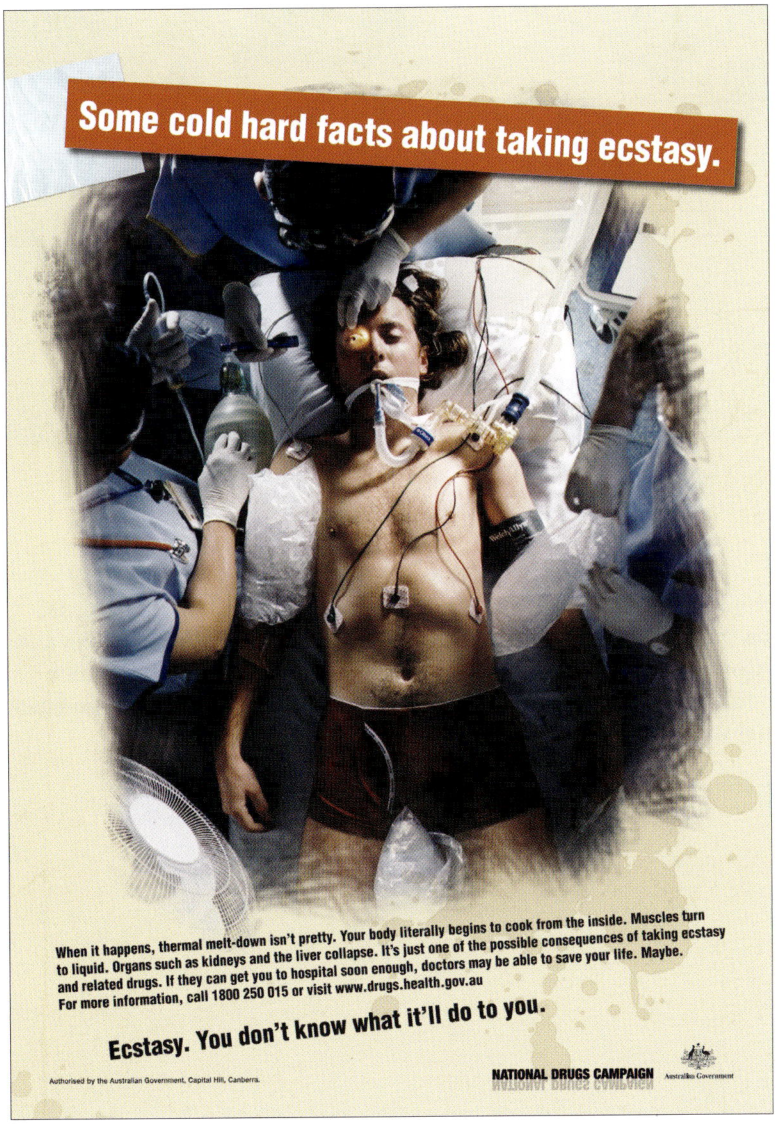

Figure 8.20: 'Some cold hard facts about taking ecstasy'

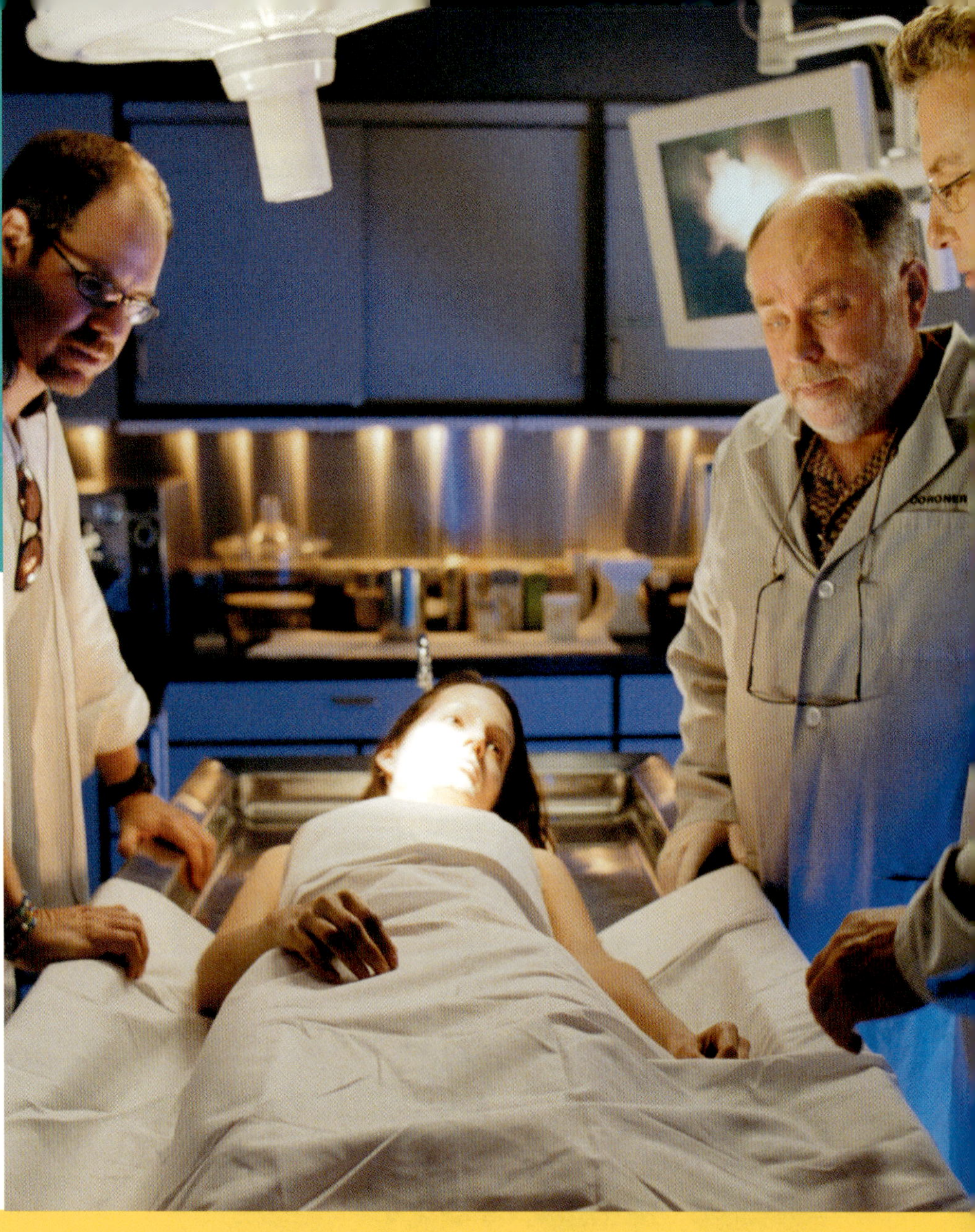

9

Crime squad

The enormous popularity of crime shows such as *CSI* and *Silent Witness*, as well as the crime fiction by authors like Jeffrey Deaver, Kathy Reichs and Patricia Cornwell, are testament to the fascination that the general public has with forensic science. However, just how well does forensic fiction stack up against the realities of the crime laboratory?

Forensic science 101

Fictitious sleuths of the past such as Sam Spade, Miss Marple and Sherlock Holmes relied more upon their keen observation of human nature, rather than forensic tools, to solve crimes. Today, it seems that as soon as a forensic technique has been devised, it is included in television dramas or in the works of crime writers. While some would argue that this inclusion makes it easier for the average person to understand forensics, there are just as many who insist that the representations of forensic techniques in fiction give people a misleading impression. Before we can really judge either way, we need to look at the basics of forensics — Forensics 101!

Eye on the media

Crime scene investigation

Television's troika of CSI shows — *CSI: Crime Scene Investigation*, *CSI: Miami* and *CSI: NY* — arguably presents popular culture's most positive view of scientists since the Professor was engaged in his unfunded better-living-through-coconut-chemistry project on *Gilligan's Island*. In February, at the annual meeting of the American Association for the Advancement of Science in Washington, DC, a group of real forensic scientists put the CSIs under the microscope.

The fictional series have inadvertently put pressure on real-life prosecutors. 'The CSI effect' is a term that came into use around 2003, when the show really started to become popular,' says trace evidence analyst Max Houck, director of West Virginia University's Forensic Science Initiative. 'It represents the impossibly high expectations jurors may have for physical evidence.' Prosecutors worry that without having the ironclad physical evidence jurors see on TV, the reasonable-doubt line may be shifting.

Houck pointed to a case in Los Angeles last year featuring a bloody coat. 'Jurors were alarmed,' Houck says, 'because no DNA testing had been done on the coat. Well, the wearer of the coat admitted to being at the murder scene trying to help the victim, so the lab had said there was no reason to test it — he said he was there.' According to Houck, the judge made a statement along the lines that 'TV has taught jurors about DNA tests but not about when to use them.'

Indeed, many people still don't know the ABCs of DNA. A lab may request a sample of a missing person's clothing in order to compare DNA on that clothing to unidentified remains. Dress shirts are particularly good at grabbing skin cells at the tight collar. 'We asked for the family to send in dress shirts,' recalls Demris Lee of the Armed Forces DNA Identification Laboratory about one case. 'And the family sent in his new shirts that were still in packaging. They couldn't believe we wanted his dirty clothes.'

Criminals may be feeling CSI's heat — and taking notes. 'What I've heard is that it's closely watched in prisons,' remarks Richard Ernest, a forensic firearms expert in Fort Worth, Texas. 'And prisons become almost like a crime school for certain individuals. They'll look at a particular segment and say, "So that's how they caught me. Well, I won't make those mistakes again."'

Instead they'll probably make new ones. 'When they try to escape detection from what they see on *CSI*, they're actually leaving more evidence,' Houck contends.

'A good example of that is instead of licking an envelope [for fear of providing DNA in their saliva] they'll use adhesive tape. Well, they'll probably leave fingerprints on the tape, and it'll pick up hairs and fibres from the surroundings. So the more effort you put into trying to evade detection, honestly, the more evidence you leave behind.'

Another CSI effect is that college kids think it's cool. In 1999 Houck's institution graduated four students with a concentration in forensics. 'We're now the largest major on campus,' he declares. 'If you add all four years together, we have over 400 students.' Perhaps their most important lesson is that real life doesn't look like a TV show. Houck tells his students that 'it's less about wearing leather pants and driving Hummers than it is about wearing Tyvek jumpsuits and crawling under people's front porches looking for body parts. Honest. I've never worn leather pants in my life.'

Houck also has a tough time watching his TV counterparts use analytical tools that don't quite really exist. 'We joke that we need to get one of those — that's a damn fine instrument,' he says.

(The amazing databases employed on some episodes prompted a friend of mine to ask, 'Why don't they just ask the computer who did it?') Another show convention that annoys Houck is investigators wandering around dark indoor crime scenes. 'They always use flashlights,' Houck notes. 'I don't know why. I usually just turn the lights on.'

Source: Steve Mirsky, *Scientific American*, May 2005, p. 80.

Clarification questions

1. What do you think that the word 'troika' means in the context of the first sentence of the article?
2. Describe the positive and the negative effects that CSI has had on the public perception of forensic science according to this article.
3. Explain what 'The CSI effect' is, as described by Max Houck.
4. What forensic science organisations are mentioned in this article?

What is forensic science?

Forensic means 'of the law court'.

Many people hear the word '**forensic**' and immediately start thinking of bodies, blood and autopsies. In fact, the word *forensic* means 'of the law court' and has its origins in the word *forum*, which was the place where the ancient Romans held their law court proceedings. So forensic science is concerned literally with the science used in the law courts.

A forensic scientist is a specialist in a particular field whose expertise allows information about a crime to be uncovered. These fields of specialty range widely, embracing disciplines as diverse as medicine, pharmacology, physics, pathology, entomology, structural engineering, chemistry, textiles, art, paleontology, podiatry, computer science, graphic animation, dentistry, psychology, graphology and metallurgy. All of these scientific disciplines and many more may be called upon to successfully solve a crime.

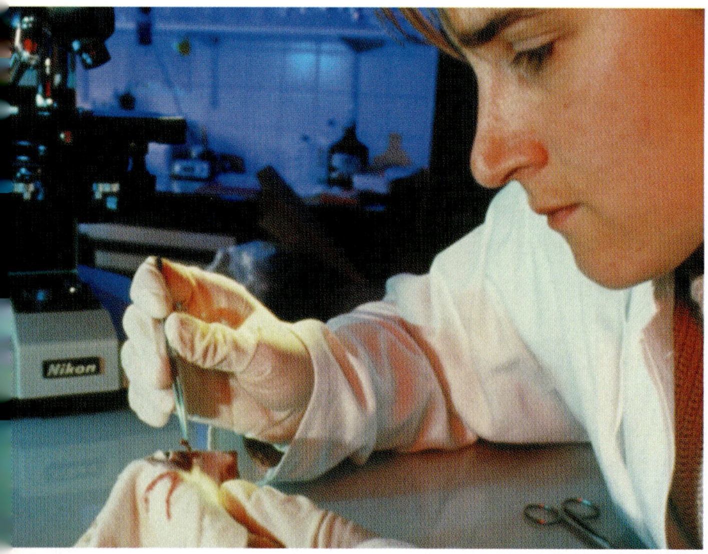

Figure 9.1:
A criminal investigation involves a wide range of scientific disciplines.

Role of the forensic scientist

The forensic scientist will have a number of special roles to play as a member of the investigating team. They will help to establish the type of crime committed, assist in the reconstruction of the crime, ensure that evidence is collected carefully, propose investigations to test the various hypotheses proposed and perform these investigations. Finally, if the case reaches the courts, the forensic scientist may be called on to act as a witness. A scientist whose usual work is not forensic science but who has a particular field of expertise or interest will often be referred to as an expert witness.

An important responsibility of the forensic scientist is to ensure that all the evidence is dealt with objectively. Since much forensic science work involves comparing one thing with another, it is crucial that all variables be controlled. For example, the scientist may want to know if the blood on the knife is the same as that on the body. Do the blue fibres on the rape victim match those taken from the jacket of the suspect? Does the bullet found at the scene of the crime match one fired from the sawn-off shotgun found in the boot of the suspect's car?

Locard Exchange Principle

The **Locard Exchange Principle** indicates that, when a crime is committed, the perpetrator will leave traces behind and will carry traces of the crime scene or victim away with them.

The basic principle of all forensic science is known as the **Locard Exchange Principle**.

Edmond Locard, a professor of criminal law in the late 19th century, postulated that when a crime is committed, the criminals always leave something at the scene of the crime that was not there before, and carry away something when they leave that was not on them when they arrived:

> Wherever he steps, whatever he touches, whatever he leaves, even unconsciously, will serve as a silent witness against him. Not only his fingerprints or his footprints, but his hair, the fibres from his clothes, the glass he breaks, the tool mark he leaves, the paint he scratches, the blood or semen he deposits or collects. All of these and more, bear mute witness against him. This is evidence that does not forget. It is not absent because human witnesses are. It is factual evidence. Physical evidence cannot be wrong, it cannot perjure itself, it cannot be wholly absent. Only human failure to find it, study and understand it, can diminish its value.
>
> Source: Edmond Locard (1877–1966).

Figure 9.2:
Edmond Locard, the father of modern forensic-science technique

Evidence describes material that may support a theory as to how, where, when, why and by whom a crime has been committed.

In other words, when a crime is committed, 'clues' or **evidence** will be left behind. Writers of crime fiction and scripts for crime shows delight in leading (and misleading) their readers and viewers through a maze of clues before the story detective explains how it all fits together at the end. In real life, the solution is rarely so neatly tied up, and it is the work of dozens of people that leads to the apprehension of the criminal, not just one enlightened supercop.

Some of the more common types of evidence found in forensic investigations include:

- footprints, shoe prints and tyre marks
- fingerprints
- dents, scratches or marks resulting from the action of tools or teeth
- soil, pollen and organic matter on clothes and shoes
- paint, glass, hair (see figure 9.3), fibre and fabric fragments
- oils, dyes and inks
- stains from body fluids (blood, saliva, semen)
- bullets and other ballistic evidence.

In the pages that follow we will be looking closely at many of these items and how they are helpful in identifying the criminal. First, you need to appreciate what happens at the scene of a crime and the role of the forensic scientist in these events.

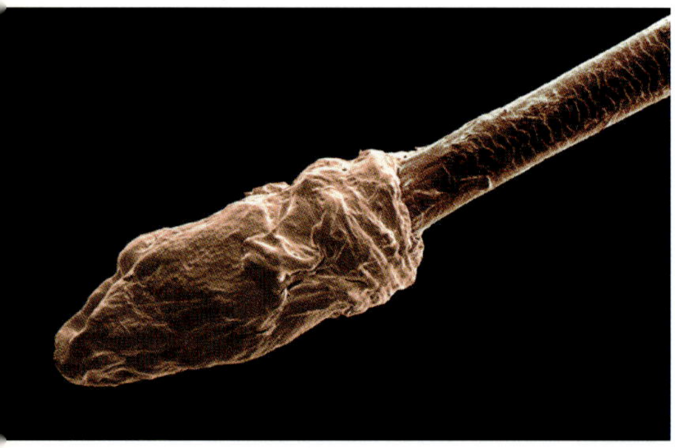

Figure 9.3: A human hair that has been pulled out

Scene of the crime

A crime has been committed. Perhaps an early morning jogger has found a body, or a couple has returned from a party to find their house burgled and their valuables gone. The police are called. What happens next?

Participants in a criminal investigation

Whenever a crime is reported, local police are usually the first to arrive at the scene of the crime. If the case is a serious one, requiring additional help, police with special skills will be called in. The scene-of-crime team will usually include some or all of the following:

- the officer in charge
- police officers to assist with inquiries
- a forensic pathologist or doctor in the case of assault or homicides
- a photographer and/or fingerprint expert
- consultant scientists where specialised knowledge is required.

These people are often referred to as **scene-of-crime officers (SOCOs)**.

Scene-of-crime officers (SOCOs) are specialist investigators who attend the crime scene.

The scientific method

During your other studies, you have probably already looked at the steps involved in the scientific method. Not surprisingly, criminal investigators follow a very similar pattern.

1. Making observations

On arriving at the scene of a crime, the first priority of those involved is to observe accurately all details that might be useful. This will involve recording data, taking statements from witnesses, and carefully collecting items that will be examined more closely later and may even be used as exhibits when the case is prosecuted. Great care must be taken, especially at the initial stages of the investigation, to ensure that evidence is not moved, contaminated or unwittingly destroyed.

The crime scene will usually be cordoned off to prevent accidental interference by the public. Even a pencil or glass at the scene of the crime may have important fingerprints that could be smudged by an unthinking person. It is particularly important that a body not be moved before the area has been marked and photographs have been taken from a variety of different angles. This stage of the investigation should not be rushed and every effort should be made to label exhibits clearly and to note anything that could be important.

2. Forming a hypothesis

A **hypothesis** is a possible explanation for a series of observations.

In criminal investigations, the process of forming a **hypothesis** occurs at two levels. First, the police try to reconstruct the events and circumstances of the crime. Second, on the basis of evidence collected, a suspect or suspects may be identified.

3. Testing the hypothesis

The reconstruction process will often lead to the identification of possible suspects whose alibis can be tested against the evidence collected.

4. Recording the results

Accurate recording of evidence is crucial. Extensive notes are taken along with statements from witnesses. Photographs and sketches are kept as well.

5. Altering the hypothesis if necessary

The initial suspect, when checked, may have a perfectly legitimate alibi. If so, the investigators must then reconsider their hypothesis and alter it in accordance with the evidence or lack thereof.

6. Reporting the results

Ultimately in a criminal investigation, the evidence will be reported in front of a jury who will judge the case on its merits.

Identifying the deceased

In homicide cases, the positive identification of the body is of primary concern. Knowing the person's identity can lead to many fruitful avenues of investigation. Apart from this, the body must be identified so that a death certificate can be completed. Identification can be done in a number of different ways.

By family or friend

This is the most common means of identification of an unknown body. However, it is not always possible if significant disfigurement of the body has occurred as a result of burning or bashing, or if the body is in an advanced state of decomposition.

By personal belongings

Identification may be difficult if there are no personal belongings, or such items may be misleading if they belong to someone else. Examples of these items include watches, credit cards, rings, jewellery, passports, drivers licences and so on.

By dental records and fingerprints

Fingerprints and dental records can be used to a limited extent. For example, the victim would need to have a criminal record for fingerprints to be helpful. Dental records are really useful only when the police have some idea as to who the person could be. It is impractical to ask all the dentists in a geographical area to check all their records for every case. What can be done, though, is to check the dental records of missing persons for a possible match. Dental records are particularly useful in cases of fire where visual identification is impossible.

Eye on the media

3D animation gives evidence more bite

Animated 3D reconstructions of a crime suspect's teeth biting a victim could make forensic bite-mark matching more accurate. Error rates are still too high for definite identification, but the technique could help jurors make their minds up.

Forensic bite-mark matching has come in for severe criticism. A handful of recent murder convictions in the US that relied on bite-mark evidence were overturned after DNA evidence came to light. And previous research has shown that even under ideal conditions matching bite marks is fallible (*New Scientist*, 13 March 2004, p. 6).

One major problem is the way the evidence is collected. Odontologists usually have to try to match photographs of bite marks to 'overlays' of the suspect's teeth — 2D images of the pattern that teeth make when they bite onto a flat surface. But bite marks that curve around arms and fingers look different.

'More than half of the photos we receive get sent back because they are too distorted,' says David Sweet, director of the Bureau of Legal Dentistry at the University of British Columbia in Vancouver, Canada. 'It's not a reflection on the police photographer, it's a reflection of the difficulty of the job.' But, says Sweet, bite marks are present in eight out of 10 sexual assault and homicide cases and can be valuable evidence.

The solution might be to use 3D imaging to match teeth to a bite. That's how John Clement and Sherie Blackwell at the University of Melbourne, Australia, set out to overcome the distortion problems.

Blackwell asked 42 dental students to make a mould of their own teeth. The students then chomped onto wax blocks to mimic a bite mark on skin. Both the teeth moulds and sample bites were digitised with 3D laser scanning. Blackwell then used a statistical technique to estimate the probability of a match.

Worryingly, the system proved to have a 15 per cent chance of matching a bite to the wrong person. Only four bites matched their corresponding set of teeth and no other. Twenty-seven bites matched their owners' teeth and at least one other person's, while seven matched the wrong teeth altogether, and four matched none of the teeth.

However, the system should do better in real cases, says Blackwell, who presented the results at the American Academy of Forensic Sciences meeting in New Orleans, Louisiana, last month. The students' teeth were all straight and unbroken, giving few of the distinguishing characteristics that forensic odontologists look for. 'Often the people involved in criminal activity aren't that fortunate,' says Sweet. 'They produce bite marks that are easier to identify than 'piano-key' teeth.'

But even without definite matches, the technology could be useful in courtrooms. Clement suggests that experts could show the animations to jurors, letting them see with their own eyes the likelihood of a bite-mark match. At present, jurors have to rely on expert testimony to assess the reliability of bite-mark evidence.

Sweet agrees. 'Bite-mark evidence is complex and difficult to explain. Anything you can do to illustrate the scenario is advantageous,' he says.

Source: *New Scientist*, 12 March 2005, p. 8.

Clarification questions

1. Why is there a problem with the currently used methods of using bite marks on victims to identify culprits?
2. Describe the process in which a 3D model of the teeth is produced.
3. Why would this technology be useful in court?
4. What do you think the term 'piano-key' teeth means in this article's context?
5. What is the advantage of using a 3D model rather than a 2D model of a bite?

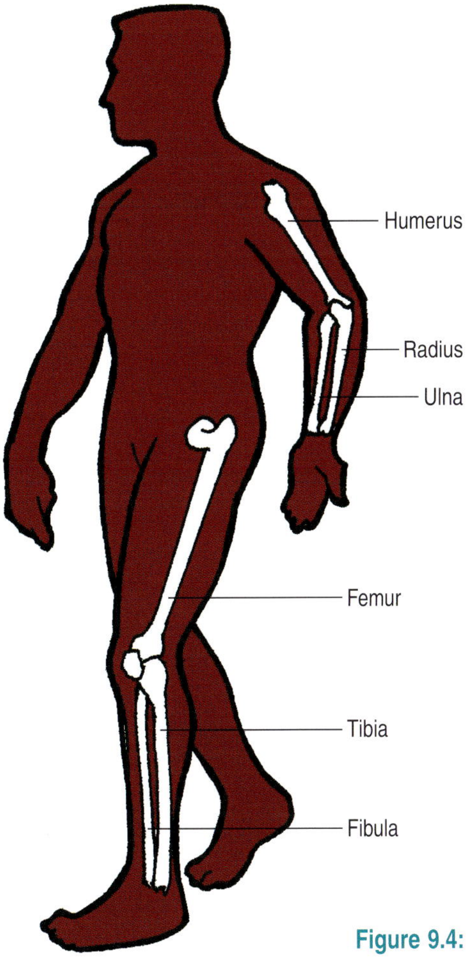

Figure 9.4: The long bones of the body are used to estimate the height of victims.

By unusual body characteristics

There may be scars or deformities on the body that can be matched with hospital records. Tattoos are often helpful in making a positive identification. X-rays of bones may reveal previous bone fractures or metal plates that can be similarly matched, and there have even been cases where the serial number on a heart pacemaker has been used to trace the identity of the victim.

By skeletal clues

The most difficult cases of identification occur when only the skeleton, or even only part of the skeleton, remains. However, the skeleton provides clues to the height, sex, race and age of the person (see figure 9.4). Measurements made on the pelvis are often used with a high degree of certainty to determine gender. The following activity illustrates how limb measurements can be used to estimate body height.

Body height can be estimated using the equations shown in table 9.1, which are based on the length of major bones of the human body. These equations were developed using the average human proportions for males and females. The more long bones that are present in the skeletonised remains, the more accurate the height estimation can be. The study of skeletonised remains is part of the relatively new science of **forensic anthropology**.

Forensic anthropology is the study of skeletal remains.

Study table 9.1 and then answer the questions that follow.

Table 9.1: Body height equations	
Height (cm): Male	**Height (cm): Female**
(2.894 × length of humerus) + 70.640	(2.754 × length of humerus) + 71.476
(3.271 × length of radius) + 85.925	(3.343 × length of radius) + 81.224
(1.880 × length of femur) + 81.305	(1.945 × length of femur) + 72.845
(2.376 × length of tibia) + 78.663	(2.352 × length of tibia) + 74.775

Note: After the age of 30, the height of a person begins to decrease at a rate of approximately 0.06 cm per year. This shrinkage must be considered when the age of the victim is known.

1. A 32-centimetre radius bone from a male body was found. How tall was the victim?
2. The skeleton of a middle-aged woman was discovered by the police. A driver's licence found near the body indicates that she may have been born in 1940 and other evidence points to her having been dead for about 12 months. Given that the tibia is 33.5 centimetres long, what is the deceased's estimated height?
3. Measure your own height and calculate the length of your humerus, radius, tibia and fibula.
4. Look at a model of a human skeleton. Make the appropriate measurements and confirm the accuracy of the table. If there are any discrepancies, can you suggest reasons?

By facial reconstruction

One interesting technique, used when only skeletal remains including the skull are available, is facial reconstruction. This procedure relies on there being a mathematical relationship between the bones of the face and the depth and position of the covering tissues.

First, the skeleton is initially examined for indications of ethnic origin, gender and age. Based on this, the average skin thicknesses and depth of muscle layers are marked at key points on the skull (see figure 9.5). From these indicators, a clay likeness can be developed. Hopefully, the likeness makes identification possible. While facial reconstruction was initially used by only forensic investigators, the technique has achieved enormous popularity among historians and has even been used on television programs such as *Meet the Ancestors.*

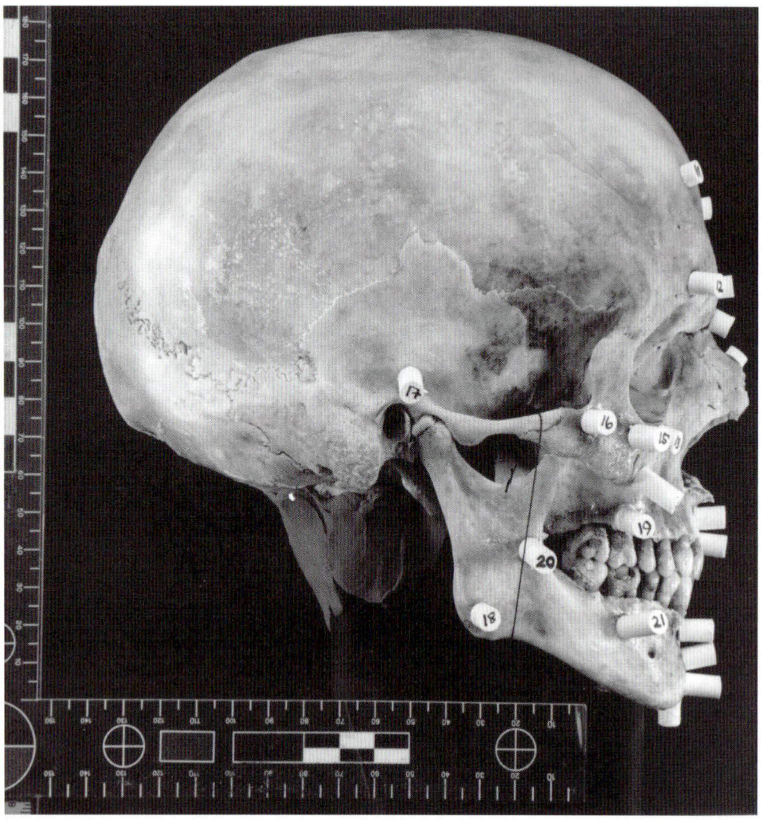

Figure 9.5:
A side view of a skull with tissue-depth indicators in place. Based on these markers, clay will be used to build a likeness of the deceased.

Estimating the time of death

It is the responsibility of the forensic pathologist to estimate the time of death. An accurate assessment is of great assistance to police in narrowing the list of suspects. However, this is not an easy task! Three main factors help the forensic pathologist to estimate the time of death: body temperature, rigor mortis and lividity.

By body temperature

Algor mortis is the process by which the human body cools after death until it reaches the same temperature as its surroundings.

The **Glaister Equation** allows an approximate time of death to be calculated based on the extent to which a body has lost heat.

At death, the heat-generating processes of the body cease and the body starts to cool until it reaches the same temperature as its surroundings (a process often referred to as **algor mortis**). The rate at which this cooling occurs depends on a range of factors. A naked body will cool faster than a clothed one; a large adult body will cool slower than that of an infant; a body in a prone, spreadeagled position will cool faster than a body slumped in a corner; and a body exposed to air currents will cool faster than one in a protected area. Forensic pathologists use general rules such as the **Glaister Equation** to estimate the time of death and then make adjustments for particular circumstances:

$$\text{Number of hours since death} \approx (37.5 - \text{current body temperature}) \div 0.83$$

Note that it is the core body temperature that is measured in the Glaister Equation rather than skin temperature. Core body temperature is taken either rectally or by inserting the thermometer into the body cavity through an incision. For example, a pathologist at a murder scene finds that the body of the victim has a core temperature of 31°C. When substituted into the Glaister equation, this allows the pathologist to determine that death had occurred approximately eight hours earlier.

The limitation of using the Glaister Equation is that it is effective for only short periods of time, as the rate at which the body cools gets slower as the ambient temperature is reached.

Another method that can be used to determine time of death from body cooling uses a graphical approach. Readings are taken at timed intervals and the results are plotted on a graph of temperature against time. The graph then can be extrapolated to give an estimate of the time of death.

By rigor mortis

Rigor mortis is the process by which a body, after death, starts to become stiff due to the accumulation of lactic acid in the larger skeletal muscles.

Dead bodies are sometimes called 'stiffs' in reference to the stiffness that occurs in body joints after death. This stiffness is caused by a condition known as **rigor mortis**. Rigor mortis occurs because the cells of the body can no longer obtain oxygen from the blood in order to produce energy (*aerobic respiration*) and so must use a process called *anaerobic respiration*, which does not need oxygen. This is the same process that the muscles use when doing very fast short-lived actions such as sprinting for the bus. One of the by-products of anaerobic respiration is *lactic acid*. In a living human, the lactic acid is broken down and removed from the body. In a deceased human, it instead builds up in muscle tissues and causes the muscle fibres to shorten and stiffen. Small muscles such as the jawbone are usually affected first, while the larger muscles in the limbs are affected last. Rigor mortis begins at the time of death but may take up to six hours to be noticeable. It is generally gone within 36 hours of death because the muscles degenerate as decomposition begins and so regain their flexibility.

By examining the extent of rigor mortis in the body, a pathologist can approximate the time of death. However, just as with cooling, the estimated time can be affected by external conditions such as the air temperature, the deceased's state of health and whether they were physically active before death.

The environmental temperature can be a crucial factor in the decomposition rate. In 1978, a case in a remote part of Australia illustrated this rather dramatically. The local telephone operator suspected foul play when she could not connect a call to a remote farmhouse after several attempts. Eventually a young child answered. After some questioning, the child explained to the operator that she didn't 'like Mummy anymore because she was turning black'. When the police arrived, they discovered two bodies in quite different states of decay. The father had been shot in the kitchen and, although dead, was not badly decomposed. The mother, however, was found shot in bed and was in a far more advanced state of decomposition because the electric blanket had been left on, thus dramatically increasing the rate of decomposition.

By lividity

Lividity is the draining of blood to the lower portions of the body due to the influence of gravity.

Lividity, sometimes called 'liver mortis' or hypostasis, is the draining of blood to the lower portions of the body due to the influence of gravity. The body develops a patchy discolouration within one to two hours of death and the process is complete within six to 12 hours.

On its own, lividity is not very helpful in establishing the time of death. It can, however, be useful in other ways. For example, areas of the body that experience continual pressure do not usually show lividity. Consequently, if areas of the body do not show lividity in expected areas, it may indicate that the body has been moved. Also, the colouring of the lividity can indicate certain types of death. Carbon monoxide poisoning, for example, results in a cherry-pink colour while cyanide poisoning gives a bright red colour.

By ocular changes

The condition of the deceased's eyes can give indications as to the time of death, with variation dependent upon whether the eyes were open or closed when death occurred. This variation is due to the drying out of the outer layer of the eye (the cornea). When the eyes are open, moisture is lost more rapidly from the eye surface, accelerating the rate of change in the eyes' appearance (see table 9.2).

Table 9.2: Ocular changes after death

	Time elapsed after death	
Eye appearance	**Eyes open**	**Eyes closed**
Cornea develops a film	Within 10 minutes	In 2–3 hours
Discolouration of the 'white' of the eye (*tachy noir*)	Up to an hour	Within 12 hours

Cornea becomes cloudy	Within 2 hours	In 12–24 hours
Cornea becomes opaque	After 2 hours	Within 3 days
Eyes bulge	Gas formation in the body	
Eyes sink in the sockets	Advanced decomposition	

As you can see from table 9.2, changes in eye appearance are very general and so cannot be used alone as precise indicators of the time of death. However, a pathologist may use ocular change observations to support timeframes suggested by other physical changes such as cooling, rigor mortis and lividity (see table 9.3).

Table 9.3: Summary of body changes within 48 hours of death

Time since death	Changes observed
1–2 hours	Early signs of lividity
2–5 hours	Clear signs of lividity throughout body
5–7 hours	Rigor mortis begins in the face
8–12 hours	Rigor mortis established throughout the body, extending to the arms and legs
12 hours	Body has cooled to about 25°C internally
20–24 hours	Body has cooled to surrounding temperature
24 hours	Rigor mortis begins to disappear from the body in roughly the same order as it appeared
36 hours	Rigor mortis completely disappeared
48 hours	Body discolouration shows decomposition beginning

Fingerprinting

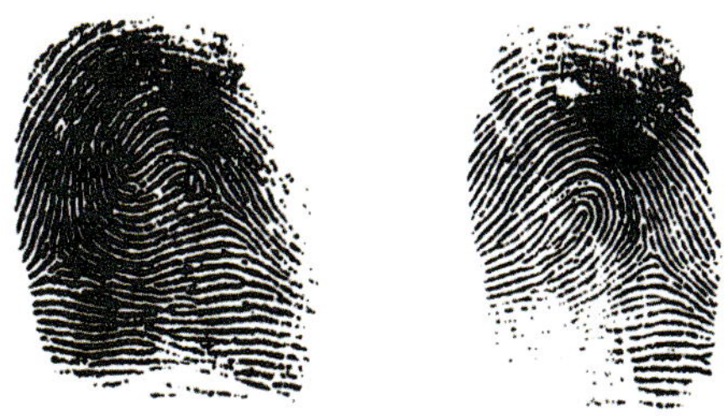

Figure 9.6: Even identical twins do not have the same fingerprints. These are the left index fingers of identical twin males.

It can be argued that fingerprinting has been the single most useful advance in forensic science. The discovery that no two people — not even identical twins (see figure 9.6) — have exactly the same fingerprints revolutionised criminal investigation, and gave police a powerful investigative tool. The presence of a person's fingerprints at the scene of a crime indicates that the person must have been at the scene at some time, though they may not have committed the crime itself. Combined with other evidence, however, fingerprints can be very convincing evidence for a jury to consider.

The surface of skin has an oiliness due to secretions from sweat glands that release salts, urea and water, and sebaceous glands that release oils. These substances remain on the surface of the skin and, when objects are touched, an impression of the fingerprint is left on the surface of the object.

Fingerprinting is so incriminating that many criminals wear gloves to avoid detection. One criminal who adopted this method was nevertheless caught. When the police found his discarded gloves, they turned them inside out and developed the fingerprints!

Latent fingerprints are fingerprints that are not readily observable on a surface.

Latent fingerprints need to be developed in some way to show up clearly so that a photographic record can be made. When blood or grease has been on the hands, the fingerprints are quite visible and can be photographed directly.

Fingerprint classification and analysis

There are four basic patterns of fingerprints: arches, loops, whorls and composites (a combination of the first three). These are shown in figure 9.7.

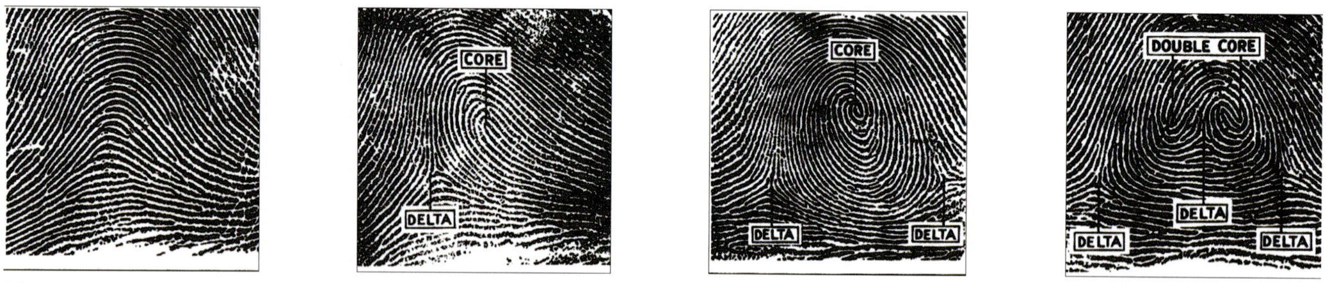

(a) arches, 5 per cent of all prints (b) loops, 65 per cent (c) whorls, 15 per cent (d) composites, 15 per cent

Figure 9.7: The four main fingerprint types

Apart from the general shape of the fingerprints, another characteristic that helps distinguish one print from another is a small triangular space called a *delta*. Deltas occur when either a single ridge separates into two ridges or parallel ridges diverge sharply. Fingerprint experts use these basic patterns to construct classification systems designed to make the matching of prints easier.

Ridge counting is the method of classifying fingerprint features by measuring the number of ridges crossed by a line drawn from a delta to a core in a loop or whorl.

Ridge tracing is the method of classifying fingerprint features such as whorls according to their alignment relative to adjacent ridges.

Loops and whorls can be further classified using systems known as **ridge counting** and **ridge tracing**. Ridge counting is the easier method. It is done by measuring the number of ridges crossed by a line drawn from a delta to a core in a loop or whorl (see figure 9.8).

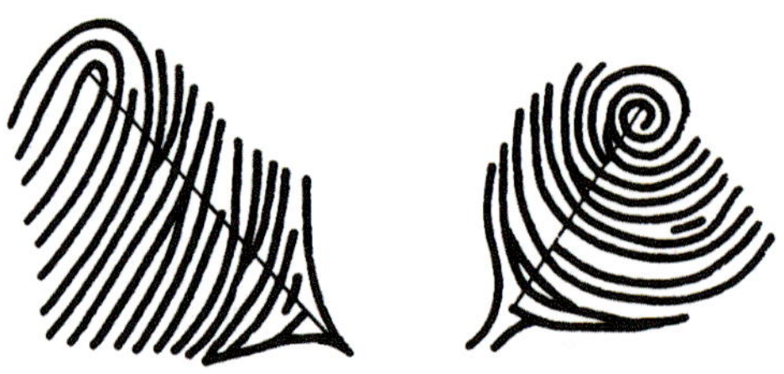

Figure 9.8: Examples of ridge counting. On the left is a loop with a ridge count of 17. On the right is a whorl with a ridge count of 13.

Ridge tracing is more complicated. Using your own fingerprints, find the delta at the bottom left-hand side of a whorl and follow a ridgeline towards the right-hand delta (see figure 9.9):

- If the line runs straight through the delta or within two ridges on either side, it is referred to as a 'middle'.
- If it passes more than two lines above the right-hand delta, it is called an 'inner'.
- If it passes more than two lines below the right-hand delta, it is called an 'outer'.

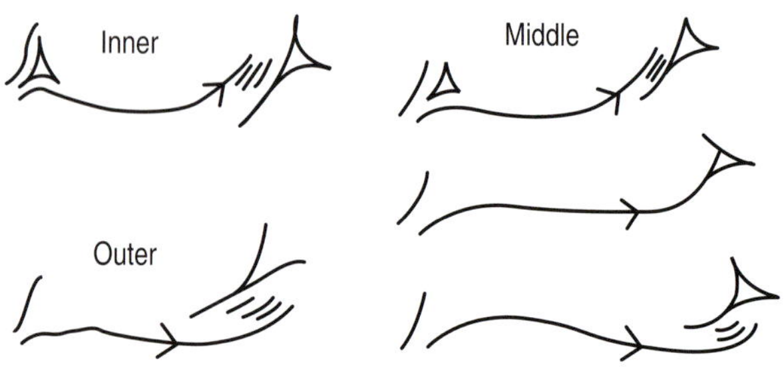

Figure 9.9: Whorls can be subclassified using ridge tracing. Categories are inner, middle and outer.

Investigation

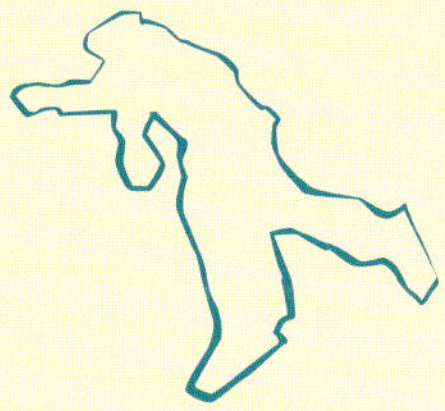

Making and classifying your own fingerprints

Aim: To prepare samples of your own fingerprints and classify them according to type

Equipment: Inkpad; copy of the chart shown in figure 9.10 (actual size); cleaning fluid; absorbent paper; hand lens

1. Right thumb	2. R. index finger	3. R. middle finger	4. R. ring finger	5. R. little finger
10. L. little finger	9. L. ring finger	8. L. middle finger	7. L. index finger	6. Left thumb

Figure 9.10:
A fingerprint chart of the type used by police

Method:

1. Examine your fingers under a hand lens. Can you detect different patterns?
2. Carefully roll your fingers, one at a time, across the inkpad. Once inked, each finger should be rolled carefully from left to right in the appropriate space on the chart to leave a clear impression of the fingerprint. You may need to experiment a little to find the best method. For example, making one print on a piece of scrap paper to remove excess ink may give a better result. Make sure that you roll each finger in the correct space.
3. Allow the ink to dry and then classify each of your fingerprints as an arch, a loop or a whorl.
4. Examine each print closely and identify the location of deltas.
5. Do some ridge counting on the loops and whorls on your fingerprints.
6. Now try some ridge tracing. Examine any whorls present and subclassify them as inner, middle or outer.

Discussion questions

1. Collect data from the whole class on numbers of arches, loops, whorls and composite fingerprints. Calculate the percentage of each type and compare these figures to those in figure 9.6. How can you account for any differences?
2. Compare your results for ridge counting with those of other class members. What is the most common ridge count for loops? Is there any difference between the numbers for males and females?
3. Compare your subclassification of whorls. What percentage of these is: inner? middle? outer? Are there any significant differences between the numbers for males and females?
4. Prepare some bar graphs summarising the data that you have gathered in this investigation.

Investigation

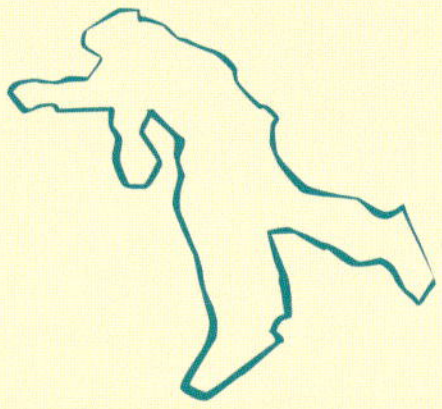

Matching fingerprints

Aim: To test your ability to match prints. Fingerprint experts are commonly required to find at least 16 points of similarity between two prints for them to be considered a match.

Equipment: Inkpad; cleaning fluid; absorbent paper; hand lens or magnifying glass; sheets of white paper

Method:

1. Ink your left-hand thumb on the inkpad and then leave an impression of your thumbprint on a small piece of paper.
2. Pass a larger piece of paper around the room for all other class members to leave their left-hand thumbprints on it, then secretly add your thumbprint to this paper too.
3. Exchange your small piece of paper from step 1 with a friend. Each of you should try to match each other's prints on the small piece of paper with the correct print on the large sheet containing all the classroom fingerprints. A magnifying glass or hand lens will be of assistance.
4. Make this activity more difficult by collecting only prints that match the basic pattern of your own print. For example, if your finger has a loop pattern, collect only those prints with loops from other class members and then try to match the prints to the people.

Developing latent fingerprints

Generally, fingerprints left behind at a crime scene are not readily seen unless they are processed or developed in some manner. Searching for objects on which fingerprints may have been left is one of the first things that crime scene investigators do.

Read the following extract from the novel *The Twelfth Card* and answer the questions after it.

> Cooper and Sachs used a variety of techniques to raise prints on the evidence: fluorescent powder, Ardrox spray and Superglue fumes on slick surfaces, iodine vapour and ninhydrin on porous, some of which raised prints by themselves, while others displayed the results under an alternative light source.
>
> Looking up at the team through his large orange goggles, the tech reported, 'Prints on the receipt, prints on the merchandise. They're all the same. Only, the thing is, they're small, too small to be from a six-foot-tall man. A petite woman or a teenage girl, the clerk's, I'd say. I see smudges too. I'd guess the unsub [unkown subject] wiped his own off.'
>
> . . .
>
> Cooper lifted copies of the prints and scanned them. Ten minutes later the FBI's integrated automated fingerprint identification system had verified that the prints did not belong to anyone on file in the major databases, city, state and federal.

Source: Extract from Jeffrey Deaver; *The Twelfth Card*, Hodder & Staughton, London, 2005, pp. 45–6.

Clarification question

List the techniques used in this extract to raise and identify fingerprints. Use the information found on the next few pages to determine how realistic these techniques are.

Prints on non-porous or glossy, hard surfaces can be detected and developed quite easily using the dusting technique. These prints can then be photographed to give a permanent record more suitable for detailed examination.

Investigation

Dusting for fingerprints

Aim: To examine a number of methods of dusting for fingerprints. (Your teacher can demonstrate dusting for fingerprints on an overhead projector by placing a print on the projection surface, dusting the print and projecting the image onto a screen.)

Equipment: Fine powder (aluminium, zinc oxide, carbon or even talcum powder can be used); soft-hair brush; variety of objects to test (e.g. glass beaker, plastic cup, garbage bag or magazine cover); absorbent paper; sticky tape; black and white cardboard; camera (optional)

Method:

1. Place some absorbent paper on a bench to protect the surface and to facilitate cleaning up.
2. Select one of the testing objects and press a finger firmly onto it to leave a good print.
3. Sprinkle a fine powder onto the surface and carefully brush off the excess powder. (Use carbon powder for light-coloured surfaces.) The powder will adhere to the oils in the print.
4. You can preserve the fingerprint by 'lifting' it with sticky tape and backing with black or white cardboard to highlight the print. If possible, photograph the prints.
5. Repeat this process for other surfaces.

Discussion questions

1. Which surfaces produced the clearest prints?
2. Which powders produced the best prints?
3. What ways can you suggest for improving this investigation?

Sometimes, dusting does not raise enough detail for clear fingerprint analysis. Another method that can be used on non-porous surfaces is to expose the surface to a vapour of ethyl cyanacrylate (contained in Superglue). The ethyl cyanacrylate polymerises when it comes into contact with the moisture in the fingerprints, forming a white deposit.

Investigation

Superglue fingerprinting

Aim: To use the vapours from Superglue (ethyl cyanacrylate) to raise fingerprints from a non-porous surface.

Equipment: Tube of Superglue; aluminium foil; 60-watt lamp; fish tank with a lid; cup of hot water; a glass that has been handled (on which fingerprints will be raised).

Method:

1. Place the lamp inside the fish tank (see figure 9.11).
2. Make a small dish from a single layer of the aluminium foil and place it on top of the light bulb.
3. Squeeze Superglue into the aluminium dish (about as much as would cover a 10-cent piece).

Safety notes

- Do not let the Superglue come into contact with bare skin.
- Do your fuming in a well-ventilated area or in a fume cupboard, as the ethyl cyanacrylate vapours are strong irritants.
- Do not let the light bulb touch a flammable material such as paper in the chamber because it might catch fire.

4. Place the cup of hot water in one corner of the fish tank. This will increase the humidity in the tank and aid the polymerisation process.
5. Place the test glass in the tank.
6. Make a control sample by rubbing your finger or thumb on the side of your nose and then pressing it onto a small piece of aluminium foil. This sample will indicate whether the Superglue has fumed adequately for results.
7. Place the control sample in the tank where you can see it clearly from the outside. Close the tank and turn on the light bulb for 10 minutes.
8. When the control sample has turned white, the fingerprints on the test glass should be ready for observation. Turn off the heat, open the tank and carefully remove your test glass. (*Note:* If the test sample is not even slightly white after some time, turn off the heat, open the tank and add some more Superglue to the alfoil bowl. Close the tank, turn the heat back on and allow the Superglue to fume until the test sample becomes white.)

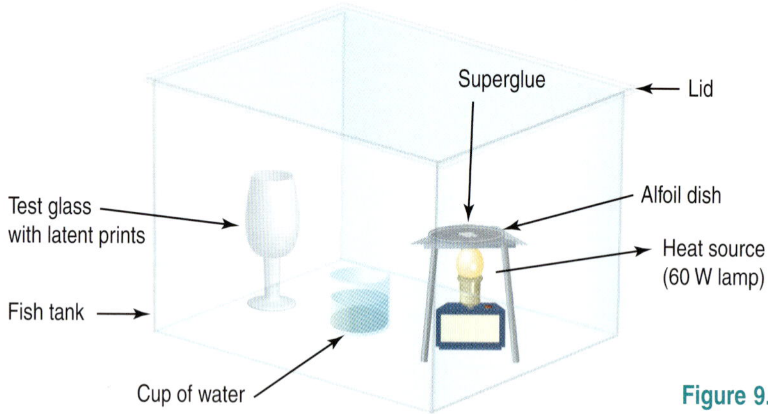

Figure 9.11: Using Superglue to raise latent prints

Often, fingerprints are present on absorbent surfaces such as paper or clothing materials. Dusting or using Superglue vapour on these surfaces produces poor results, so other methods must be used. Exposing the sample to iodine vapour is one way in which prints can be raised from porous surfaces. Iodine is absorbed by the fats and oils in the fingerprint and gives an obvious purple colour.

Investigation

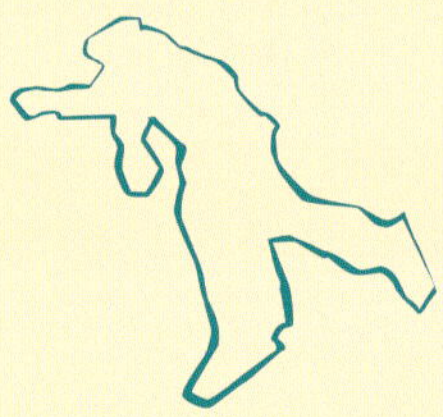

Raising fingerprints with iodine

Aim: To use iodine vapour to develop fingerprints from a porous material
Materials: A piece of paper that has been handled (to leave fingerprints on it); iodine crystals; zip-lock plastic bag; metal spatula
Method:

1. Place the paper in the zip-lock bag.
2. Add three spatulas of iodine crystals and seal the bag.
3. Gently shake the bag back and forth horizontally (this way the crystals stay at the bottom of the bag). The crystals will start to sublimate (turn from a solid into a gas without turning into a liquid along the way).
4. Observe the fingerprints that have developed on the paper.

Discussion questions

1. When the bag is opened and the paper removed, how permanent are the fingerprints raised in this method?
2. Has the iodine had any effect on the rest of the paper?

The advantage of using iodine on porous materials is that it does not damage the material or any writing on the item being tested, but fingerprints raised using this method tend to fade quite quickly. Therefore the fingerprints must be photographed to keep them for examination. Another of iodine's disadvantages is that the reaction agent it bonds to — fat and oil — is usually the first material to break down with exposure.

The chemical ninhydrin (triketohydrinden hydrate) will be used on porous materials instead of iodine if it is necessary for the prints raised to remain permanent. The ninhydrin reacts to the amino acids that are exuded in sweat from the hands. Unlike the fats left behind in fingerprints, which break down very quickly, the amino acids will remain on porous materials such as paper for long periods of time. When the ninhydrin combines with the amino acids, the fingerprints will appear as purple-blue impressions on the material.

Investigation

Developing prints using ninhydrin

Aim: To develop latent fingerprints on a porous surface using the chemical ninhydrin

Equipment: 1-gram triketohydrinden hydrate crystals; 100 mL methylated spirits; spray bottle; piece of paper with fingerprints on it, 250 mL beaker

Method:

1. Place the methylated spirits in the beaker.
2. Add the crystals and stir until dissolved completely.
3. Pour the ninhydrin solution into the spray bottle.
4. In well-ventilated area (e.g. outdoors), spray the paper liberally with the solution.
5. Allow the paper to dry for at least 24 hours until the prints start to appear.

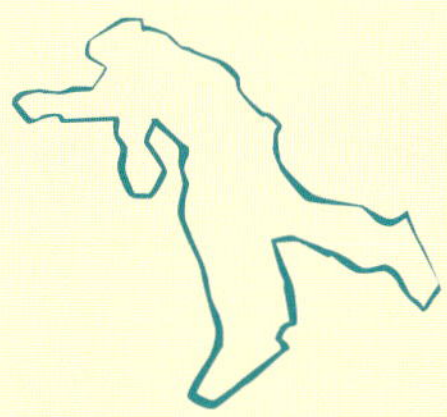

Safety notes:

- Methylated spirits is very flammable, so keep it away from possible heat sources.
- Avoid getting the solution on your skin, perhaps by holding the paper with tongs while you spray it.

Methods of developing fingerprints are constantly improving. One of the latest developments is a laser. It is reputed to give excellent results for very faint fingerprints on surfaces that do not yield prints with other methods. The laser works by causing certain chemicals in the fingerprints to fluoresce. These images are then photographed for closer examination.

Bloodwork

Blood at a crime scene can reveal enormous amounts of information about both the victim and the perpetrator. Blood is studied from two different perspectives — biological and physical. The study of the blood itself from a biological point of view is called **serology** and looks at aspects such as blood type and DNA, while the patterns formed by blood and other fluids at a crime scene are considered more from a physics point of view.

Serology is the study of the structure of the blood.

Bloodstain pattern analysis

The shape and pattern of blood spots may yield additional information helpful to the police in reconstructing a crime. Like all fluids, the way in which blood moves and spreads depends on the droplet size, the height from which it falls, the speed at which it travels and the characteristics of the surfaces with which it comes into contact. By examining the shape and size of blood droplets and smears at a crime scene, investigators can find indications of how a crime scene developed.

In the following extract from *The Last Precinct*, the character Kay Scarpetta, a forensic scientist, describes to a companion how important information can be found from the blood stains:

> I know experts who have spent their entire careers affixing medicine droppers of blood to ring stands, using plumb lines, squeezing or dripping or pouring or projecting blood onto a variety of target surfaces from a variety of angles and heights, and walking through puddles and stamping and slapping, and experimenting. Then, of course, there is the math, the straight-line geometry and trigonometry for figuring out point of origin.
>
> The blood in Diane Bray's bedroom, at a glance, is a videotape of what happened, but it is in a format that is unreadable unless we use science, experience and deductive reasoning to sort it out.

Source: Extract from Patricia Cornwell, *The Last Precinct*, Warner Books, London, 2000: pp. 398–99.

Clarification question

A number of 'facts' are given by the character in the first part of this extract. Use research techniques to determine the accuracy of these statements.

Transfer patterns are blood patterns that are formed when a bloodied object such as a hand, a weapon, shoes or even hair come into contact with a surface.

Blood patterns formed when a bloodied object such as a hand, a weapon, shoes or even hair come into contact with a surface are referred to as **transfer patterns**.

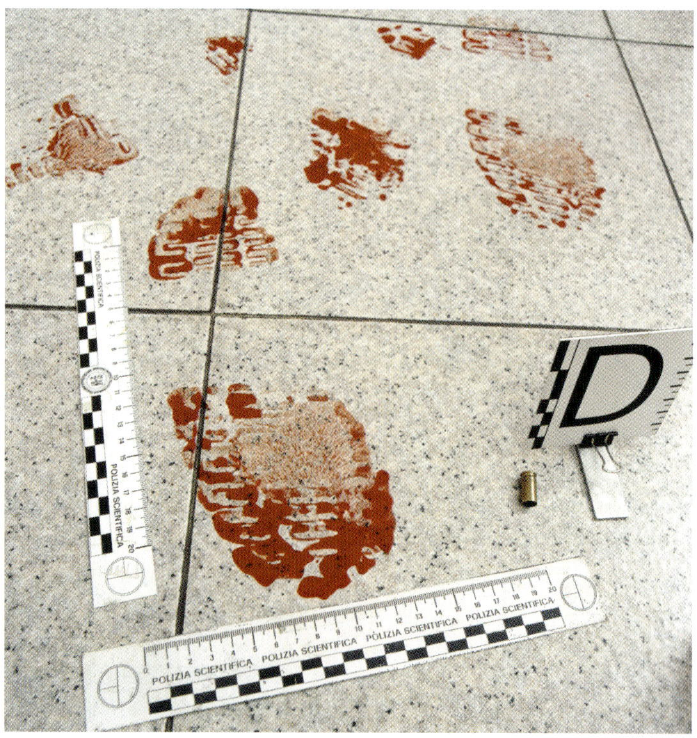

Figure 9.12: Blood pattern, formed by a bloodied shoe, at a murder scene

Other blood spots may be caused by splashing, indicating a crime of a particularly violent nature where blood has spattered from a severe wound. Blood spatters can generally be classified according to the velocity with which the blood leaves the impact site.

- *Low-velocity spatter* results when the source of the blood spots is struck by an object moving at less than 1.5 metres per second. This is common for blood dripping freely or the cast-off spatter from knife wounds. Low-velocity spatter is made up of droplets that are usually 4 to 8 millimetres in diameter.
- *Medium-velocity spatter* is indicated by blood spots that are between 1 and 4 millimetres in diameter, and is caused by the blood source being struck by an object travelling up to 10 metres per second. This is typical of spatter produced by hammers and cricket bats.
- *High-velocity spatter* is produced by gunshots and high-speed objects. The droplets are very fine, less than 1 millimetre in diameter, and leave the impact site at velocities over 10 metres per second.

The size and shape of blood splashes are directly related to the height from which the blood has fallen and the angle of collision. This is because, unlike water droplets that separate into smaller droplets as they fall, blood droplets do not separate. When blood strikes a surface, the outer rim of the blood spot forms a characteristic jagged edge. This jagged shape is referred to as **crenation** (see figure 9.13). The greater the height from which the blood has fallen, the greater the degree of crenation.

Crenation refers to the jagged edging around the rim of a blood spot that forms when the blood droplet strikes a surface.

When blood strikes at an angle to the surface, other patterns can be observed. As the collision angle diminishes, the blood spot will become increasingly elongated and eventually separate into two spots. Measuring the length and width of the spatter allows scientists to determine the angle at which a particular blood droplet struck the surface, because there is a mathematical relationship between these factors:

$$\sin \theta = \frac{\text{width of droplet}}{\text{length of droplet}}$$

For example, by measuring the width and length of the blood spatter in figure 9.14, we can determine that the blood struck the surface at an angle of 30°.

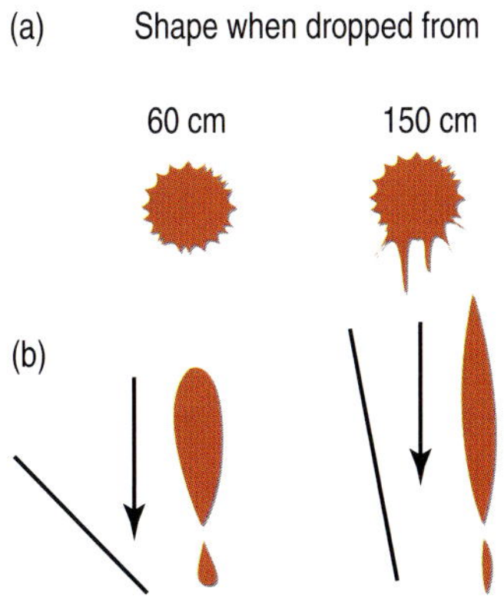

Figure 9.13:
(a) Blood spot showing crenation
(b) Shapes of blood splashes related to variations in striking angle

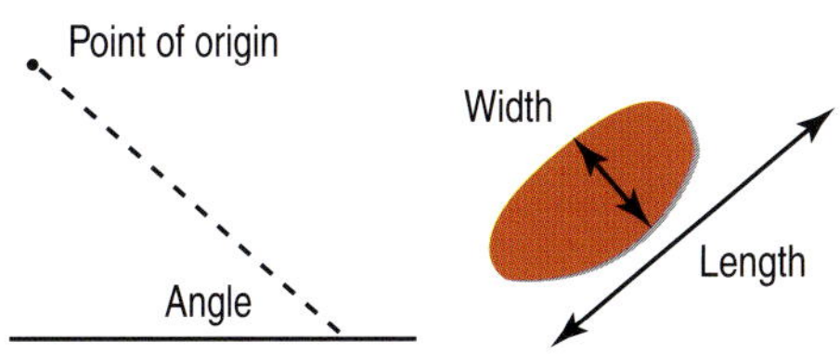

Figure 9.14:
Determining the angle of impact for blood spatter

Investigation

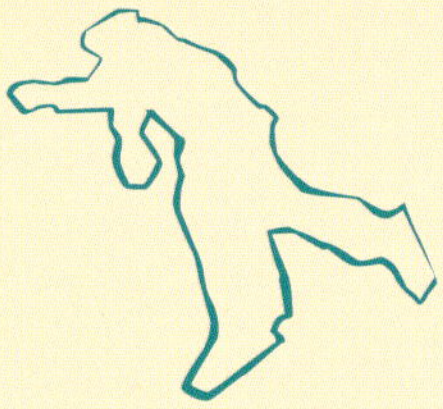

Effect of height on blood spatters

Aim: To see how blood dropped from differing heights makes different spatter patterns on impact

Equipment: 100 mL warm water mixed with 2 teaspoons of cornflour and some red food colouring (blood substitute); metre ruler; eyedropper; a small stepladder (optional); 30 cm ruler; large sheet of glossy paper; an area of flooring that is easily cleaned

Method:

1. Place the glossy paper on the floor.
2. Fill the eyedropper with the 'blood' and hold it so that the tip of the dropper is 0.5 metres above the paper on the floor.
3. Release a single drop onto the paper.
4. Measure the diameter of the drop and note any crenation.
5. Repeat steps 2 to 4 for heights of 1, 1.5, 2 and 2.5 metres.

Discussion questions

1. Was there a noticeable relationship between the drop height and the diameter of the spatter? If so, what is it?
2. Did the amount of crenation vary when the height was changed?
3. How would you test the effect of the surface onto which blood was dropped?

Blood typing

Having determined that a suspicious stain is of human origin, further testing will reveal the sex and blood type of the person in question. This information can be just as helpful in eliminating suspects as it is in leading to a final conviction. The fact that a suspect has a matching blood type is not conclusive; it simply proves that the person may have been at the scene of the crime. However, when combined with other evidence, it may be incriminating.

Human blood contains substances such as dissolved salts, red and white blood cells and platelets, all within a watery fluid called plasma. Since many characteristics of blood are inherited and therefore differ from one individual to another, the ability to distinguish between different blood types can be of great assistance in paternity suits and rape cases.

Secretors are people whose body fluids contain type A or type B antigens dissolved in them.

Another significant inherited factor is that nearly 80 per cent of the population are **secretors**. This means that the antigens A and B will dissolve in other body fluids such as saliva and semen, allowing blood grouping for these people to be determined. For example, a blackmailer sending a threatening note may lick the envelope before mailing it. If the blackmailer is a secretor, forensic tests will be able to determine the person's blood type. Because the antigens will also dissolve in semen, similar methods can be used in rape cases.

The most commonly used method of blood grouping is known as the ABO system. This system differentiates between blood types on the basis of the presence of special proteins, or antigens, called antigen A and antigen B. Table 9.4 summarises the four blood groups possible under this scheme.

Table 9.4: Antigens and antibodies present in blood		
Blood type	**Antigens present**	**Antibodies present**
A	Contains antigen A only	Contains anti-B
B	Contains antigen B only	Contains anti-A
AB	Contains antigens A and B	Contains neither anti-A nor anti-B
O	Contains neither A nor B antigens	Contains both anti-A and anti-B

Agglutination the process in which red blood cells are clumped together by hostile blood antigens.

You will notice that table 9.4 also mentions substances called *antibodies*. These substances, produced by the body, cause clumping of red blood cells in a process known as **agglutination**. For instance, the antibody A will cause clumping in blood that contains red blood cells with antigen A in them. This is why doctors must be careful when giving blood transfusions. A person with blood type A cannot tolerate blood from a person with blood type B: clumping of the recipient's red blood cells will result in capillaries, causing death. Clumping of the red blood cells in the donated blood is not such a problem because there are fewer of these cells.

Table 9.5 summarises the possible exchanges of blood using the ABO system. Can you see why people with group AB blood are referred to as universal recipients and people with group O blood are referred to as universal donors?

Table 9.5: Possible blood donors and recipients					
		Blood type of donor			
	Group	A	B	AB	O
Blood type of recipient	A	Yes	No	No	Yes
	B	No	Yes	No	Yes
	AB	Yes	Yes	Yes	Yes
	O	No	No	No	Yes

In the ABO blood system, the gene responsible for determining blood type comes in three different forms called alleles: A, B and O. Since only two of these alleles (one from each parent) combine to determine blood type, the possible combinations are AA, AB, AO, BB, BO and OO. Genes are referred to as either dominant or recessive. In the ABO blood system, genes A and B exhibit equal dominance towards each other, and both are dominant over O, O being recessive. Consequently, a person with blood group A carries the A gene in the form of either AA or AO.

As mentioned earlier, the fact that blood type is an inherited characteristic is most useful in paternity and rape cases. The diagrams in figure 9.15 show possible combinations. You will see that if one parent has type A blood (as either AA or AO) and the other has type AB blood, the offspring can be type A, B or AB, but not O. A full list of the possible combinations that may occur is shown in table 9.6.

Parent X (type AO)

Parent Y (type AB)

×	A	O
A	AA	AO
B	AB	BO

Parent X (type AA)

Parent Y (type AB)

×	A	A
A	AA	AA
B	AB	AB

Figure 9.15:
Possible offspring from AB × A parents

Table 9.6: Possible offspring using ABO blood groupings

Blood groups of parents	Possible groups of offspring	Impossible groups of offspring
O × O	O	A, B, AB
O × A	O, A	B, AB
O × B	O,B	A, AB
A × A	O, A	B, AB
A × B	O, A, B, AB	None
B × B	O, B	A, AB
O × AB	A, B	O, AB
A × AB	A, B, AB	O
B × AB	A, B, AB	O
AB × AB	A, B, AB	O

Blood typing as done by the Red Cross Blood Bank depends on observing how blood samples react with special chemicals called anti-A serum and anti-B serum. If antigen A is present in a blood sample, then it will react with anti-A serum to produce agglutination. This blood sample would produce no reaction with anti-B serum since no antigen B is present.

Table 9.7: Blood typing using anti-A and anti-B serum

If you observe ...	It indicates the blood type of the person is ...
Both samples show clumping	Type AB
Only anti-B clumps	Type B
Only anti-A clumps	Type A
No clumping	Type O

Blood types are not distributed evenly throughout the population. In Australia, 49 per cent of the population is type O, 38 per cent is type A, 10 per cent is type B and only 3 per cent is type AB. Forensically, this can be very significant. For instance, if type AB blood is found at the scene of the crime and a key suspect also has type AB blood, police will certainly pay close attention to the suspect's activities.

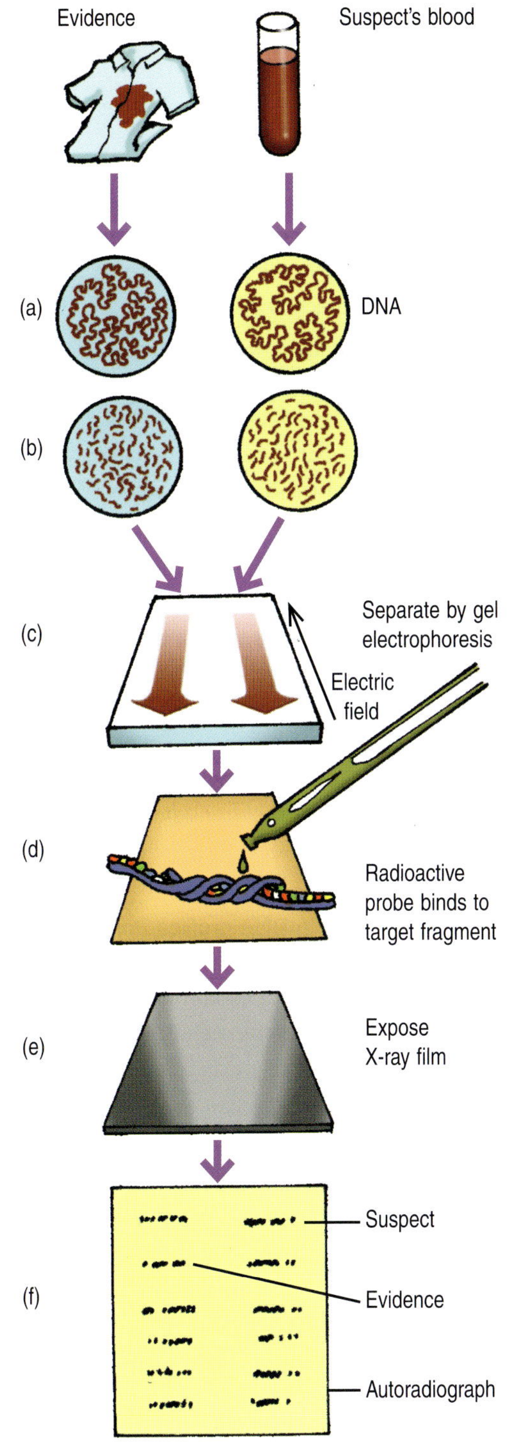

Figure 9.16: The process of DNA identification involves these key steps: DNA molecules (a) are extracted from the cells of a suspect's blood then (b) chopped up into smaller fragments using enzymes, (c) separated from one another by a process called gel electrophoresis (d) tagged with a radioactive isotope so that, (e) they will show up on an X-ray film. (f) If the bars on the pattern from the evidence and suspect match, calculations are then done that indicate the chances of such a match occurring randomly.

Source: Adapted from *Scientific American*, May 1990, p. 20.

DNA analysis

Arguably the most exciting recent advance in the world of forensic science has been the development of DNA analysis. DNA (deoxyribonucleic acid) is a complex chain of chemicals present in every cell of the human body. The DNA molecule contains all the genetic information necessary to make a human being, including details such as the number of fingers and toes, hair and eye colour, and even some personality traits. Since each person is unique, each DNA pattern is also unique. Identical twins are the exception. As they both came from the same fertilised egg, they will have the same DNA.

DNA samples for analysis can be obtained from any body cell including blood, saliva, skin, hair, bones or semen. All these are materials commonly found in criminal investigations. DNA is collected from suspects by using mouth swabs.

The steps in the process of DNA identification are shown in figure 9.16. Most developments in this field have taken place in the US and Japan.

DNA testing is helpful because each person's DNA 'fingerprint' is a combination of characteristics drawn from each parent. This means that, apart from helping to solve criminal cases, DNA testing is also useful in solving paternity cases and immigration disputes. (A person's chances of immigrating are often improved if they can prove a close relationship with a current Australian resident.)

While DNA analysis is seen by many viewers of crime shows as a magic process that points the finger at the culprit with 100 per cent certainty, the reality is very different. The analysis can be done only if there is enough good-quality DNA in the sample to test. Like all biological material, DNA degrades over time so a very old sample may not give reasonable results. Similarly, a very small sample may not provide enough to test.

While DNA is contained in nearly every cell of the body, not all the cells contain the full set of DNA that may be needed for a positive match. Skin cells, for example, contain a complete set of DNA in their nuclei (called *nucleic DNA*) while blood cells, which have no nucleus, can provide only a partial set (called *mitochondrial DNA*).

Another problem is that a DNA sample can be compared only to a known sample in the tester's possession, such as samples from a suspect. Some countries are moving towards establishing a database that contains samples from people who have been convicted of crimes. In this way, DNA evidence from new crimes can be compared to known offenders. This is considered by many people to be a violation of human rights. Furthermore, cross-contamination of a crime scene or mishandling of a piece of evidence can corrupt the evidence so that the DNA extracted from a sample may have nothing to do with either the culprit or the crime. False matches and inaccuracy have become less common in DNA analysis and techniques are improving all the time but, at this stage, DNA evidence on its own is not accepted in court as the basis for conviction.

Eye on the media

Your DNA in their hands

The boundary between the desires of the state and the rights of the individual requires constant surveillance. Last December, the American Civil Liberties Union filed a suit to stop a Californian law that allows DNA testing of anyone arrested for a felony but not charged or convicted. The ACLU argues that the law, known as Proposition 69, violates due process and is a 'vicious attack' on privacy.

The contrast could not be starker in the UK, where a series of laws passed over recent years go further than Proposition 69. In England and Wales, the state retains DNA not only from innocent people who have had a brush with the law, but even from those who volunteer a DNA sample during a police sweep. Four years ago, this magazine warned of the potential dangers of these laws (*New Scientist*, 5 May 2001, p. 3), yet public interest in the issue has remained close to zero. In the interim, things have got worse. The retained DNA is being put to uses never originally envisaged and without public approval or ethical constraints.

There is no doubt that DNA profiling is a useful tool. For some crimes it can nearly double clear-up rates. Yet DNA profiles are still obtained from fewer than 1 per cent of all reported crimes, so how important profiling is in the grand scheme of things is not clear. Last month, the House of Commons committee on science and technology called for a review to spell out the database's value. This is a good idea. Most people would agree that it is worth surrendering some personal freedom if serious criminals are caught, but they need to know where to strike the balance.

But people will still need convincing. The decision to retain DNA profiles from anyone who is arrested — even if not charged — lacks any sensible rationale. As the inventor of DNA profiling Alec Jeffreys argues, it would make more sense to test everyone than to continue with such a haphazard, discriminatory system.

It discriminates in more ways than one. Using the incomplete information available about the National DNA Database (NDNAD), *New Scientist* calculates that it contains DNA profiles of about 8 per cent of adult white males and 32 per cent of adult black males. Does this reflect the real balance of criminals in the UK, or merely who the police think commit crimes? Many will see these figures as a tangible reflection of the institutional racism the police are so anxious to shake off.

The problems go deeper. The UK keeps not only DNA profiles but also the samples from which they came. These contain a huge amount of information about their owners. It is already possibly to tell if a sample comes from somebody with red hair, for example, and genes for other characteristics will emerge in time. Should that information be unlocked? That decision rests solely with the police and forensic scientists on the NDNAD Board.

The board also decides what research can be carried out on the contents of the database. Already it has let researchers test ways to tell a person's ethnicity from their profile. In any other area of science such a project would have to have been approved by an ethics committee. Likewise, it would be unthinkable to conduct research on a sample taken from a person without their consent. Yet those on the NDNAD have no option. If the samples were all from criminals, one might argue that they had forfeited their right to withhold consent. But many on the database are not criminals.

The NDNAD Board also approved familial testing in cases when DNA from a crime scene partially matches a profile in the database. The relatives of the person on the database are then tested for a match. This raises the possibility of sons and daughters inadvertently discovering that they are not in fact related to the people they thought were their parents. Such implications have never been debated.

It is unacceptable that the NDNAD board is answerable only to itself. The database should be overseen by an accountable, independent body that includes lay people. And it needs a new ethical framework. California beware!

Source: *New Scientist*, 9 April 2005, p. 3.

Clarification questions

1. What do the following abbreviations stand for?

   (a) NDNAD (b) ACLU (c) DNA (d) UK

2. Why does Alec Jeffreys describe the current UK system of retaining DNA samples from anyone who is arrested as 'haphazard' and 'discriminatory'?
3. Where do you think the author of this article stands with regard to collecting DNA samples? Justify your answer.

Ballistics

Ballistics is the study of the interaction between firearms and their projectiles.

An increasing number of crimes involves the use of firearms, making the study of **ballistics** one of growing importance. The most common role of the ballistics department is in identifying an offending weapon but much other information can be advanced by a sound knowledge of ballistics.

Identifying the weapon

Criminals may use a wide variety of firearms, including rifles, revolvers, pistols and the much-favoured sawn-off shotgun. Each leaves telltale signs after being fired that assist in the identification of the weapon used — a sort of 'ballistic fingerprint'.

The first job of the ballistics expert is to identify the type of weapon used in a crime. A bullet collected at the scene of the crime, or perhaps taken from the victim's body during an autopsy, is weighed and measured to determine the calibre of the ammunition.

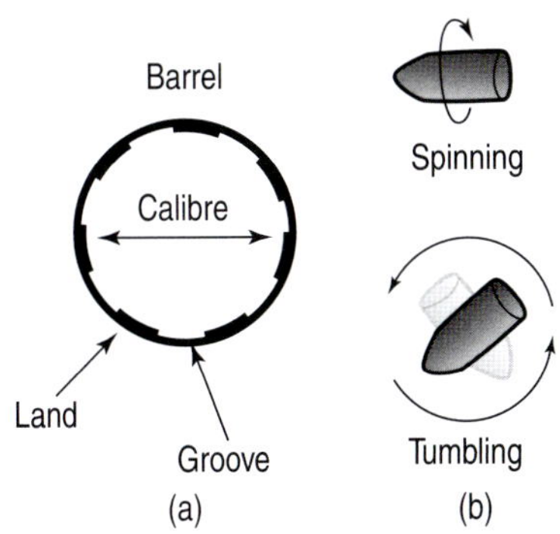

Figure 9.17: (a) Cross-section of rifle barrel showing lands and grooves (b) Spinning motion of a bullet compared to tumbling motion

The bullet is then examined microscopically for rifling marks. Many weapons have spiralled ribs on the inside of the gun barrel. These are called *lands* and *grooves*, and they cause the bullet to spin on its axis as it leaves the gun instead of tumbling end over end. This gives the weapon much greater accuracy and range. Different manufacturers cut these spiral grooves to different specifications in terms of (a) the depth of the lands and grooves, (b) the direction of spiral — left or right, and (c) the number of spirals per given length (called the *pitch*). The importance of this forensically is that each weapon cuts a unique pattern of grooves, or *striations*, into the soft outer coating of the bullet (see figure 9.17).

Indentations are also left on the base of each bullet or shotgun cartridge as a result of the action of the weapon's firing pin. Ballistics experts use these markings not only to determine the type of weapon but also to track an individual weapon, since each individual firearm has characteristics that result in slight variations in markings on the spent ammunition.

To make the examination, the expert needs both the offending bullet and a 'control' or test bullet. Once a possible weapon has been found, a test bullet is fired into water, cotton wool or ballistics gel (a jelly-like substance that has similar properties to human flesh) to minimise damage so that the comparison will be a fair one.

Both bullets are then examined under a *comparison microscope* (see figure 9.18). This instrument consists of two microscopes linked in such a way as to allow the user to view two objects side by side using a single eyepiece. Bullets from the same weapon will have identical striations and firing pin indentations (see figure 9.19). Since shotguns are smooth bored, (they have no rifling in the barrel) other methods of identification are used. An examination can be made of the markings on the base of spent cartridge cases that were caused by the action of the firing pin. Shotgun cartridges contain material known as *wadding*. The wadding separates the propellant from the lead pellets (shot), and is made of cardboard, plastic or felt. An examination of this material can be used to determine the type of weapon used.

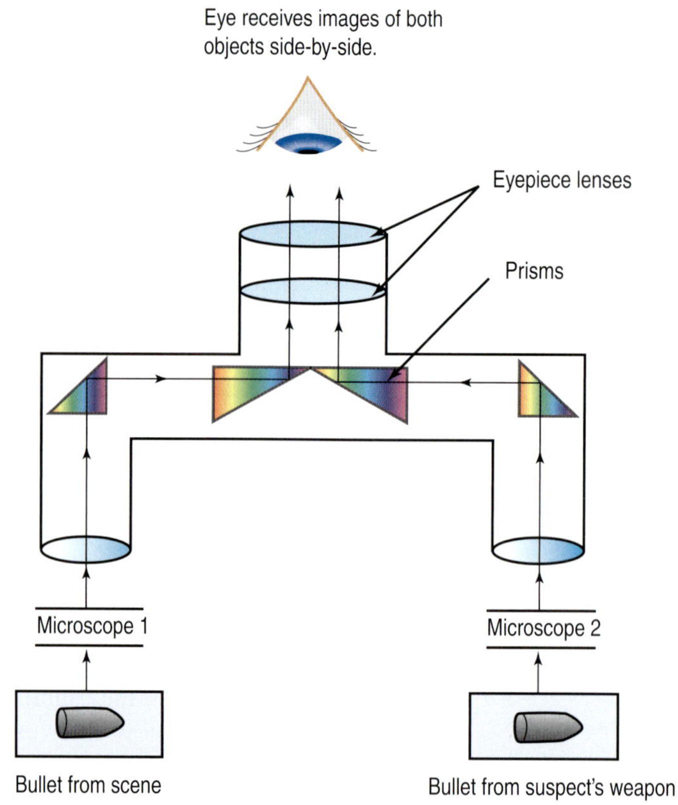

Figure 9.18:
A comparison microscope

Figure 9.19:
Photograph from a comparison microscope showing a perfect match between the features of the left and right bullet. The join between the two images is just visible.

Collect a variety of spent bullets and cartridges. A shooting club is a good source. Try to get six to eight spent bullets from a variety of weapons of the same gauge. Group the bullets so that you have two bullets from one weapon and a single bullet from the others.

These should then be numbered for identification purposes.

Once you have a number of sets ready (do not mix them up!), get class members to examine each set carefully to see if they can identify those that were fired from the same weapon.

Other evidence from ballistic analysis

Ballistics experts can also help with the following types of analysis:

- Chemical traces of propellant may be left on the hand of the firer. Chemical analysis of swabs of this material may be matched with propellant from other ammunition of the same type.
- The presence or lack of the wadding can indicate firing range. Usually the wad falls to the ground after a few metres so that its presence in or on the body will usually suggest that firing occurred at close range.
- Measurements of the force required to pull the trigger of a weapon (the trigger-pull) helps to determine whether or not accidental discharge of the weapon was possible.
- Examination of the gunshot wound itself can help in the estimation of the firing direction and distance. The direction of firing can be determined by examining the entry and exit wounds, and the shape of the wound. A neat, circular wound may indicate a direct shot, whereas an oval wound would suggest firing from an oblique angle.

Firing at close range results in bullet holes larger than the bullet itself, leaving an imprint of the gun muzzle and scorching around the wound or on clothing — the result of the discharge of the explosive material. Small, neat wounds generally result from firing at ranges of greater than a metre.

Shotguns do not usually produce an exit wound. Since the shot spreads out as it moves away from the gun, the extent of the effect (size of the area on the body) helps to determine firing distance.

How to make a good impression!

Any impression left in soft material at the scene of a crime can be an important source of evidence. The most common impressions are foot/shoe prints and tyre tracks (see figure 9.20) left in soft material such as soil. Significant impressions can also be left by metal tools used for breaking into buildings. If the tool has an unusual scratch or shape, it may be possible to match up the impression with the tool if the tool is later found. If these impressions are to be used as evidence in court, a permanent record must be made, either a photograph or some sort of casting material.

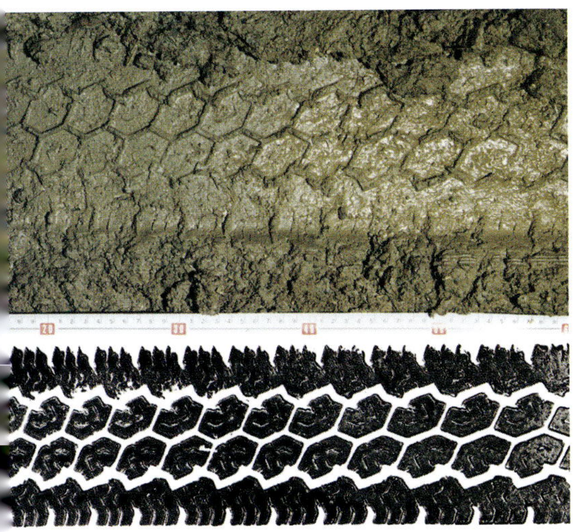

Figure 9.20: Tyre prints matched to tyre tracks

The properties of the casting material are very important. It should be cheap (if large quantities are needed) and easy to prepare and use under all sorts of conditions. It must set quickly and result in an accurate and detailed imprint. On the other hand, the material should not shrink after setting, adhere to the original material or damage the impression when it is removed.

The choice of casting material will depend on such factors as the size of the impression, the material in which the impression has been made (soil, wood, metal), whether or not it matters if the impression is destroyed by the process, whether or not the impression is on a vertical or horizontal surface, and the amount of detail that is required.

Investigation

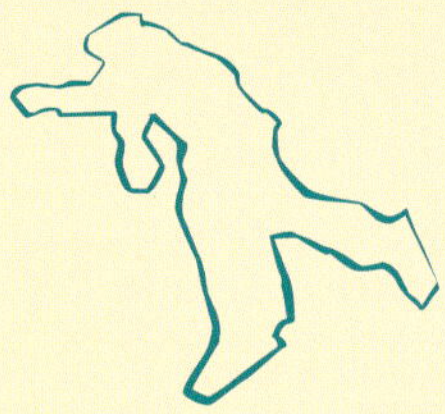

Making a plaster cast of a footprint

Aim: To make a plaster cast of a shoe print using plaster of Paris. It might be a good idea to practice first on impressions from leaves, shells, tyre treads and so on.

Equipment: An old shoe (preferably with an interesting or distinct pattern on the sole); a mould large enough to surround the shoe (ice-cream container or cardboard; see figure 9.21); plaster of Paris; a container to prepare the plaster of Paris; a soft brush or toothpick; Paddlepop sticks or old cloth; hairspray

Method:

1. Find a suitable area, preferably outdoors because this can be quite a messy activity. The area should be protected, however, as you will need to leave the cast to set overnight and you will not want your work interfered with.
2. Make an impression in some moist sand or soil with your shoe. Spray the impression with hair spray (this helps to hold the sand/soil together).

Figure 9.21: Box mould around a footprint impression

3. Use the mould to create a border around the impression you have created (see figure 9.21). If the impression is too large, confine yourself to the heel or ball of the shoe.
4. Prepare a slurry by combining about seven parts plaster of Paris to four parts of water and mix thoroughly, being sure to remove all bubbles.
5. Carefully pour the plaster of Paris into the mould, trying not to disturb the impression. Some cloth or Paddlepop sticks placed gently across the top of the slurry will give the final product greater rigidity. Allow the plaster to set, preferably overnight.
6. Once set, gently remove the cast and clean the surface carefully, using a soft brush or toothpick.
7. Display all the casts and the shoes randomly and see who can match them up.

Extension

8. Compare shoe prints made under a variety of circumstances such as running, jumping, carrying a heavy load, turning, and so on. See if other students can identify the circumstances under which the impression was made.
9. Develop a scale for comparing and matching the length of print with common shoe sizes (for example, a size-11 man's shoe is about 33 centimetres long).

Hairs and fibres

Hairs and fibres can easily become detached in crimes involving violence. Chemically, they are very stable materials and have a wide variety of other properties, so they can give quite conclusive results.

Hairs can be identified as animal or human, and fibres as natural or synthetic. Under a microscope, natural fibres tend to have a rough, irregular external surface whereas synthetic fibres have a much smoother surface due to the way in which they have been manufactured. Hairs have characteristic **scale patterns** on them that can be used for identification purposes. Wool has a particularly pronounced scale pattern.

Scale patterns are the characteristic shapes visible along the length of a natural fibre such as hair or wool when viewed under a microscope.

Investigation

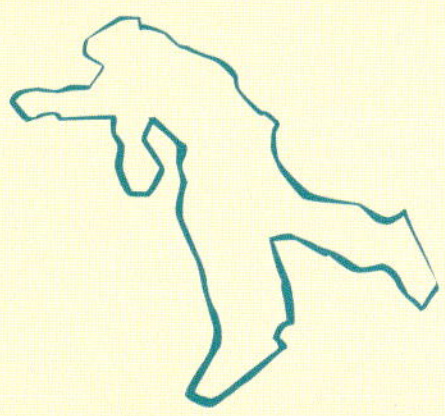

Examining hairs and fibres

Aim: To make an examination of a range of animal hairs and synthetic fibres, and to try to confirm the general characteristics of each as described in the text

Materials: Samples of animal and human hairs, natural and synthetic fibres (or prepared slides if available); microscope, glass slides and coverslips; mounting fluid; sticky tape; prepared slides of some 'unknown' fibres or hairs; acetone

Method:

1. Prepare slides for viewing by taking the hairs and/or fibres provided and attach them to a microscope slide with some sticky tape. (Animal hairs should first be cleaned by washing them in a suitable solvent such as acetone.) More permanent slides can be prepared by placing a 0.5 centimetre length of the material on a slide with a couple of drops of mounting fluid over it. Lower a glass coverslip onto the specimen and wait for the fluid to harden.
2. Using a microscope, carefully examine each of the slides in turn. In your notebook make careful notes about each of the fibres. Use diagrams and comment on features such as scale patterns, colour and surface characteristics.
3. Take some of the 'unknown' slides and try to identify them.

Extension

4. Try to distinguish between hair that has been forcibly removed and hair that has fallen out naturally. Look at the base of the hair strand. What use could be made of this difference?
5. Compare hair samples from different parts of the body (body hair, scalp hair and eyelashes).
6. Compare dyed hair with naturally coloured hair.
7. Compare hair samples from people with different ethnic origins.

Forensic toxicology

Toxicology the study of substances that are harmful to the human body.

The study of substances that are harmful to the human body (poisons) is called **toxicology**. Forensic toxicologists analyse blood and urine samples for the presence of alcohol (although this is now done routinely by police using breathalysers), they identify other drugs such as heroin and cannabis that are found in the possession of criminals, and analyse cases of suspected

Thin layer chromatography is a process by which the components of a mixture to be analysed are separated using a thin layer of material or film.

Gas chromatography is a process by which the mixture to be analysed is vaporised and mixed with a gas that then carries it through a column of absorbent material.

poisoning, whether accidental or deliberate. Two methods commonly used in the detection or identification of drugs are known as **thin layer chromatography** and **gas chromatography**. Both methods are used to separate and identify mixtures of drugs (see figure 9.22).

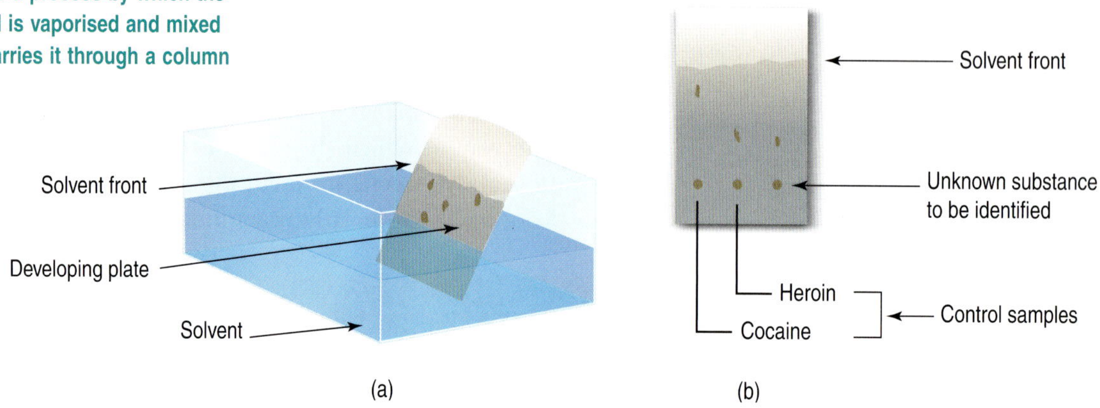

Figure 9.22:
(a) Thin layer chromatography (b) Sample chromatogram of a mixed drug sample

Thin layer chromatography

In thin layer chromatography, the mixture is spotted onto special plates that are coated with silica gel or alumina (*stationary phase*). These plates are placed in chromatography tanks containing organic solvents (*mobile phase*). The solvents slowly move up the plate by capillary action in much the same way as water rises up inside a tree trunk. As the solvent moves, the various drugs are moved up the plate as well, but at a rate that depends on their relative solubilities in the solvent and the strength of their attraction with the stationary material. Different solvents will separate mixtures at different (known) rates.

Since each drug moves a characteristic distance depending on the solvent used, identification can take place once separation has occurred. Often, known standard drugs are run alongside the unknowns to give an easier comparison. The separated spots are sprayed with special solutions that enable them to be seen as coloured spots under ultraviolet light.

Investigation

Chromatographic analysis of inks

Aim: To use a simple method of thin layer chromatography to separate the various dyes from an ink sample

Materials: Glass rod; paperclip; 250 mL beaker; various solvents (water, methanol, ethanol, diethyl ether); filter paper; black or brown inks from either bottles or felt pens of different brands

Method:

1. Cut a suitably sized piece of filter paper and mark a pencil line 1 centimetre from the base of the paper.
2. Place a spot of each ink sample on this pencil line.
3. Place the filter paper in the solvent and observe the results. You may need to experiment with the solvent you use because different inks may not be soluble in certain solvents (such as water).

4. Record the results carefully. Note any differences in position and colour intensity.

Discussion questions

1. Why were you advised to mark the base line in pencil?
2. What is the name given to the resulting filter paper?
3. How many different colours were you able to separate? How clear was the separation? Were some solvents better than others? Prepare a report on your results.

Gas chromatography

Gas chromatography is a process that works on very similar principles to thin layer chromatography except that, as the name suggests, the mobile phase is a gas. A common gas used is nitrogen since it is inert (unreactive) even at the high temperatures used.

A constant flow of the inert gas is passed through a column containing the stationary material, which is coated on the inner walls. Different materials adhere to the stationary material to different degrees and so take a longer or shorter period to pass through the column. The arrival of the different materials is measured by a detector that displays the information as a series of peaks on a chromatogram. This chromatogram can then be compared with the chromatograms of known substances.

Gas chromatography has two advantages: (1) only a small amount of material is required, and (2) the results obtained are quantitative. The gas chromatography process is also an excellent way of comparing two samples of material. For example, a paint sample left behind in a hit-and-run accident can be compared with a sample from a suspect's vehicle. Identical displays would provide conclusive evidence.

End of chapter questions

1. When would an 'expert witness' be called on to give evidence? Give some examples.
2. (a) What are the causes and symptoms of the conditions known as rigor mortis and lividity?
   (b) How is an understanding of these conditions helpful in establishing the time of death?
3. List the factors that are likely to affect the rate at which body temperature falls following death.
4. Explain what is meant by the following terms when used in relation to fingerprinting:
   (a) latent
   (b) ridge counting
   (c) ridge tracing
   (d) delta.
5. (a) Using diagrams, outline the main steps involved in the process of DNA fingerprinting.
   (b) Why is DNA fingerprinting such a powerful forensic technique?
6. Explain each of the following terms as they relate to ballistics:
   (a) rifling
   (b) lands
   (c) grooves
   (d) wadding
   (e) trigger-pull
   (f) striations.
7. (a) Explain how a ballistics expert would go about proving that a bullet found at the scene of the crime came from a particular weapon.
   (b) What special piece of scientific equipment is likely to be used? How is it helpful?
8. Apart from examining the bullet, what other evidence might a ballistics expert or pathologist look for in a case of murder with a firearm?
9. (a) What are the properties of a good casting material?
   (b) What factors should be taken into account when selecting a substance to use in making a cast?
10. How would you be able to differentiate between an animal fibre and a synthetic fibre?

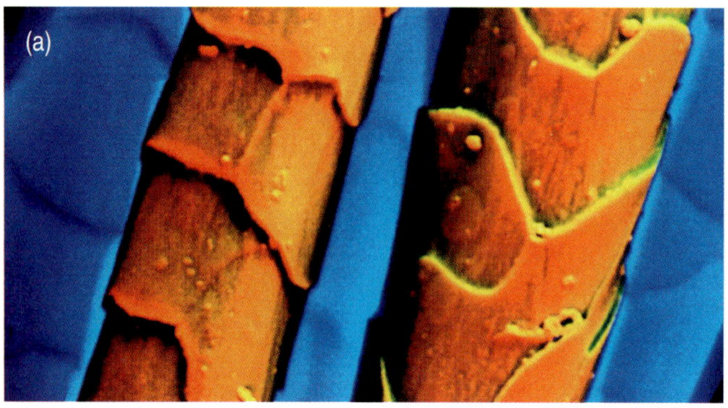

Figure 9.23:
How can you tell an animal fibre from a synthetic fibre?

11. Explain what is meant by the ABO blood-grouping system.
12. Explain why it is possible for parents who are both type A blood carriers to produce a child with blood type O.
13. Explain how accurate observations of blood spattering patterns can yield helpful forensic evidence.
14. Explain how the process of thin layer chromatography could be used to help compare the ink used to forge a signature on a cheque with ink from a bottle found at a suspect's home.
15. A body is found lying face down in the centre of a freshly-ploughed field on a farm. There are no footprints, tyre tracks or marks of any kind in the field. The body has lividity colouration on the back.
    (a) Is there any evidence of foul play?
    (b) What is the most likely way the body came to be in the middle of the field?
16. Describe the ways in which police can ensure that no evidence is overlooked, accidentally disturbed or destroyed at the scene of a crime. Explain why these actions are essential.
17. 'The charging of a suspect in a criminal investigation is like the forming of a hypothesis in a scientific investigation.' Comment critically on this statement.
18. The skeleton of a woman believed to have been 60 years old is found. The tibia is measured at 20.6 centimetres. Work out what the woman's height would have been.
19. A naked body is found in the bush. Describe the types of evidence that a forensic pathologist might look for to assist in the identification of the body.
20. Read the article below and answer the questions that follow.

Gunshot traces similar: expert

Similar gunshot particles were found in a suspect's car and on the body of murdered police chief Colin Winchester, the ACT Coroners Court was told yesterday.

Crime scene expert Det-Sgt Peter Nelipa said gunshot particles of the same composition were found at the scene of the murder and in the car of a man whose name was suppressed by the Coroner, Mr Ron Cahill.

The man is expected to appear at the inquiry on Monday.

The court was told the man's flat was raided by police and his car seized on January 18, eight days after Australian Federal Police assistant commissioner Winchester was shot outside his Canberra home.

Nelipa told the inquest into Winchester's death that a Victorian forensic expert found gunshot particles on the door handle, indicator stalk and rear-vision mirror of the man's blue Mazda sedan.

He said similar particles were found on the ground near Winchester's car, on his body and in the car where he was found murdered on January 10.

Nelipa said he believed the particles found at the murder scene were from PMC-brand ammunition. He said similar particles were also in cartridges found on the ground near Winchester's body.

He said he believed Winchester had been shot twice in the head as he attempted to leave his car.

Nelipa said a police officer found green particles in the boot of the suspect's car which looked similar to green particles which he vacuumed from the ground near Winchester's body.

He said he compared the particles with unexploded PMC .22 ammunition and found they were similar.

Nelipa said blue paint particles found at the scene and blue paint taken from the man's car did not match. The paint found at the scene matched Winchester's trailer, he said.

Nelipa told the inquiry that police movements around the crime scene on the night of the murder had damaged his chances of gathering important evidence.

One of the first police officers on the scene the night of the murder told the inquiry the crime scene was cordoned off about 20 minutes after he arrived.

Det-Constable Karl Pattenden said he arrived at the scene at 9.30pm and then conducted a visual search of the immediate area.

He said he conducted the search instead of immediately blocking off the area to determine exactly what had happened and discover if an offender or weapon was nearby. Pattenden said a guard was placed near the crime scene and he did not see any police officers walking on the grass near the body or touch the body.

He said the immediate crime scene and the area around the Winchesters' home was cordoned off by 9.50pm.

Pattenden also revealed yesterday that the first suspect in the murder was a former police officer.

He said police had difficulty finding the former officer only hours after the murder but the man was later cleared of any involvement.

Source: David Washington, *The Courier-Mail*, 1 September 1989.

(a) How many examples can you identify of the Locard Exchange Principle being used by detectives?

(b) Did the police waste their time in investigating the blue paint found at the scene of the crime since it did not match the blue paint on the suspect's car? Explain your opinion.

(c) What evidence is there that the police were negligent at the scene of the crime? What precautions could have been taken?

(d) How did the police justify their delay in cordoning-off the scene of the crime?

(e) Imagine that the police later find a weapon they suspect fired the PMC-brand ammunition. Explain how ballistics experts could go about determining if it was in fact the weapon used.

21. A forensic biologist takes temperature measurements on a body every 30 minutes to help determine the time of death. Using the data in the table below, estimate the time of death.

Time	1.00 pm	1.30 pm	2.00 pm	2.30 pm	3.00 pm
Temperature (°C)	29.6	29.2	28.6	28.0	27.8

What adjustments to the estimated time of death would need to be made under the following circumstances?

(a) The body is found unclothed in August in Tasmania.

(b) The body is found lightly clothed next to an electric heater that is still operating.

22. Explain why a tense bank robber is likely to leave better fingerprints at the scene of the crime than a composed one.
23. Give two reasons why fingerprints, rather than signatures, would be used in some countries for identification purposes.
24. How might a person's occupation affect their fingerprint or handprint pattern? Give examples of such occupations.
25. Ballistics experts refer to the 'trigger-pull' of a weapon. This is the force required to cause the trigger to discharge.
    (a) Design an experiment that could be used to measure the trigger-pull of a rifle.
    (b) Explain why trigger-pull is often measured in cases where the accidental discharge of a firearm is claimed.
26. An interesting mark was found in the woodwork around a window at the site of a burglary. The officer in charge asked a young officer to make a cast of the mark. Why did the officer in charge burst into laughter when he saw the young officer stirring a plaster of Paris slurry? What advice would he have given the young officer?
27. Describe the properties you would need in a casting material if you were preparing a cast of:
    (a) an indentation in a timber window frame caused by a screwdriver
    (b) a footprint in moist, clay garden soil
    (c) a tyre mark on a rainy day.
28. In a homicide investigation it is found that type AB blood is found on the clothing of the victim, whose blood type is O. The main suspect has type AB blood. Comment on the strength of the evidence.
29. A sample of blood is found to agglutinate with anti-A serum. Without any additional information, what are the possible blood types of the sample?
30. Can a mother of type A blood produce a child of type O if the father of the child is type B? Explain.

Appendix A: Investigation and research methods

During this course, you will be working for all intents and purposes in the same way that any scientist would. This means that you will conduct and take part in investigations, inquiries and research into a wide range of different areas. You may even end up discovering something quite unique.

However, a scientist's role is not just to find things out. They must also communicate their findings effectively to others. In this chapter, we will look at how you can carry out effective investigations and research by going through a few different ways in which your work can be communicated. We will focus on four means of communication:

- an extended investigation
- a research paper
- a media file
- a scientific journal.

Extended investigation

An extended investigation is one in which a detailed series of experiments are conducted over a number of weeks, sometimes even over a number of months. Often, these investigations arise out of smaller class experiments. Depending on the nature of the Science 21 course that your class is developing, it may be left to you to decide on what topic, problem or line of inquiry your extended investigation will be about.

Selecting a topic

Selecting a topic is the first and one of the most important steps in conducting a practical investigation. An inappropriate choice of topic can lead to frustration and disappointment.

The selection of a topic should begin with a search for a suitable general topic area. A simple topic is almost always better than a complex one that requires advanced background knowledge. It is wise to choose a topic that builds on your current knowledge of science and your laboratory skills.

A topic may come to mind immediately, but sometimes it takes a few weeks to select one that you are happy with. The following list of 'springboards' for generating ideas might help:

- List your hobbies and other interests. For example, an interest in music could lead to topics involving the way that musical instruments produce different frequencies.
- List the topics that have interested you. For example, if you enjoyed an 'electric power' area of study, you might want to investigate the efficiency of an alternative electricity-generation method.
- Brainstorm with a group of friends and generate a list of possible topics. You might be surprised at the value of group brainpower in generating topics. Each suggested topic might generate many others when you work in a group.

- Browse through some scientific journals (like *New Scientist* or *Australasian Science*), other textbooks and topic books.
- Find out what other students have investigated in the past. The titles of their investigations might help you think of some topics of your own.
- Consider which ideas can *realistically* be investigated with the resources and time available to you

Practical investigations in science can take several forms. Your topic should fall into one of the following four categories.

1. *Designing, constructing and testing a device.* The purpose of the device must be clearly defined. This type of investigation should emphasise the ability of the device to fulfil its purpose and how the device can be modified to achieve its purpose more effectively.
2. *Investigating the operation of a device.* This type of investigation involves measuring the performance of a device that has already been designed and manufactured. The purpose of the device needs to be clearly defined. Any measurable aspect of the performance of the device under different conditions could be investigated.
3. *Solving a scientific or technological problem.* The problem to be solved must be clearly defined. Care needs to be taken to ensure that the problem can be solved with the equipment and within the time you have available.
4. *Investigating a physical phenomenon.* This type of investigation involves measuring a limited number of physical properties of objects or substances under different conditions. Examples of this include investigating the Greenhouse Effect in a glasshouse, or the resistance of materials to penetration by a sharp object.

Refining the topic

It will be necessary to define the topic in detail. You need to decide exactly what problem you are trying to solve or what questions you are asking so that the aim of your investigation is clear and achievable.

It is better to investigate one aspect of the topic in detail rather than a broad topic area superficially. For example, the topic 'Standing waves in metal wires' is quite broad, and there are many variables that could be investigated. You could not examine them all successfully in the available time. Instead, you would need to make the topic narrower so that you could examine the effect of changing two variables on a third dependent variable. One possible question to ask is: 'What is the effect of the tension and length on the fundamental frequency of a steel wire of constant cross sectional area?'

Once you have defined your topic in detail, you should be able to answer 'yes' to all of the questions listed in table A.1.

Table A.1: Refining the topic — a checklist	
Item to be checked	**Checked?**
Does my topic fall into one of the four categories described above?	
Is my topic one that takes account of my own interests?	
Does my topic involve the use of the techniques of experimental physics?	
Do I know where to look for any background information that might be needed?	
Do I have a clear aim (that could be modified as the investigation proceeds)?	
Is the equipment that I am likely to need readily available or easy to construct? (Be realistic in assessing this!)	
Will I be able to make preliminary observations and measurements within a period of about one hour?	
Will I be able to collect all necessary data within the time that has been allocated to the investigation?	
Is my investigation likely to produce meaningful conclusions?	
Have I discussed my choice of topic with my teacher?	

Getting organised

Time spent on detailed planning is a good investment. An extended practical investigation needs careful planning. Without it, you could find yourself lacking direction, wasting time or even having to start again at a time when you should have almost completed your investigation.

Devising a timeline

A timeline allows you to break an extended investigation into a set of manageable tasks. You can then organise dates and tasks to make sure that the whole investigation, including the report, is completed by the due date. A table like table A.2, listing the key tasks and deadlines, can be very helpful.

Table A.2: Timeline for the investigation	
Task	**Due date (set your own if your teacher has not done so)**
Selection of topic	
Submission of detailed plan	
Final check of apparatus and preliminary measurements	
Collection of data	
Analysis of data	
Completion of first draft	
Submission of final report	

Maintaining a logbook

A logbook is an essential element of an extended practical investigation. It provides you with a complete record of your investigation and is very much like a diary. You should record every aspect of your investigation into your logbook as it is undertaken. The logbook will assist you to prepare your report. An exercise book would make an ideal logbook.

All good scientists keep logbooks when they are experimenting and will often refer back to earlier logbooks, for example, if they have forgotten how a particular problem with equipment was solved. Sometimes these logbooks end up being historical documents in their own right. Figure A.1 shows some of the pages from Thomas Edison's notebooks when he was developing the electric light.

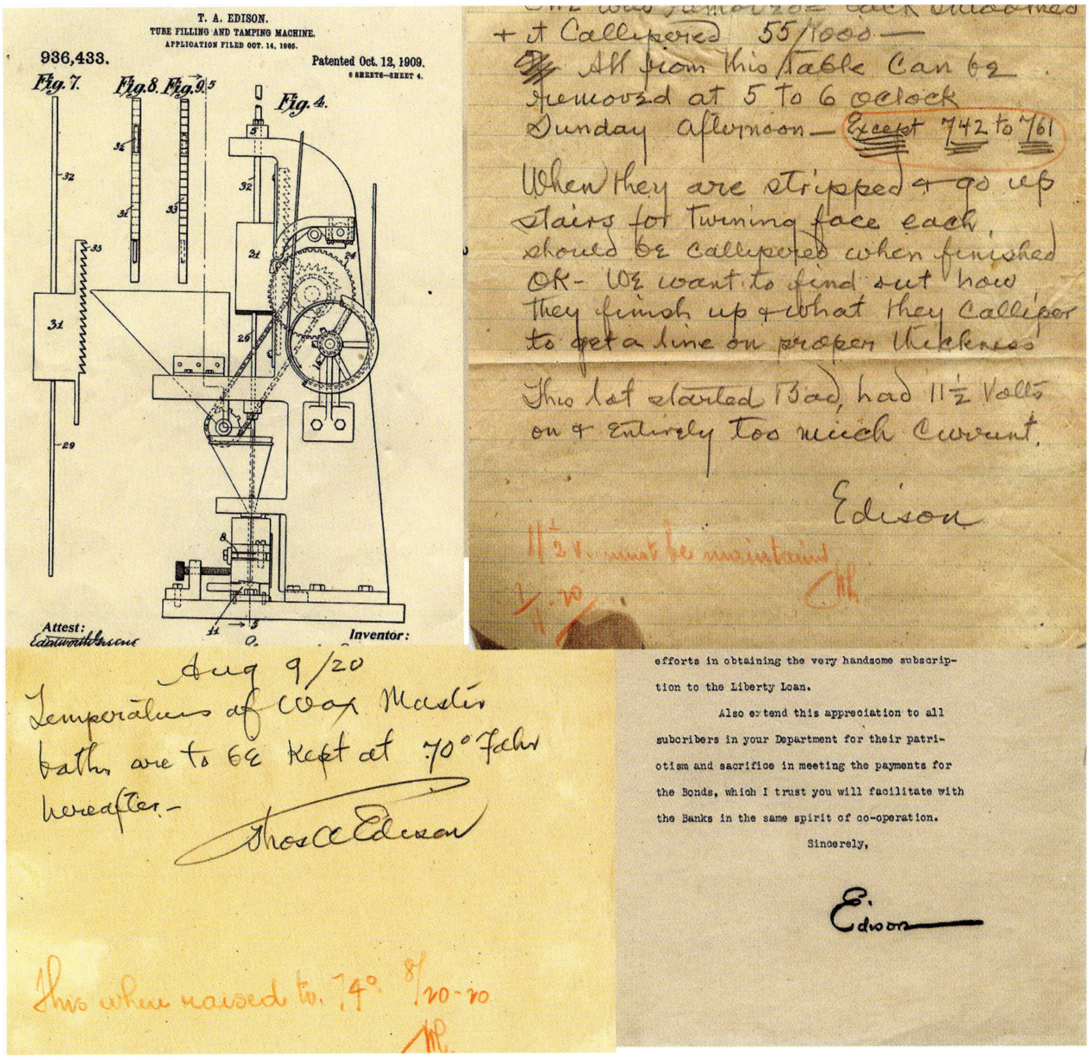

Figure A.1: Pages from Thomas Edison's laboratory notes

The logbook should include the following information:

- A record of the time spent on each aspect of your investigation. An example of some entries in such a record is presented in table A.3.
- Lists and notes taken during the process of selecting your topic and deciding on your aim.

- Lists of resources as they are found and used. Record author, title and other details needed for the bibliography of your report. If the resource is from a library, record the Dewey number so that you can find it again if you need to.
- Notes made during your search for background information
- Lists and diagrams of equipment needed for your investigation.
- A record of people consulted for advice, including your teacher.
- A record of any changes made to your selected topic, your aim or procedures.
- An outline of any difficulties encountered and how they were overcome.
- All of your raw data. Wherever possible, data should be recorded in table form, which is easier to read. These records can sometimes be a little messy because they are recorded as you gather the data, so make sure your data is readable and understandable both to you and your teacher.
- Remember that the data recorded in your logbook will be used to help you to construct tables and graphs in your final report.
- Preliminary drafts of your report, including the development of conclusions and any other thoughts that might help in the discussion section of your report.

Table A.3: Record of progress — some sample entries

Date	Time and location	Task	Comments
16 March	2 pm to 3 pm School library	Researched possible topics	Found lots of resources about parachute design (see list).
4 May	Periods 4 and 5 Laboratory	Set up equipment for the first time	Need more height. Need video camera or much smaller weights.
4 May	7.30 pm to 8.30 pm Home	Remade model parachutes	No problems — lighter plastic, strings and will use smaller weights.

Researching your topic

Before you start planning your investigation in detail, it is a good idea to find out more about your topic. Some background information about your topic might help you in the design of your experiment, the selection of appropriate equipment and in explaining your results later on.

The best place to start your search for background information is your own textbook. Following that, it is definitely worth a visit to your school library. Use the library catalogue too, to find books and journal articles with information about your topic. Other textbooks can also be useful as can the index of a good encyclopaedia, which is often a good place to start looking for general information about your topic.

In your search for background information, it is tempting to spend time looking for resources beyond home and your school library. Keep in mind, however, the constraints of time and assessment criteria before spending too much time and energy seeking background information.

The Internet provides a wealth of information about almost every topic imaginable. However, it can also be a trap in that you can spend a large amount of your time wading through huge amounts of irrelevant (but interesting!) information. You should not need to use it to find background information unless you are unable to find the required information in print at home or at school. In order to use your time on the Internet efficiently, you need to make sure that you know exactly what you are looking for.

Planning ahead

An investigation plan provides you and your teacher with a clear outline of your aim and how you are going to achieve it. It is also necessary to ensure that the equipment that you need is appropriate, available and safe. It is usually a good idea to write this plan in your logbook. The *minimum* information in your written plan should include:

- title
- clear statement of your aim or hypothesis
- list of variables involved in your investigation (include those that you will need to keep constant, those that you will deliberately change and any that you are not able to control), as well as an estimate of the range of values that you expect changing variables to take
- list of the ideas and relationships that are relevant to your investigation
- list of the equipment that you will need for your investigation and an estimate of how long it will take you to set up the investigation
- 'plan of attack' for your investigation, including a statement about what you will measure, how many measurements need to be made and an estimate of how long it will take to make the measurements
- list of resources, other than apparatus, that you will use for background information, and their location.

Experimental design

Variables are the factors that can affect the outcome of an experiment.

The **independent variable** (or **manipulated variable**) is the variable that is manipulated by the experimenter.

The **dependent variable** (or **measured variable**) is the variable that changes in response to the manipulation of the independent variable and is measured by the experimenter.

Controlled variables are variables that are kept the same throughout the investigation.

In order to reach sound conclusions, it is necessary to ensure that your experiments are well controlled. The factors that change in an experiment are called **variables**. In an investigation at this level you are likely to examine the effect of two variables on another. The variables that you deliberately change are called **independent variables** (or **manipulated variables**). The variable that is affected by changing the independent variables is called the **dependent variable** (or **measured variables**). You need to design your investigation so that you can separately examine the effect of each independent variable on the dependent variable. All other variables should be kept constant throughout the investigation. Variables that are kept the same throughout the investigation are called **controlled variables**. Sometimes it is not possible to keep all other variables constant. If this is the case, an explanation needs to be provided in your report.

For example, if you are examining the transmission of sound through polystyrene foam sheets, the dependent variable might be the sound intensity level (in dB). The independent variables might be the frequency of the sound and the thickness of the polystyrene foam. You would need to ensure that you gather enough data to examine the effect of frequency alone and the effect of thickness alone. All other variables should be kept constant (e.g. the sound level intensity in dB on the same side of the sheet as the sound source, the distance between the source and the sheets, and the distance between the sound level meter and the sheets). You might argue that although variables like air pressure and temperature are not kept constant they have little effect on the result. On the other hand, you might be able to minimise the changes in such variables. The reflection of sound from surrounding objects is a variable that could change when you alter the frequency of your sound source. The effect of that variable can be minimised by choosing the location of your measurements wisely. Background noise is a variable that is very difficult to keep constant. However, changes in background noise can be minimised with a wise choice of time and location for your measurements.

All of these aspects of variable control need to be considered during the planning of your investigation and discussed in your report. You should make notes on your thoughts about variable control in your logbook.

Consider this example: Helen has decided that her experiment will determine if the colour of the light that a plant receives affects its growing rate. Her experiment will involve growing a number of plants in separate cupboards, each of which contains a light of a particular colour. So one cupboard will contain a red light, the next a green light, the next blue and the next white. She will measure the height of each of the plants every day for two weeks and compare their growing rates according to the colour of light they were exposed to. Here are the variables in her experiment:

- The independent variable (the factor that she will purposely change) will be the colour of the light a plant is exposed to.
- The dependent variable (the thing that she will measure) is the height of the plant.
- The controlled variables will be the things that she keeps the same for every plant. This will include the amount of water the plants receive, the plant species, the type of soil they are planted in, the intensity (brightness) of the lights, the length of time that the plants are exposed to light and even how far the plants are placed from the light in the cupboard.

Selecting the right equipment

In order to collect reliable data, you need to select or construct equipment that is appropriate. Your choice of equipment should be discussed with your teacher because you need to make sure that the equipment is available. In selecting your equipment, you should consider the questions listed below.

- Is the equipment safe to use?
- If the equipment is to be constructed, is time available to do so?
- Is the equipment appropriate for the task?
- Do I know how to use the equipment, or can I learn quickly?
- Do I understand how the equipment works?
- Will the equipment allow me to measure with sufficient accuracy?
- Can I modify or replace the equipment if something goes wrong or if the investigation changes direction?
- Will I be able to critically evaluate my choice of equipment in my report?

Some events that you might want to observe occur either too quickly or too slowly to record with simple laboratory equipment. You might need more advanced equipment, such as a cathode ray oscilloscope or a video camera to record such events.

A *cathode ray oscilloscope* (CRO) allows you measure the voltage and frequency of output signals from electronic devices such as microphones and photoelectric cells. If the pulses are not displayed on the screen long enough to observe, you can use a video camera to record the signal. Ensure that the camera, CRO and lights are positioned to give you a clear image.

Video cameras are very helpful in recording the motion of objects or people for later analysis. You can record an event and then play it back frame by frame to record position and time intervals. This will allow you to analyse motion in detail. A great deal of care needs to be taken to ensure that your results are reliable:

- You must ensure that the camera remains still. Use a tripod and avoid looking through the viewfinder as you record to ensure that you do not accidentally move the camera while recording.
- Put small, brightly coloured markers on the object or person's clothing so that you can observe the motion of a particular point.
- Place a large ruler (self-made if necessary) near or behind the object or person so that measurements can be made from the TV screen during play-back, and then scaled up to the correct size. The ruler should be as close to the plane of motion as possible so that a parallax error does not occur when the object or person moves from directly in front of the camera.
- When you play back the motion, use a clear plastic sheet and a fine-point marking pen to record the position of the marker in each frame. If there is more than one marker on the object or person, use a different colour for each marker. View the marker from directly in front of the screen when you record its position to avoid a parallax error. The image that you view is behind glass up to 3 centimetres thick.

If you want to use a video camera to record motion involving large distances (like a long jump or a parachute drop), you will need to find a way to solve the problem caused by parallax errors.

Data loggers (produced by both Casio and Texas Instruments) are devices that can be used to record quantities like light intensity, temperature, force, voltage and many other physical quantities over short or long time intervals. This information is transmitted from the appropriate sensor attached to the data logger. For example, a voltage probe attached to a data logger could be used to record the voltage induced in a solenoid when a magnet falls through; such a measurement would be more easily and accurately recorded with a data logger than with a CRO. The data logger can then transfer the data to another device such as a graphics calculator or a computer and then the data can be graphed, manipulated or transferred to another application such as a spreadsheet (see figure A.2).

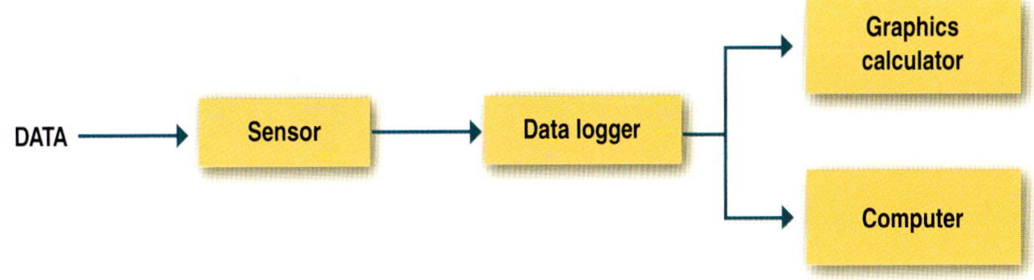

Figure A.2: How data loggers collect data

Data loggers are particularly useful when a large number of events needs to be recorded over a long period (e.g. a voltage every 10 minutes for eight hours) or over very short periods of time (e.g. a change in position measured every 0.001 second for five seconds). They can also be used to control devices. For example, in an investigation to examine the output voltage of a battery used in an appliance that is used periodically, the data logger could be used to switch the appliance on for fixed intervals and a predetermined frequency. The same interface could be used to monitor the voltage output of the battery (see figure A.3).

Figure A.3: A data logger and temperature probe will be used here to measure the temperature every second as ammonium nitrate dissolves in water.

Experimental error

All physical measurements have a degree of uncertainty. The uncertainty can be due to human error or due to the limitations of the measuring instrument. When you make measurements in the laboratory, you can indicate the degree of uncertainty by recording quantities in the appropriate number of significant figures. For example, a time interval measurement of 42.5 seconds suggests that there is a degree of uncertainty in the last figure quoted. However, it does not indicate whether the time interval is accurate to within 0.1 seconds, 0.2 seconds or even 0.5 seconds.

A more precise way of expressing the quantity and its uncertainty would be:

42.5 ± 0.2 s

This expression indicates that the real measurement could be up to 0.2 seconds higher or lower than the measurement that we have come up with, and that the time interval is most likely to be between 42.3 seconds and 42.7 seconds. Similarly, if you express a measurement as:

10.5 ± 0.3 s

you are stating with confidence that the measurement lies between 10.2 seconds and 10.8 seconds. The uncertainty expressed in this example is $\pm$ 0.3 seconds.

Uncertainty can be expressed as a percentage. For example, the percentage uncertainty of the measurement of the time interval above is:

$$\frac{0.2 \text{ s}}{42.5 \text{ s}} \times 100\% = 0.5\%$$

It is important to remember that uncertainty in measurement is not a mistake. One of the challenges of a scientific investigation is to make the uncertainty as small as possible; many experiments routinely have an accuracy of about 10 per cent. The uncertainty in measurements in the laboratory can be minimised if proper care is taken. Some of the precautions to be taken are described below:

- Use the appropriate measuring instrument. For example, you would not attempt to use a sundial to measure the time taken to run 100 metres! Nor should you try to measure the diameter of a copper wire with a ruler; a micrometer should be used.
- Measure multiples of quantities where necessary. For example, if you need to measure the thickness of a sheet of paper, measure the thickness of 50 identical sheets and divide by 50. If you need to measure the time taken for a pendulum to complete one swing, measure the time taken for 10 complete swings and divide by 10. If you wish to work out how many peas are in the pods from a particular pea plant, you would open five pods, count the number of peas in each and then calculate the average. Usually, the more measurements taken, the more reliable your value.
- Take care when reading scales. You should be able to read a set of scales to about one-tenth of the smallest division marked on it. Make sure that you read the scale from directly in front or above and not from the side. For example, because of the distance between the needle of an ammeter and its scale, if you view it from the left you will get a higher reading than if you view it correctly from directly in front. If you consistently view a scale from one side, all of your measurements will be either too high or too low. This type of error is called a parallax error.

Of course, no matter how much care you take with your measurements, some uncertainty will occur owing to the limitations of the equipment itself. Errors such as an incorrect placement of the zero mark or an uneven scale will lead to measurements that are consistently high or low. Errors that are consistently high or consistently low as a result of either the incorrect use of equipment or limitations in the equipment itself are called **systematic errors**. For example, a poorly calibrated thermometer that *consistently* reads two degrees higher than the actual temperature produces a systematic error.

Systematic errors occur as the result of either incorrect equipment use or limitations with the equipment itself. These errors produce values that are consistently higher or lower than the accepted value.

Random errors are caused by unpredictable or unnoticed variations in measurements that occur randomly throughout the course of an experiment.

Errors that occur due to estimation when reading scales or random variation in the quantity being measured due to unnoticed factors (such as a passing breeze tilting the balance scales a little in one of the measurements) are called **random errors**. Random errors can cause higher readings or lower readings irregularly throughout an experiment. As a result, they are often difficult to detect.

Estimating uncertainty

The uncertainty of a quantity due to random errors can be estimated by repeating measurements. The quantity is then specified as the mean (or average) of the measurements. The difference between the highest measurement and the mean will not always be the same as the difference between the lowest measurement and the mean. The uncertainty can be expressed as the larger of the two differences.

Do not waste time looking for equipment that is more accurate than necessary. For example, there is not much point using an electronic timer that is accurate to one-thousandth of a second to record the falling time of a parachute when a stopwatch would be quite adequate. When using digital electronic equipment, think carefully about how many significant figures should be recorded.

In making a decision, take into account the uncertainty (discussed earlier) due to other variables in the physical quantity being measured. Also, consider whether it is worth measuring and recording one variable to many significant figures when other variables are measured to only two, three or four significant figures.

Analysing your results

The quality of the analysis of results ultimately determines the validity of your conclusions. You need to choose appropriate mathematical tools to analyse your data and the errors associated with the data. In most cases, analysis of data and errors will involve the use of graphs.

Displaying uncertainty on a graph

The uncertainty of measured quantities can be displayed on a graph using error bars as illustrated in figure A.4, which shows the part of a graph of period versus length of a pendulum. A line of best fit has been drawn. Even though it does not pass through all of the points plotted, it does pass through all of the error bars. The easiest way to draw a line of best fit is to place a transparent plastic ruler just under the points and reposition it so that you can draw a line with the points distributed evenly above and below the points. Lines of best fit are not always straight. They can be curved in shapes like parabolas and hyperbolas.

Strange results

It is not uncommon to obtain some results that do not quite fit when repeating measurements. When calculating the mean and the uncertainty, it is reasonable to discard such results if a mistake or error of judgement can be identified. However, such apparently strange results should still be recorded in your logbook along with an explanation.

When things go wrong

It is very likely that during the course of your practical investigation, not everything will go according to plan. You might face one or more of the following problems:

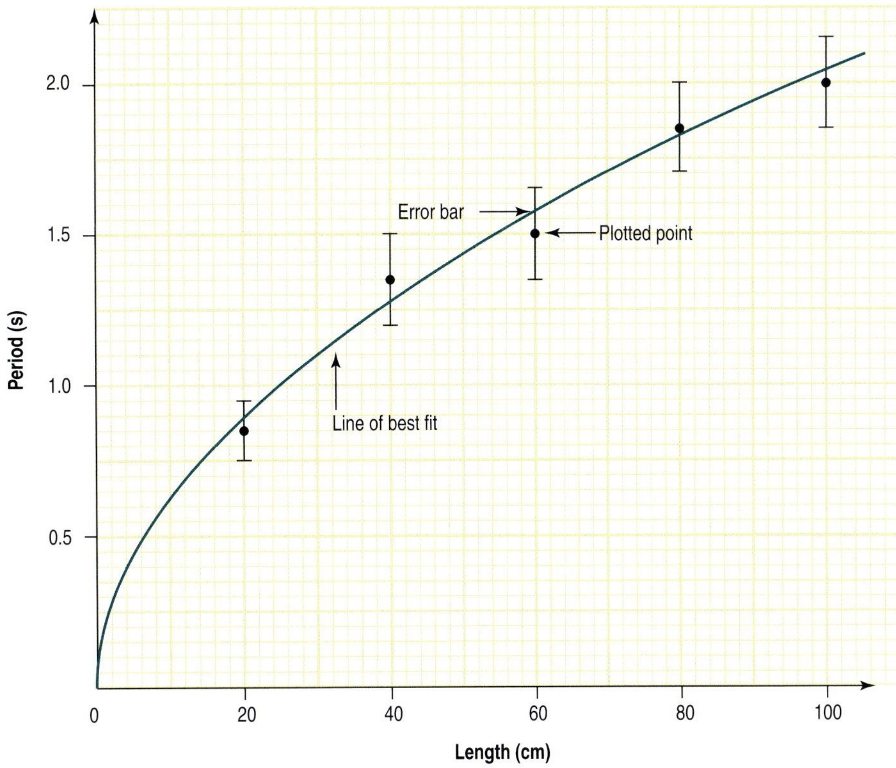

Figure A.4:
A graph showing period versus the length of a pendulum

- your equipment does not work correctly
- it takes you far too long to gather data
- you cannot measure as accurately as you had expected
- you get too many strange or anomalous results
- your results do not seem to make sense
- unexpected variables interfere with your investigation.

These sorts of problems are often encountered by experienced scientists, so do not panic. Problems such as these form part of the process of scientific inquiry.

The way that you respond to problems obviously depends on the nature of the problem. However, the list of suggestions given below might help.

- Reconsider your procedure to check that it is appropriate.
- Ensure that you are using your equipment correctly.
- Check each item of your equipment, and modify or replace it if necessary.
- Check your calculations.
- Discuss your problem with other students.
- Consult textbooks or other references.
- Seek assistance from your teacher.

You need to record the details of the problems and how you responded to them in your logbook.

The effects of problems encountered during your investigation can be minimised if they are detected early. By analysing your data as you go, rather than waiting until all data have been gathered, problems can be detected early enough to allow time for reflection and an effective response. Thorough planning and early checking of equipment by making preliminary observations and measurements should reduce the likelihood of encountering serious problems.

Writing your report

The only way that others can evaluate your investigation is by reading your report. It must describe your investigation completely, concisely and within the prescribed word limit. Your report should be written so that other students are able to read and understand the content.

Your report should include:

- a statement of the purpose and context of the investigation
- a discussion of the physics on which the investigation is based
- a list and description of the equipment used in your investigation
- a description of your procedure
- observations and measurements
- an analysis and interpretation of your data
- a description of the problems encountered and how you responded to them
- an evaluation of the success of your investigation
- conclusions
- a list of resources used and acknowledgement of any assistance or advice given.

Before you start writing your report, ensure that you know the:

- criteria used for assessment
- word limit
- format required by your teacher.

Profile of a report

There is no single universally accepted format for writing reports of practical investigations. The headings used below are typical. Your teacher will indicate exactly which headings should be used.

Abstract. An abstract comprises a brief summary of your investigation and its main conclusions, and is usually included at the beginning of the report. It defines the scope and nature of the investigation. You should also include in the abstract a statement of the aim of your investigation. Even though it appears at the beginning of your report, it is best not to write the abstract until after you have completed the rest of your report.

Introduction. In the introduction you can describe the context of your investigation. Explain briefly why the topic is relevant and discuss the science on which the investigation is based. List the variables that are relevant to your investigation. Identify those that you will deliberately change, those that you will control and those that you are either unable to control or do not need to control.

Aim. A clear statement of the specific purpose of your investigation is needed. If a hypothesis is proposed, it should be included immediately after the statement of your purpose. Reasons for your hypothesis should be stated.

Method and materials. A detailed description of how you carried out your investigation is essential. An incomplete or poor description of your equipment and procedures makes it difficult for the reader to interpret your data and conclusions. A complete list of the equipment used should also be included, along with an explanation of how it was set up and why it was used. Include labelled diagrams where necessary. The description should include enough detail so that the reader does not have to guess how any of the steps were undertaken. It should demonstrate to the reader that data have been collected carefully and accurately, and that variables have been controlled. Describe any modifications made to your planned procedure during the course of your investigation.

Results. Data should be presented clearly, neatly and in such a way that patterns and relationships can be easily seen. Wherever possible, observations and measurements should be presented in tables and the appropriate graphs should be drawn. Calculations or sample calculations should be included with your data. Use the appropriate number of significant figures and include uncertainties in your tables and graphs wherever possible. Use error bars and lines of best fit on graphs. If you have used a spreadsheet for data organisation and analysis, think carefully about whether a hand-drawn graph would provide more information than one generated by the spreadsheet software. Figure A.5 illustrates a sample table and graph.

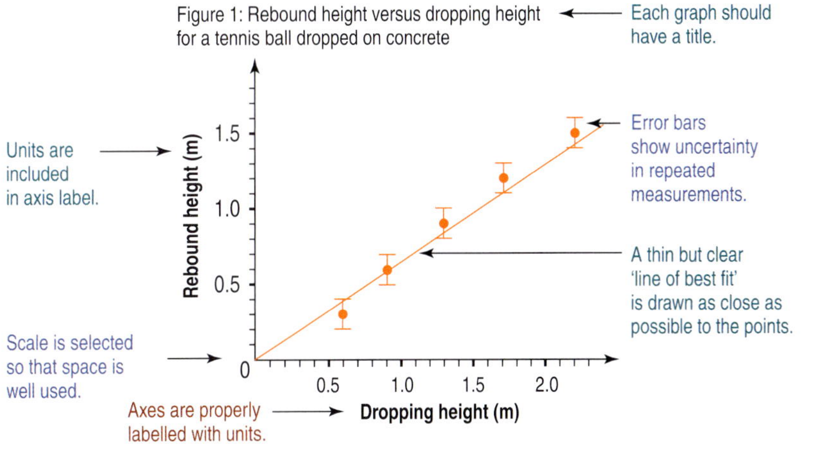

Table 1: Voltage and electric current across a resistor ← Each table should have a title.

Include units in column headings.

Trial	Voltage (V)	Electric current (mA)
1	2.4	423
2	2.3	425
3	2.5	427
4	2.4	428
5	2.4	426
Mean	2.4 ± 0.1	426 ± 3

Include clear headings.

Rule rows and columns to make data in tables clear.

Units are not needed in the body of the table, because they are in the headings.

Include uncertainties for repeated measurements.

Include mean value of repeated measurements.

Figure A.5: A sample graph and table

Note the following requirements when drawing a graph:

- Give each graph a title.
- Show the units on the graph.
- Properly label the axes with units.
- Use error bars to show uncertainty in repeated measurements.
- Draw a thin but clear line of best fit as close as possible to the points.
- Select a scale to best use the space.

Similarly, for tables:

- Give each table a title.
- Include units in column headings
- Rule rows and columns to make data in tables clear.
- Include clear headings.
- Units are not needed in the body of the table as they are in the headings.
- Include uncertainties for repeated measurements.
- Include the mean value of repeated measurements.

Discussion. A critical analysis of your data needs to appear in your report. You should discuss the meaning of your results and how they relate to the aim of your investigation. You should also discuss the degree of success of the investigation and the limitations of your equipment and procedures. Any weaknesses in the design of the investigation should be pointed out along with suggestions for improvements. Identify sources of errors and summarise their effect on your investigation.

Conclusions. A clear statement of what you found out is written under this heading. It is important that you read your statement of the purpose of your investigation again before you write your conclusions. Your conclusions must be relevant to the purpose of the investigation. They must also be consistent with the data obtained and the limitations of your investigation. If you included a hypothesis, you must state whether or not it was supported.

Bibliography. A list of books, journals and other printed or audiovisual material resources to which you have referred must be included. For general references, list the relevant chapters. For specific references, list the page numbers used. The list should include enough information to allow the reader to locate the resource. The information should be included in the following order:

Books:

Author; book title (in italics); place of publication; publisher; date of publication; chapter or page numbers referred to.

Journal articles and web pages:

Author (surname, first name); title of article (in quote marks); name of journal (in italics) or URL (in angle brackets); date of issue or publication; page numbers referred to.

Here are some examples:

> Adams, Douglas, *The Hitch Hiker's Guide to the Galaxy: A Trilogy in Four Parts,* London, Heinemann, 1989.
>
> Williams, Caroline, 'Elephants on the Edge', *New Scientist*, 18 February 2006, pp.39–41.

NIDA, 'NIDA InfoFacts: High School and Youth Trends', <*www.nida.nih.gov/infofacts/hsyouthtrends.html*>, accessed 22 May 2006.

Acknowledgements. This is a list of people and organisations who gave you assistance or advice.
Ensure that you state the nature of the assistance or advice given.

Appendices. The appendices are part of your report and are provided so that readers can choose to look at them if they need to. Your logbook should be included as an appendix. Hard copies of data recorded in spreadsheets should also be included, along with any other necessary attachments that are not needed in the main body of your report.

Research paper

A research paper is an essay describing your own ideas or opinions of a topic supported by the information and findings of other sources. These other sources are generally publications of works written by experts in their particular field.

Purpose

Research papers can be in the form of analytical essays or argumentative essays. The purpose of an analytical research essay is to use evidence to consider facets of a particular issue, while an argumentative research essay uses evidence to convince the reader of your opinion with regard to a particular issue or argument.

Topic

The topic is a general area of inquiry that forms the basis of your initial research. It may be dictated by the assessment item or by the subject you are studying. Your topic will be quite broad, for example 'Species extinction' or 'Road safety'. Preliminary information on the topic should be gathered and read as soon as possible, so that you may determine just how much is available. If a topic you have chosen for yourself has very little written about it, finding this out at the beginning allows you to pursue another topic instead before it is too late.

Thesis statement / research question

Once you have an idea of what others have written regarding your topic, you should now begin to formulate a more specific idea as to what your essay is going to be about. The thesis statement for an argumentative essay (or the research question for an analytical essay) is a one- or two-sentence statement that outlines the purpose of the essay. This statement should signpost the direction that your essay will take.

It is important that a thesis statement or research question is specific. A statement such as 'Forensic science is really useful' is too general; 'How much information can be gained from a study of blood patterns at a crime scene?' is better.

The thesis statement or research question is usually found near the end of the introduction.

Research

Based on your preliminary research into the topic, you should have a good idea of what sources are available that are likely to be useful. For scientific research essays, you should try looking at the following sources:

- textbooks
- general science journals such as *New Scientist, Scientific American, Australasian Science, Australian Geographic, Popular Science*
- for more specialised interests, look for specialist magazines and journals (e.g. researching different car designs may lead you to magazines such as *Wheels* or *The Road Ahead*)
- newspapers
- Internet
- transcripts of radio and television programs (e.g. *The Science Show* and *Catalyst*), which are often available on the appropriate station's website.
- podcasts.

Most of these will be available through your local library or, if nearby, a university library.

It is essential that you keep track of your sources because you must reference any works that you use, whether directly or indirectly. Some students find it useful to create a file in which they place photocopies of the relevant pages, each source clearly marked with the title, author and publishing details.

Thinking about it

This is a very important part of the writing process and should occur throughout the research and structuring phases. Once you have read the research material, reflect on what you have learned:

- What are my own immediate responses to the thesis statement / research question?
- What evidence/quotes/works have I found that support my response?
- Do I know enough to answer the question? If not, what do I still need to find out?

Structuring the essay

The way in which you organise your essay is important so that the reader can follow your line of reasoning easily. The building block of the essay is the paragraph. A paragraph in a research essay is usually between 100 and 200 words in length. This allows you to plan roughly how many points you can make for an essay of a particular length. This, in turn, gives you an idea as to how much detail your research needs to go into. For example, if your essay has a limit of 1000 words, you can allocate roughly 200 to 300 words for the introduction and conclusion combined, leaving 700 to 800 words to be made up in the body. This in turn indicates that the body will contain three or four paragraphs. You can then reason that you should limit yourself to making three or four points or arguments in an essay of this length.

Essays are organised into three main parts:

1. introduction (one paragraph)
2. body (one paragraph for every new idea or argument)
3. conclusion (one paragraph).

Introduction. This contains background information or ideas that lead to the posing of the research question or thesis statement. It may consider the general topic to start with and then narrow the focus to the specific purpose of this essay. Avoid including the words 'In this essay, I will ...' or 'The purpose of this essay is to ...'. The purpose of your essay should be clear, and should not need overt explanation! Strangely, many people (including the author) find it easier to write the introduction last, but this varies between individual writers.

Body. This is where you make each of your individual points, backed up by evidence from your research. Just as the thesis statement indicates the direction in which your essay will go, so the opening sentence for each paragraph should indicate the content of that paragraph. Remember that each new idea should have a new paragraph and a topic sentence.

Conclusion. This should briefly remind the reader of the points that have been made in the essay and then present an evaluation of what these points indicate. In doing this, you should ensure that your summations are consistent with your original thesis.

Drafting and revising

Keeping to the previous guidelines, you should come up with a first draft for your essay. This is not the version that you hand in for assessment! Once your first draft is written, you should leave it for a day or two and work on something else. That way, when you come back to read it again, you are far more likely to spot mistakes such as clumsy grammar, poor sentence construction or unsupported arguments. Be ruthless when you look at your draft!

- Check your assessment criteria and, if your essay is missing something vital, fix it immediately.
- Is your meaning completely clear in your arguments?
- Do not be afraid to cross out whole paragraphs if necessary.
- One of the best things that you can do to check for grammatical errors is to get someone else to read it out loud to you. Problems with grammar and sentence construction are often easier to pick up when you hear them rather than when you see them written down.
- If your teacher offers to mark a draft before submission, take advantage of it. It is better to find out what the problems are before the work is formally assessed. Take notice of any suggestions that are made and act on them.
- Do not leave your first draft until the night before it is due. It will not be your best work!

The drafting and revision process can often take the longest time, but it is absolutely essential that it be done. While many writers revise only twice, some spend months working on revisions.

Media file

A media file is a collection of articles and television and radio program transcripts covering a particular issue, and is usually best suited to current controversial topics. This collection is accompanied by a summary of their content (usually in data table form) and may also include a critical analysis (a brief analytical research essay based on the media file contents). It takes at least a month to collect a good range of articles for a media file.

Collecting articles

Once you have established the issue on which you are basing your media file, collecting articles is simply a matter of going through every newspaper and magazine that you can get access to and finding appropriate items. You may either tear the page out (if it belongs to you) or take a photocopy of the article. Make sure that every item has the name of the publication, the date of publication and the page number, as well as the article title and the author's name noted on it. You might also be able to find transcripts for some television programs such as *Catalyst* and *Four Corners* on their websites.

Collecting should start as soon as possible because a good media file cannot be created the day before it is due.

Organising articles

Usually you will be directed to have a specific number of items in your file. This could be anywhere from eight to twenty articles, depending on the assessment conditions. By the time you come to organise your file, you should have been collecting for weeks and have a folder full of items. This will give you a lot of choice as to what you include. The items you select should reflect the file's purpose, variety of viewpoints and range of media.

Purpose

What you select depends on what you want your file to do. The two most common selections are conflicting viewpoints of an issue or the timeline of an event as it unfolds.

Variety of viewpoints

Try to select articles that are written from different perspectives. For example, a media file on reproductive technology could include a human-interest item about a woman who had an IVF baby after many years of trying, a scientific article about the dangers of IVF and a treatise on the ethics of artificial insemination written from a religious perspective.

Range

A media file is most useful when it covers more than one type of media, so try to avoid having it full of nothing but newspaper articles or only full of magazine articles. Remember that media includes art and music. Sometimes a cartoon or a protest song says just as much as an essay, so these may also be added to your file if appropriate.

File summary

Once you have selected the items for your file, you will need to read them in order to evaluate what they are saying. Your file summary will usually be in table form and the headings will depend upon the requirements of your course. The example provided in table A.4 is fairly typical.

Table A.4: Example of a media file summary				
Title	**Author**	**Source**	**Genre**	**Emphasis**
'Lost in Space?'	JJ Hart	*New Scientist;* 14/7/05 pp. 34–5	Scientific article	The dangers of sending humans into space
'Space waste'	LHD	*The Courier-Mail;* 12/3/05 p. 23	Letter to the editor	Money should be spent on health care instead of space travel
'Diary of an astrogirl'	Sue Gale	*Women's Weekly;* May 2005; pp. 20–22	Feature article	Life on board the International Space Station

Scientific journal

A journal is to the scientist what a visual diary is to the artist. It is a tool in which ideas, interpretations, research notes and reflections can be recorded for future reference. Depending on the nature of your Science 21 course, a journal may span as little time as a term or as much as an entire two-year course. As a result, the physical size of the journal itself will vary. However, exercise books are usually a good place to start, although some people who collect a variety of different media for their journal actually have a 'journal box'.

Journals are particularly useful when you are gathering ideas and information for research work. Your journal may include:

- personal writings or opinions
- interpretations of research findings, articles, references or news clippings
- reflections on a topic
- critiques of scientific journals or texts
- drawings, photos, videos, recordings and transcripts
- records of and reflections on conversations, interviews, discussions, group work
- teacher notes, draft feedback and clarification of points.

Appendix B: The periodic table of elements

Key

Atomic number	79	
	Au	Symbol of element
Atomic weight	197.0	
	Gold	Name of element

1 H 1.008 Hydrogen																	2 He 4.003 Helium
3 Li 6.941 Lithium	4 Be 9.012 Beryllium											5 B 10.81 Boron	6 C 12.01 Carbon	7 N 14.01 Nitrogen	8 O 16.00 Oxygen	9 F 19.00 Fluorine	10 Ne 20.18 Neon
11 Na 22.99 Sodium	12 Mg 24.31 Magnesium											13 Al 26.98 Aluminium	14 Si 28.09 Silicon	15 P 30.97 Phosphorus	16 S 32.07 Sulfur	17 Cl 35.45 Chlorine	18 Ar 39.95 Argon
19 K 39.10 Potassium	20 Ca 40.08 Calcium	21 Sc 44.96 Scandium	22 Ti 47.87 Titanium	23 V 50.94 Vanadium	24 Cr 52.00 Chromium	25 Mn 54.94 Manganese	26 Fe 55.85 Iron	27 Co 58.93 Cobalt	28 Ni 58.69 Nickel	29 Cu 63.55 Copper	30 Zn 65.41 Zinc	31 Ga 69.72 Gallium	32 Ge 72.64 Germanium	33 As 74.92 Arsenic	34 Se 78.96 Selenium	35 Br 79.90 Bromine	36 Kr 83.80 Krypton
37 Rb 85.47 Rubidium	38 Sr 87.62 Strontium	39 Y 88.91 Yttrium	40 Zr 91.22 Zirconium	41 Nb 92.91 Niobium	42 Mo 95.94 Molybdenum	43 Tc [97.91] Technetium	44 Ru 101.1 Ruthenium	45 Rh 102.9 Rhodium	46 Pd 106.4 Palladium	47 Ag 107.9 Silver	48 Cd 112.4 Cadmium	49 In 114.8 Indium	50 Sn 118.7 Tin	51 Sb 121.8 Antimony	52 Te 127.6 Tellurium	53 I 126.9 Iodine	54 Xe 131.3 Xenon
55 Cs 132.9 Caesium	56 Ba 137.3 Barium	57–71 * Lanthanides	72 Hf 178.5 Hafnium	73 Ta 180.9 Tantalum	74 W 183.8 Tungsten	75 Re 186.2 Rhenium	76 Os 190.2 Osmium	77 Ir 192.2 Iridium	78 Pt 195.1 Platinum	79 Au 197.0 Gold	80 Hg 200.6 Mercury	81 Tl 204.4 Thallium	82 Pb 207.2 Lead	83 Bi 209.0 Bismuth	84 Po [209.0] Polonium	85 At [210.0] Astatine	86 Rn [222.0] Radon
87 Fr [223.0] Francium	88 Ra [226.0] Radium	89–103 ** Actinides	104 Rf [261.1] Rutherfordium	105 Db [262.1] Dubnium	106 Sg [266.1] Seaborgium	107 Bh [264.1] Bohrium	108 Hs [277] Hassium	109 Mt [268] Meitnerium	110 Ds [281] Darmstadtium	111 Rg [272] Roentgenium	112 Uub [285] Ununbium	113 Uut [284] Ununtrium	114 Uuq [289] Ununquadium	115 Uup [288] Ununpentium	116 Uuh [292] Ununhexium	117 Uus Ununseptium	118 Uuo Ununoctium

* Lanthanides

57 La 138.9 Lanthanum	58 Ce 140.1 Cerium	59 Pr 140.9 Praseodymium	60 Nd 144.2 Neodymium	61 Pm [144.9] Promethium	62 Sm 150.4 Samarium	63 Eu 152.0 Europium	64 Gd 157.3 Gadolinium	65 Tb 158.9 Terbium	66 Dy 162.5 Dysprosium	67 Ho 164.9 Holmium	68 Er 167.3 Erbium	69 Tm 168.9 Thulium	70 Yb 173.0 Ytterbium	71 Lu 175.0 Lutetium

** Actinides

89 Ac [227.0] Actinium	90 Th 232.0 Thorium	91 Pa 231.0 Protactinium	92 U 238.0 Uranium	93 Np [237.0] Neptunium	94 Pu [244.1] Plutonium	95 Am [243] Americium	96 Cm [247.1] Curium	97 Bk [247] Berkelium	98 Cf [251.1] Californium	99 Es [252.1] Einsteinium	100 Fm [257.1] Fermium	101 Md [258.1] Mendelevium	102 No [259.1] Nobelium	103 Lr [262.1] Lawrencium

The names for elements 111–118 are temporary.
For elements with no stable nuclides, the mass of the longest living isotope is given in square brackets.

Figure B.1: The periodic table of elements

Appendix C: Physical constants and values

Table C.1: Physical constants and values

Quantity	Value
One light year	9.5×10^{15} m
Absolute zero	–273.15°C = 0 K
Acceleration due to gravity	9.801 m/s^2
Speed of light in a vacuum	2.998×10^8 m/s
Boiling point of water	100°C
Freezing point of water	0°C
Speed of sound in air (20°C)	343 m/s
Radius of Earth (average)	6.38×10^3 km
Earth–Moon distance (average)	3.84×10^5 km
Earth–Sun distance (average)	1.496×10^8 km
Mass of Sun	1.99×10^{30} kg
Mass of Earth	5.97×10^{24} kg
Mass of Moon	7.36×10^{22} kg

Appendix D: Answers to selected end of chapter questions

Chapter 1: Life, the universe and everything

13. (a) E
    (b) D
14. (a) $\frac{1}{4}$
    (b) $\frac{1}{16}$
20. (c) (i) ~ 500 000 years (ii) ~ 70 years
    (d) 600 000–700 000 years ago

Chapter 2: The restless Earth

10. (a) 1.5
    (b) 15 s
    (c) ~ 175 km
15. (a) 145°C
    (b) 6.5 km

Chapter 3: Colonising other worlds

6. (a) 12.8 m/s^2
   (b) 39.27 s
   (c) One revolution every 62.8 s
15. (a) 109 light years
    (b) 1.03×10^{15} km
17. (a) F, white
    (b) A, blue–white
    (c) M, red
18. (b) 8 solar masses
20. (b) Approximately 90 km

Chapter 4: Nuclear energy — Australia's energy future?

6. (a) 92
   (b) 235 – 92 = 143
   (c) 235
11. 7 500 (These X-rays would need to be taken over a very short period of time for acute symptoms to occur.)

Chapter 5: Weapons through the ages

6. (a), (d), (e) and (f)
9. 1 272 m
10. 39.3°
11. (b) About 0.15 m/s
18. (a) (i) 449 649 people
        (ii) 27 950 people
19. 160 000 000 Joules (or 160 MJ)
20. The B-53 is 600 times more powerful than the Hiroshima bomb.

Chapter 6: The brave new world of genetic engineering

4. (a) 32
   (b) 31
   (c) 63
25. (c) The Zebu has a breathing rate of 35/min while the Holstein has a breathing rate of around 95/min. Therefore, the breathing rate of the Zebu is about 36 per cent that of the Holstein at this temperature.

Chapter 7: Survive that drive!

3. 89 km/h
4. (a) 80 km/h
   (b) 90 km/h
   (c) 52 km/h
5. 2.4 m/s^2
13. (b) 33.4 m
14. (a) 12 m
    (b) 24.4 m
    (c) 36.4 m
15. 6.7 m/s
16. 17.6 m/s
17. (a) 5.56 m/s
    (b) 27.8 m/s
    (c) 54 km/h
    (d) 180 km/h
18. The dog catches up with the car after 17.6 seconds.

Chapter 8: What's your poison?

11. (a) BAP = $(0.016 \times N) - (0.015 \times H)$
    (b) Bill will need to wait a further 3.2 hours without alcohol before he is legally able to drive home.

Chapter 9: Crime squad

18. The woman would have been about 144.9 cm tall.
21. Around 4.00 am

Glossary

a

absolute temperature scale: devised so that absolute zero is equal to –273.15°C, and a difference of one degree in the absolute temperature scale is equivalent to a one-degree difference in the Celsius (Centigrade) scale 92

absolute zero: the lowest temperature that is theoretically possible. It is equal to –273.15°C or 0 K 92

absorbed dose: the amount of radiation energy that a person has absorbed per kilogram of body mass. It is measured in grays (Gy). 110

acceleration: the rate at which an object changes its velocity 201

accelerator mass spectrometers (AMSs): devices that allow the number of carbon-14 atoms in a sample to be counted with high accuracy 28

accretion: the process by which a star or planet gains mass when material is pulled into it by gravitational attraction 4

acute radiation syndrome: caused when a person is exposed to high levels of radiation over a short period of time 111

aerofoil: any surface that extends into the air so as to create lift, such as an aircraft wing 129

agglutination: the process in which red blood cells are clumped together by hostile blood antigens 263

algor mortis: the process by which the human body cools after death until it reaches the same temperature as its surroundings 251

alleles: are possible gene variations 166

alloy: a combination of metals or a metal mixed with non-metals. For example, brass is an alloy of copper and zinc. 195

amber: a transparent yellow, orange or greenish material that has formed from petrified tree sap. It is often used in jewellery. 21

amine neurotransmitters: a family of neuro-transmitters made up of compounds that have nitrogen molecules attached to them. This group includes the neurotransmitters acetylcholine, dopamine, serotonin and norepinephrine. 218

amphometer (or Digitector): a device that measures the time elapsed between the activation of two sensor cables 210

amplitude: the height of a wave 64

anaerobic respiration: the process by which food is turned into energy without the use of oxygen 10

animals: multicellular organisms that use other organisms as a food source that they digest internally 11

asteroid belt: the belt of rocky debris that surrounds the sun between the orbits of the planets Mars and Jupiter 4

asthenosphere: a section of the mantle on which the lithosphere plates move 39

astrophysicist: a scientist who uses physics to explain the forces acting on and between objects in space 92

atherosclerosis: the narrowing and hardening of arteries 230

atomic mass: equivalent to the number of nucleons (protons and neutrons) in the nucleus of an atom 108

atomic number: the number of protons in the nucleus of an atom 108

axon: a long extension from the soma that transmits electrochemical impulses to other neurons 217

b

background radiation: radiation that is always present, coming from sources such as the sun, space, soil, building materials and medical equipment 110

ballistics: the study of the interaction between firearms and their projectiles 267

Big Bang Theory: proposes that all of the matter and energy presently in the universe began as a single super-dense mass that exploded and expanded outwards 2

billion years: equal to 1 000 000 000 years 2

blood alcohol concentration (BAC): a measure of the amount of alcohol in grams that is present in a 100 mL sample of blood 232

blue-shifted: describes the change in light frequency of stars towards the blue end of the spectrum when they are moving towards us 92

braking distance: the distance required for the brakes to bring the car to a halt 197

c

carrier: an organism that carries the gene for a specific trait in its genotype but does not show that trait in its phenotype 167

centripetal force: the inward-acting force that acts on an object moving in a circle 82

chemosynthesis: the process by which organisms make organic compounds from elements other than oxygen 10

chromosomes: long thread-like structures made of DNA that are found in the nucleus of cells 163

circumstellar habitable zone (CHZ): the region around a star where a planet would need to be located in order to have liquid water on its surface. Also known as a habitable zone 94

cloning: the process by which an exact genetic copy of an organism is made 172

coefficient of friction (μ): a measure of the friction between two surfaces in contact with each other 205

complementary base-pairs: pairs of nucleotide bases that form the cross-links of the DNA molecule 162

compression wave (or longitudinal wave): a wave in which the disturbance moves in the same direction as the wave 46

continental crust: the thicker regions of the Earth's crust, made of a variety of less dense rock types. 38

controlled chain reaction: not all of the neutrons released cause the fission of further nuclei 113

controlled variables: variables that are kept the same throughout the investigation 283

convection: the cycle in which hot, less dense material rises and cooler, denser material sinks 41

convergence: when two plates are moving towards each other 43

crenation: the jagged edging around the rim of a blood spot that forms when the blood droplet strikes a surface 261

critical mass: the smallest spherical mass that will sustain an uncontrolled chain reaction 112

crumple zones: sections placed at the front and rear of cars to increase the collision time and so reduce the forces acting on the passengers in a collision 192

d

daughter nucleus: the nucleus formed when an unstable nucleus undergoes decay 109

decay chain or decay series: the sequence of decay that an isotope undergoes until it forms a stable nucleus 109

dendrites: spiky extensions from the soma that receive impulses from other neurons 217

dependent variable (or measured variable): the variable that changes in response to the manipulation of the independent variable, and is measured by the experimenter 283

dilemma zone: a region before a set of traffic lights where a driver's course of action (to either accelerate or brake) on seeing a yellow light is not clear 202

divergence: when two plates are moving away from each other 43

DNA (deoxyribonucleic acid): a long molecule that is the carrier of genetic information 161

dominant allele: the characteristic that will be expressed whenever it appears in a genotype 166

Doppler effect: the apparent change in frequency observed when there is relative movement between a wave source and an observer 91

dormant: volcanoes that are not currently active but have the potential to erupt 60

dose equivalent: the biological effect of radiation that has been absorbed by living tissue, and is measured in sieverts (Sv) 111

e

earthquake: occurs when the stresses between locked plates and rocks are suddenly relieved, causing sudden land movement and the release of energy 46

electrolysis: electric current is passed through a liquid to cause a chemical change 78

electromagnetic pulse: very strong electromagnetic fields formed by an atomic blast suddenly collapse; this can cause electrical and electronic systems to burn out 149

electrons: negatively charged particles orbiting the nuclei of atoms 107

enrichment: the process by which the percentage of uranium-235 in a uranium sample is increased so that it can sustain a chain reaction. Rather than the 0.7 per cent uranium-235 found in normal uranium ore, enriched uranium may contain 1–4 per cent uranium-235 114

eons: the four major units into which the time that has passed between the Earth's birth and the present time is divided 5

epicentre: the point on the surface of the Earth that lies directly above the focus 46

epochs: the subdivisions of periods 5

eras: the three subunits into which the Phanerozoic eon is divided 5

evidence: material that may support a theory as to how, where, when, why and by whom a crime has been committed 246

exoplanets or extra-solar planets: planets in orbit around stars outside our own solar system 89

extinct: volcanoes that have not erupted for thousands of years and are effectively dead 60

f

fast reactors: use neutrons that have high speed and energy to cause nuclear fission 114

fibrillation: when a heart beats rapidly but ineffectively, creating an inadequate supply of oxygen to the body 231

first-class lever: one in which the fulcrum lies between the load and the effort 134

fission bombs: another name for atom bombs — devices that use an uncontrolled chain reaction to produce destructive energy and heat 147

focus: the location inside the Earth of the fracture or faulting that causes an earthquake 46

forensic: 'of the law court' 244

forensic anthropology: the study of skeletal remains 249

freefall: floating effect that objects experience when both they and their environment are falling at the same rate 80

frequency: a measure of how many waves pass a particular point in one second 91

friction: the force that acts between two contacting surfaces resisting their motion past each other; caused by interaction of surface irregularities and molecular bonds 203

fusion bombs: use the nuclear fusion of hydrogen nuclei to release energy 148

g

gas chromatography: a process by which the mixture to be analysed is vaporised and mixed with a gas that then carries it through a column of absorbent material 272

gas giants: large planets composed almost entirely of gases. Jupiter, Saturn, Uranus and Neptune are the gas giants in our solar system. 4

Geiger counter: a device that measures the rate at which a material is undergoing decay. It does this by detecting the radioactive particles produced during the decay. 28

gene sequencing: identifying the exact order in which the bases occur on the chromosomes of an organism 167

gene therapy: a form of treatment in which specific sections of DNA are introduced into a tissue or organ in order to compensate for or replace DNA that is defective 175

genes: sections of DNA carried in the chromosomes that provide instructions for the production of specific proteins 164

genetic engineering: the technology by which the genetic make-up of an organism may be manipulated to fulfil a specific purpose 170

genotype: an organism's gene combination that it will be carrying for a specific characteristic 167

geodesic domes: domes made of interlocking steel struts that are arranged in small triangles enclosing a panel (usually glass). They were first designed by the architect Richard Buckminster Fuller. 85

geological timescale: a chart that depicts the Earth's history as found in the geological and fossil records of its past 6

geology: study of the structure of the Earth 23

germline therapy: treatment of genetic diseases in embryos by replacing faulty sections of DNA with healthy DNA 182

Glaister Equation: allows an approximate time of death to be calculated based on the extent to which a body has lost heat 251

global warming: term used to describe the present increase in the average temperature of the Earth's atmosphere over time 118

ground zero: location on the ground directly under the detonation point of an atomic weapon 148

gyroscopic motion: demonstrated by any object spinning about its axis. Such an object resists any changes in the direction of the axis by giving rise to opposing forces. 131

h

Hadean: comes from the word *Hades*, which is another word for Hell 6

half-life: the time taken for half of a group of unstable nuclei to decay 23

heart attack: a biological event that occurs when the narrowing or blockage of the blood vessels that supply the heart interrupts the supply of oxygenated blood to the heart muscle 231

Human Genome project: a project in which several institutions around the world combined their efforts to sequence all of the human genetic code 167

hydroplaning: when water is not carried from underneath the tyres quickly enough by the wheel tread, causing the car to ride on a thin cushion of water rather than making solid contact with the road 208

hypothesis: a possible explanation for a series of observations 247

i

independent variables (or manipulated variables): the variable that is manipulated by the experimenter 283

inertia: the tendency of an object to resist changes to its motion. Newton's First Law of Motion (the Law of Inertia) states that an object at rest will remain at rest, and an object in motion will maintain its motion, unless an outside force is exerted on the object. 191

instantaneous speed: speed of travel at a specific moment in time 209

intra-plate earthquakes: occur when the plate flexes in response to stressing forces on the edge of the plate 46

ions: atoms that have an overall charge due to either the loss or gain of electrons 107

isotopes: different forms of the same element that have the same number of protons but a different number of neutrons 23

k

karyotype: shows the chromosomes present in an individual's body cells 163

Kuiper Belt: a region of icy bodies orbiting the sun beyond the orbit of Neptune 4

l

latent fingerprints: fingerprints that are not readily observable on a surface 254

lava: molten rock that comes out of a volcano onto the Earth's surface, where it forms igneous rock when it solidifies 58

lift: the upwards-acting force created by a wing moving through the air 129

light-year: the distance that light travels in a year. It is equivalent to 9.461×10^{15} metres. 90

lithosphere: a series of plates made up of the crust and the top layer of the mantle, forming the outermost skin of the Earth 39

lividity: the draining of blood to the lower portions of the body due to the influence of gravity 252

Locard Exchange Principle: indicates that, when a crime is committed, the perpetrator will leave traces behind and will carry traces of the crime scene or victim away with them 245

logarithmic: a relationship in which a single unit increase in the independent variable results in tenfold increases in the dependent variable 53

Lorentz transformations: a set of calculations by the mathematician Hendrik Lorentz that predict the effects of travel at speeds approaching the speed of light 99

m

magma chambers: underground cavities, usually found beneath volcanoes, that contain magma 57

magnitude: a number assigned to an earthquake that reflects the destructive energy it has released 53

Manhattan Project: the codename used in World War II for the secret US project to beat Germany in developing the first atomic bomb 146

mares: large lava plains found on the surface of the moon 84

mass extinctions: large numbers of different species die within a short period of time, often as the result of a single catastrophic event 17

Mercalli intensity scale: used to describe the effects of an earthquake in a particular region 54

microbe: an organism so small that you need a microscope to see it 10

micrometeorite: a dust-sized piece of space rock 85

millisecond: 0.001 seconds 193

Mohorovicic discontinuity ('Moho'): the region where the bottom of the Earth's crust meets the upper layers of the mantle 39

momentum: the product of an object's mass and its velocity 141

monogenic traits: the result of a single gene 166

multicellular: made up of more than one cell 11

n

natural selection: the process by which organisms exhibiting characteristics that increase their chance of survival in their environment pass these traits to their offspring, thereby increasing the incidence of that characteristic in the organism group 12

nebula (plural: nebulae): a glowing cloud of gas resulting from the debris of stars that have exploded 3

neurons: nerve cells 217

neurotransmitters (or neuromediators or transmitter substances): chemicals that travel across the synapse between the axon of one neuron and the dendrite of the next 217

neutrons: particles with no charge found in the nuclei of atoms. They are similar in size to protons.

non-renewable energy sources: are finite in amount and cannot be replaced once exhausted 106

normal force: a force that acts perpendicularly to a surface as a result of, and in the opposite direction to, the net force applied to the surface 82

nova: a star that has depleted its supplies of hydrogen fuel and then collapsed, causing its gaseous envelope to ignite in a massive explosion that scatters the gases into a cloud surrounding the remains of its core. The nova's brightness is much greater than that of the original star. 3

nuclear energy: energy emitted by a nucleus as it becomes more stable 112

nuclear fallout: the return to Earth of the irradiated material sent into the atmosphere by an atomic blast 149

nuclear fission: process by which large nuclei break up to form smaller nuclei 112

nuclear forces: the forces that arise from the interaction of the particles that make up an atom. They are called nuclear forces because the most influential of the particles are found within the nucleus (or centre) of the atom 2

nuclear fusion: smaller nuclei combine to form larger nuclei, releasing a great amount of energy 2

nuclear reactor: a device in which controlled fission is used to provide energy 113

nucleons: particles found in the nucleus of an atom. Protons and neutrons are both nucleons. 107

nucleotides: the subunits that make up the DNA molecule. Each nucleotide is made up of a sugar block, a phosphate block and a nucleotide base. 161

nucleus: the centre of an atom 107

o

oceanic crust: describes the thinner regions of the Earth's crust, made of dense igneous rock such as basalt. Oceanic crust is constantly being formed. 38

p

Pangaea: the name for the original supercontinent that contained all the present continents as a single mass. It broke up into Laurasia and Gondwanaland. 39

panspermia: theory that the earliest microbes were brought to Earth inside meteors from space 10

parabola: a symmetrical curved path. In geometric terms, it is formed by the intersection of a plane through a cone. 139

parallax error: misjudgement of distance that occurs as a result of a shift in the position of the observer 210

periodic table: a chart devised by the Russian chemist Dimitri Mendeleev that displays the known elements in rows and columns according to their physical and chemical properties 107

periods: subdivisions of an era that signal changes in the fossil or geological record 5

petrification: the process by which organic matter is replaced with minerals that are harder than the original material 20

phenotype: a characteristic that the organism expresses as a result of its genotype 167

photosynthesis: the process by which sunlight is used to convert carbon dioxide and water into oxygen and carbohydrate 10

planetismals: chunks of material formed in the accretion disc of a solar nebula that combine to form planets 4

plate boundary: where one tectonic plate meets another 43

polygenic traits: traits that depend on two or more genes 166

protons: positively charged particles found in the nuclei of atoms 107

punnet square diagram: a 3 × 3 table that allows the possible genotypes of offspring to be determined from the genotypes of the parents 167

P-waves: compression waves, which are the fastest of the waves produced in earthquakes and are the first waves to arrive at seismic recording stations 46

pyroclastic flow: hot volcanic material produced by an explosive eruption flows like a liquid 59

r

radiation: particles released when an unstable nucleus decays or a nucleus is forced to undergo a nuclear reaction 109

radioactive decay: breakdown of unstable isotopes into more stable nuclei by the release of particles 23

radioisotopes: isotopes that are unstable and likely to decay, releasing radiation 109

ram-augmented interstellar rocket (RAIR): a modified form of a Bussard ramjet that uses hydrogen atoms to allow the fusion of lithium or boron nuclei to release energy, rather than using just the fusion of the hydrogen atoms themselves as a fuel source 98

random errors: caused by unpredictable or unnoticed variations in measurements that occur randomly throughout the course of an experiment 287

range: the furthest horizontal distance that a projectile weapon is able to send a missile 136

reaction distance: the distance a car travels between the driver seeing the hazard and actually starting to brake 197

reactor core: the heart of the nuclear reactor 115

recessive allele: a characteristic that is expressed only when there is no dominant allele present in the genotype 166

red-shifted: describes the change in light frequency of stars towards the red end of the spectrum when they are moving away from us 92

renewable energy sources: can be replaced endlessly 106

Richter scale: used to determine the magnitude or size of an earthquake and reflects the amount of energy released 53

ridge counting: classifying fingerprint features by measuring the number of ridges crossed by a line drawn from a delta to a core in a loop or whorl 254

ridge tracing: classifying fingerprint features such as whorls according to their alignment relative to adjacent ridges 254

rigor mortis: the process by which a body after death starts to become stiff due to the accumulation of lactic acid in the larger skeletal muscles 251

Ring of Fire: a seismically active region that follows the outline of the Pacific Plate 63

s

scale patterns: shapes visible along the length of a natural fibre such as hair or wool when viewed under a microscope 271

scene-of-crime officers (SOCOs): specialist investigators who attend the crime scene 246

secretors: people whose body fluids contain type A or type B antigens dissolved in them 262
seismic waves: the waves formed by the sudden release of energy that occurs when locked plates move suddenly 46
seismogram: the record of the ground motions detected by a seismometer 48
seismograph: made up of a seismometer, an amplifier and a system that records the readings of the seismometer 48
seismologist: a scientist who studies earthquakes 39
seismometer: a device that detects small horizontal and vertical movements of the Earth's surface 48
selective breeding: process by which organisms with favourable characteristics are bred together in the hope of obtaining offspring that have those same favourable characteristics 170
serology: the study of the structure of the blood 259
shock wave: the rapid expansion and contraction of particles in a gas or liquid caused by a sudden change in pressure or density. Shock waves travel at or above the speed of sound 149
silent errors: genetic substitution errors that occur in such a way that there is no effect on the protein formed 166
sleeper ship: a type of manned spaceship in which the crew is held in suspended animation until arriving at the destination 96
soma: the cell body that contains the nucleus 217
Special Theory of Relativity: part of a paper written by Albert Einstein called 'On the Electrodynamics of Moving Objects', which discussed the effect of an object's speed on the space around it 99
speciation: the process by which one species gives rise to another 14
spectral class: determined by a star's surface temperature, its colour and its composition 92
spectrometer: a device that measures the frequencies of light produced by a luminous object such as a star 92
speed: rate at which an object travels a distance 197
speed guns: handheld devices that use the reflection of soundwaves or light waves to determine the speed of an object at which the guns are directly pointed 210
standard drink: contains 10 g of pure alcohol in the form of ethanol 233
stopping distance: distance that a car travels between the driver seeing a hazard and the car coming to a halt 197
strata: layers of rock that have formed from the compression of sediment 20
strong nuclear force (SNF): the force that holds the nucleons together in the nucleus of an atom 109
subduction: when one plate moves downwards under the edge of another plate 44
subduction zone: the region where the subduction of one tectonic plate under another occurs 44
substitution error: a genetic error in which one of the bases is changed or miscopied, causing a different protein to be made in the cell 166
S-waves: transverse or shear waves produced during an earthquake 47
synapse: the small gap between the dendrite of one nerve cell and the corresponding axon of the next nerve cell 217
systematic errors: the result of either incorrect equipment use or limitations with the equipment itself. These errors produce values that are consistently higher or lower than the accepted value. 287

t

terminal velocity: speed at which a falling object ceases to accelerate because the upward force of air resistance is balanced by the downward force of gravity 153
terraforming: the process in which a planet is artificially adapted to have an Earth-like climate and atmosphere 88
Theory of Evolution: describes the way in which organisms change in appearance, structure and function over long periods of time 12
thermal flash: the enormous amounts of heat and radiation that rapidly spread out from the centre of an atomic blast 148
thermal neutrons: used in nuclear reactors to cause fission in uranium nuclei 114
thermal reactors: nuclear reactors in which the neutrons produced in each fission reaction are slowed down so that they may cause further fission of nuclei 114
thin layer chromatography: a process by which the components of a mixture to be analysed are separated using a thin layer of material or film 272
toxicology: the study of substances that are harmful to the human body 271
traction: a tyre's ability to grip a surface and so push the car forward 195
transfer patterns: blood patterns that are formed when a bloodied object such as a hand, a weapon, shoes or even hair come into contact with a surface 260
transform faults: areas of weakness and cracking in the Earth's crust that are caused by the forces generated by plate motion 45
transgenic organisms: contain gene sequences that originated in another organism 174
translation: occurs when two plates slide past each other either horizontally or vertically without creating or destroying crust material 45
transverse wave: one in which the disturbance caused by the transfer of energy acts perpendicularly to the direction of the wave itself 47
triangulation: the process by which the location of an earthquake's epicentre can be determined based upon the difference in arrival times of seismic waves from three different seismograph stations 50

tsunamis: giant sea waves produced when water is displaced by underwater earthquakes, landslides or volcanic eruption 64

u

uncontrolled chain reaction: occurs when all of the neutrons released during fission cause the fission of further nuclei, allowing all of the fissionable material present to be involved in the chain reaction 112

v

variables: factors that can affect the outcome of an experiment 283

viscosity: a measure of the 'stickiness' of a fluid. A fluid with high viscosity (like honey) flows slowly, while a fluid with low viscosity (like water) flows easily. 58

visible spectrum: composed of light that has frequencies between 4.3×10^{14} Hz and 7.3×10^{14} Hz 92

volcanic gases: those released by a volcanic reaction. The most common gas is superheated steam but carbon dioxide, carbon monoxide, sulfur dioxide and hydrogen sulfide can also be produced. 59

volcanic outgassing: the process by which gases are released into a planet's atmosphere from its molten interior by volcanic activity 8

volcaniclastic material: a general term for the solids, liquids and gases that are ejected from a volcano during an eruption 58

vulcanologist: a scientist who studies volcanoes 59

Weblinks

The website for this book is at: www.jaconline.com.au/science21

Chapter		Page no.	On-screen text
1	Life, the universe and everything	15	Evolution lab
1	Life, the universe and everything	22	Fossils in Australia
1	Life, the universe and everything	26	Radioactive decay
2	The restless Earth	49	Earthquake seismogram
2	The restless Earth	49	Model seismograph
3	Colonising other worlds	74	Mir modules
5	Weapons through the ages	141	Cannon simulator

Index

a

b

c

d

i

j

k

l

m

n

o

p

q

W